28.50

NUMERICAL METHODS
IN PRACTICE

INTERNATIONAL COMPUTER SCIENCE SERIES

Consulting editors   **A D McGettrick**   University of Strathclyde

J van Leeuwen   University of Utrecht

SELECTED TITLES IN THE SERIES

NUMERICAL METHODS IN PRACTICE
Using the NAG Library

Tim Hopkins
University of Kent at Canterbury

Chris Phillips
University of Newcastle-upon-Tyne

ADDISON-WESLEY
PUBLISHING
COMPANY

Wokingham, England · Reading, Massachusetts · Menlo Park, California
New York · Don Mills, Ontario · Amsterdam · Bonn
Sydney · Singapore · Tokyo · Madrid · San Juan

To Claire, Emily, Ruth, William and Zena

Cover designed by Crayon Design of Henley-on-Thames and printed by The Riverside Printing Co. (Reading) Ltd.
Typeset by the authors using LaTeX at the University of Kent at Canterbury.
Printed and bound in Great Britain by Mackays of Chatham PLC, Chatham, Kent.
First printed in 1988.

British Library Cataloguing in Publication Data
Hopkins, Tim
   Numerical methods in practice : using
   the NAG Library.
   1. Numerical methods. Applications of
   computer systems. Software packages
   I. Title   II. Phillips, Chris
   III. Series
   519.4'028'553

   ISBN 0–201–19248–9

Library of Congress Cataloging in Publication Data
Hopkins, Tim.
   Numerical methods in practice.

   (International computer science series)
   Bibliography: p.
   Includes index.
   1. NAG Library (Computer programs)   2. Numerical analysis — Data processing.   3. Subroutines (Computer programs)   4. FORTRAN (Computer program language)   I. Phillips, Chris, 1950–   II. Title.
   III. Series.
   QA297.H589   1988   519.4'028'55369   88– 24179
   ISBN 0–201–19248–9 (pbk.)

Foreword

The NAG Project began in May 1970, when a group of us from six British university computing centres decided to develop a library of numerical routines for our ICL 1906A/S machines. Mark 1 of the NAG Library, containing 98 documented routines, was released on the 1st October 1971.

The Library now contains, at Mark 12, 688 routines accessible to users. The original informal project has grown into a non-profit-making company, NAG Ltd, with a subsidiary, NAG Inc, in the USA. Numerical analysts and software specialists from all over the world have contributed, and continue to contribute, to the Library, which remains the principal software product distributed by NAG.

Quality of documentation is at least as important as the quality of the software, and requires just as much care and effort. The NAG Fortran Library Manual has expanded to fill seven volumes and contains definitive reference documentation for all the routines in the Library, as well as background material, guidance on choice and use of routines, and decision trees. However, different users need different kinds of documentation, and neither the manual, nor its machine-based version, the NAG On-line Information, can hope to satisfy all their needs.

This book is designed to fill part of the gap. It is intended as an aid to the use of the NAG Library that will be especially helpful to students of numerical analysis. The book is based on a description of the most commonly used numerical methods, and discusses their advantages and disadvantages. The book emphasizes those methods selected for use in the NAG Library, but not to the exclusion of other methods or of other sources of numerical software. Hence it can stand alone as an introductory text in numerical analysis, and is recommended for that purpose.

The book also discusses some of the issues that arise when numerical methods are implemented in software, such as the design of subroutine calling sequences and the choice of data structures. It explains some of the conventions that have been adopted in the NAG Library.

Preparation of numerical software is a topic attracting increasing interest. This volume exposes a number of the primary concerns of the topic.

For any user of the Library, one of the most daunting problems is how to select the most suitable routine to solve a particular problem, out of the many hundreds that are provided. Here again the book provides valuable assistance.

Thus it is a pleasure to welcome the appearance of this book, and to recommend it in particular to anyone who wishes to use the NAG Library as an aid to studying and applying modern numerical methods.

Brian Ford
Director, The Numerical Algorithms Group

Preface

The rapid development of the digital computer has been matched by an increase in the derivation of sophisticated numerical methods for the solution of mathematical problems, and the availability of computer codes which implement them. Many of these routines have been incorporated into the NAG (Numerical Algorithms Group) library which is widely available on a number of machines. Users of the library come from many disciplines, and their understanding of the underlying numerical techniques ranges from comprehensive to very little. Our experience suggests that many users frequently have difficulty in making correct use of the routines available, either because they are unsure which particular routine to use, or, having decided on a routine, how to set up the call sequence, or even how to interpret the results produced by the routine. In each case, some idea of the fundamental workings of individual routines can be of considerable help. The excellent NAG documentation can assist, but its comprehensive nature means that the uninitiated can find ploughing through it an extremely daunting task. Moreover, the reader may find the references to original sources unhelpful if all he wants is some brief understanding of the basic numerical method.

This volume addresses a number of important areas of numerical problem solving, outlines the facilities available in the NAG library for solving such problems, and describes the numerical methods on which these routines are based. As a deliberate policy we have omitted detailed references to methods which, although of theoretical interest, are considered to be of insufficient general use for their inclusion in the NAG library. Each chapter of this book corresponds to a topic in numerical analysis covered by one, or occasionally two, chapters of NAG. Users who have a specific problem in mind may find it useful to turn immediately to the end of the appropriate chapter, where a summary on routine selection is given with references back to earlier sections for a detailed discussion of a particular routine.

The book may also be viewed as an introductory text on numerical methods; all we expect of the reader is a reasonable competence in Fortran programming and mathematical analysis. The text is interspersed with code fragments and complete programs which illustrate implementation aspects of the development of numerical software. Rather than dwelling on the theory of numerical analysis, we feel that those new to the subject should be encouraged to obtain numerical results in order to gain greater insight. The availability of a comprehensive routine library avoids the 're-invent the wheel' syndrome, and eases the task considerably.

We emphasize that this volume does not intend to be a replacement for the NAG documentation, but a supplement which may be used as a first source of information. As such we hope that it will help to overcome the misgivings that many first users have, and avoid misuse of the routines.

Acknowledgements

We would like to thank

- Ian Gladwell, Diane Grace, Sven Hammarling, Gordon Makinson and David Sayers who read all or part of earlier drafts and who provided a wealth of improvements to the content and the spelling,

- Mike Delves for providing funds so that we could meet over a curry from time to time to sort out problems,

- Allison King who, along with others at Addison-Wesley, put up with two amateur authors, typesetters and book designers and provided a memorable lunch,

- Ian Utting for generating the runes required to transform our text files into a book,

- Sean Leviseur for maintaining LaTeX, TeX and various previewers and drivers at Kent and hacking them when necessary,

- Leslie Lamport and Don Knuth for making LaTeX and TeX available.

Any remaining mistakes are entirely the responsibility of our wives and, although our bank balances can't withstand the financial inducements offered by Don Knuth, we would be very pleased to hear from anyone who finds errors or shortcomings in the text.

Finally, Bristol Rovers and Derby County are mentioned so we can get them in the index!

Tim Hopkins *University of Kent at Canterbury, Canterbury, UK.*

Chris Phillips *University of Newcastle-upon-Tyne, Newcastle, UK.*

Contents

Chapter 1
Introduction

In this chapter we lay the foundations for a subsequent detailed study of certain of the NAG routine chapters. In particular, we

- consider the practical aspects of implementing some simple numerical algorithms;

- review certain features of Fortran;

- give simple examples of calls to NAG routines;

- review a number of mathematical and computational concepts needed elsewhere in the book.

1.1 Introduction to problem solving

It is assumed that the reader of this text wishes to solve mathematical problems. These problems, whether they originate from engineering, sociology, physics or any other discipline which needs to model or simulate some physical situation, are assumed to be in the form of mathematical descriptions of the original phenomena. This description may vary in complexity from a small well-conditioned set of linear equations, to a large coupled system of partial differential equations acting over a geometrically complicated region with nonlinear boundary conditions. For the majority of practical problems no analytic solution will exist, and hence some computed numerical solution will be sought. Even when an analytic solution does exist it may still be more efficient to solve the problem numerically. Furthermore, to obtain all but the most trivial of numerical approximations the solver will require the assistance of a computer. The overriding requirement is to obtain the 'correct' solution. In this volume we make no attempt to teach either the art of mathematical modelling, or basic computing; rather, we address the problem of developing a computer program which can be used to determine an approximate solution. Although we hope to show how this may be achieved as painlessly as possible, we do not believe that numerical problem solving can be totally successful without

1

some knowledge of the underlying techniques, their possible weaknesses, and of the problems inherent in computer arithmetic.

There are several courses of action open to our intrepid problem solver. He might

(1) Attempt to avoid writing a program altogether by using an existing *black-box package*. This type of software ranges from packages which allow little or no user interaction to highly sophisticated *expert systems*. At the lowest level data, in a package-dependent format, are transformed into a tabular set of results. More advanced packages allow users to specify their own input format, to perform a variety of computational tasks via a built-in command language, and to produce a number of different views, both textual and graphical, of the results. Some packages have command languages which differ little from high level programming languages.

    Packages of this form exist for various types of statistical analysis (e.g., GLIM, GENSTAT), for the solution of partial differential equations arising in a number of branches of engineering (e.g., finite element analysis using PAFEC), for circuit design analysis (e.g., SPICE), and for many other applications.

    Further, there are expert systems which attempt to analyse the problem as presented by the user and, possibly with user interaction, thence decide upon the best means of solution. This type of system is still in its infancy and it may be some years before complex numerical problems are solved by such means.

    Black-box software offers an excellent means of solution provided that the problem can be completely solved by the package, and provided that the user is satisfied with the algorithms which are used internally, and with their implementation. Unfortunately detailed descriptions of the internal workings of many packages are not usually made available to the end user. Major problems arise when the package will not perform all the computational tasks needed for a complete solution, or does not provide access to all the information required, or uses techniques which the user does not consider suitable. Often such packages are available only as object codes; even when sources are provided it may well be beyond the programming capabilities of the user to make any necessary changes.

(2) Split the problem into a number of well-defined mathematical steps and attempt to beg, borrow, or steal already available subroutines to solve some or all of the individual subproblems. This approach allows flexibility in the solution by giving the user control over both input and output, and the method of solution used to solve the various subproblems. The subprograms are used as building blocks, and, to be of general use, such procedures need to be able to deal with a general

class of problem. There is currently a wealth of very high quality software, in the form of collections of subroutines, available either in the *public domain* and usually obtainable for a modest handling charge, or as proprietary software for which a more substantial fee is generally charged. In the first group we have, for example, the *PACK software*, developed over the last decade or so, which includes

- LINPACK - for solving systems of linear equations,
- EISPACK - for eigenvalue/eigenvector calculations,
- QUADPACK - for numerical integration computations,
- MINPACK - for function minimization, etc.

There is also the *Collected Algorithms of the ACM*, now published in *Transactions on Mathematical Software* and available in machine readable form. The problem with software from different sources is that it lacks cohesion and consistency in implementation details, documentation and *user interface* (the way the user and the software communicate with one another).

Ideally, uniformity should be designed into, or imposed onto, the software. However, complete retesting and revalidation is required to ensure that there are no hidden problems when moving code to a new environment. A good subroutine library would thus be expected to consist of a homogeneous collection of subroutine codes, and documentation, specifically tailored to a particular machine/compiler combination. In addition, wide use and a central bug fixing organization are important in that they help to improve the quality and reliability of the code.

To be of use in many different applications such a collection of subroutines would need to implement a wide range of general purpose problem solving algorithms. The *NAG numerical subroutine library*, whose use is described in this book, is one such collection which fulfils the above requirements. If correctly used, routines from this library should allow quite difficult problems to be solved with a minimum of programming effort.

(3) Write a complete purpose built program. This approach may well require extensive knowledge of both *numerical analysis* (Section 1.2) and computing to ensure a successful implementation and is likely to be extremely time consuming. It must be considered as a last resort solution when all attempts at a type (1) or (2) solution above have failed. Unfortunately many users still appear to have phobias against using other people's software. One commonly heard excuse is that general purpose software cannot possibly be efficient enough. This may well be true if particularly large problems are to be solved many times over. However, a subroutine library can be useful for the rapid

development of a prototype solution. Only when this prototype is functioning correctly need the question of efficiency be tackled. It is often the case that a large percentage of the execution time of a program is spent in only a very small percentage of the code, typically 95% of the time in 5% of the code. The transformation of the prototype into an efficient production code may thus require changes to a small number of routines only. Furthermore, the prototype system is useful during this transformation step as it provides a means of cross checking the results.

Finally we recognize the existence of a, fortunately small, band of computer users who have the 'Not Invented Here' disease. These people only trust software written by themselves and generally turn out to be masochists!

EXERCISE 1.1   Find out what packages and subroutine libraries are available at your computer installation. Note down any software in your area of interest; it may save you some effort in the future. Look carefully at the last large program you wrote; how much code can be replaced by existing routines?

EXERCISE 1.2   What criteria can be used to measure the 'cost' of a computer program? Consider the importance of each in the context of micro, mini and mainframe environments. How would you attempt to quantify these costs?

1.2   Numerical analysis and algorithms

It is not easy to put forward a simple all-encompassing definition of the term 'numerical analysis'. The best perhaps that we can do is to say that it covers the study of methods for obtaining approximate solutions to mathematical problems. At one end of the spectrum we may consider the addition of two real numbers; here the approximation lies in the fact that certain numbers cannot be specified exactly using a finite number of decimal digits. We may thus find it necessary to work with numbers which differ, albeit only slightly, from the originals. In this context numerical analysis is concerned, amongst other things, with the effects such errors have on the computed sum and any subsequent computations. At the other end of the spectrum we have complex problems, such as partial differential equations, which are difficult, or impossible, to solve by conventional analytic techniques. Here we need to have a thorough understanding of the problem and its properties before attempting to determine an approximate solution. In particular we must be aware of relevant results from mathematical theory. Within such an environment the numerical analyst can, with some confidence, hope to derive and analyse methods which, ultimately, may be used to develop computer programs producing numerical results purporting to be a solution to the original problem. The approximation used can take many forms. It may, for example, lie in the assumption that the solution may be adequately represented, at least locally, by a low degree polynomial, or that an infinite process may be terminated after a finite number of steps without incurring a serious loss of accuracy. It will almost certainly involve the problems associated with the manipulation of approximate numbers already mentioned. In each case it is important to build a rigid mathematical framework within which we may study the effect that these approximations have on the computed solutions; without such a framework we can say little about the reliability of any numerical results. We are primarily interested in numerical methods which may be coded into general purpose computer programs; such methods are often termed *numerical algorithms* or *computational methods*. Their realizations as computer programs are generally termed *numerical software*.

An algorithm is a mathematical recipe; it consists of a finite number of *input* parameters which define a particular problem, and a set of well-defined steps which transform the input data into a set of desired *output* values. Strictly speaking (see Knuth [48]) each step of the algorithm should be sufficiently basic that it could be performed exactly and in finite time by a human using pencil and paper. Many of the methods we consider are thus not algorithms in this strict sense since we will not perform exact arithmetic; as such, they are sometimes referred to as *computational algorithms*. We distinguish between two types:

- *Direct methods* – these transform the input data into a solution in a predeterminable number of operations. Generally the number of operations required is dependent on some size parameter, e.g., the number of equations in a system of linear equations. If exact arithmetic were used, direct methods would compute an exact solution to the mathematical model as defined by the algorithm (but not, necessarily, to the original problem). Analysing the effects of the errors caused by using finite-precision arithmetic forms an important area of numerical analysis.

- *Iterative methods* – even using exact arithmetic these would theoretically require an infinite number of operations to compute an exact solution. We need to be able to terminate the process after a finite number of steps at a point where the computed results are 'close', in some sense, to the true solution. The problem of terminating iterative methods so as to guarantee a user specified accuracy is also of interest to the numerical analyst.

We are further concerned with the effectiveness of computational algorithms, especially when they are realized as computer programs. It is quite common for there to be several different algorithms available for solving the same class of problem; how do we decide which method is the most effective? We require some measure of effectiveness; this may well depend on more than just the requirement that the results be as accurate as possible. For example, we may compare algorithms from the viewpoints of the following factors:

- *Efficiency* in terms of the number of arithmetic operations required to solve a particular problem. Such an operation count may often give a good indication of which algorithm will run fastest when implemented as a computer program.

- *Storage costs* — we must bear in mind that storage may be limited and that the overheads required by algorithms which require an excessive amount of array space may be prohibitively expensive. Careful analysis of the implementation of the algorithm will be required.

- *Reliability* — does the algorithm either accurately solve all problems in a particular class, or detect that it has failed to solve a particular problem?

- *Generality* — how wide a class of problem can be solved? If we are to make the effort to produce sophisticated numerical software, we want it to be as widely applicable as possible.

Rarely do we find an algorithm which satisfies more than one or two of the requirements simultaneously; rather we have to trade one requirement for another. Thus we may, for example, increase generality at the expense of efficiency. We discuss a number of the points raised in this section in more detail in Section 1.14 and Section 1.16.

EXERCISE 1.3   Consider an iterative method which generates a convergent sequence of approximate values $x_0, x_1, x_2, \ldots$. What different termination criteria could you use to ensure a final result 'close' to the true solution? How would the magnitude of this result affect your choice of criteria?

1.3   Computer arithmetic

In this section we consider the problems associated with the use of *finite-precision arithmetic*. We are particularly concerned with *number representation* within a computer but some of what we have to say is relevant to number manipulation using a pocket calculator, or even the basic tools of pen and paper.

Computer *memory* is divided into addressable *words*, each of which is made up of a fixed number of *bits* (*binary digits*), known as the *word length* of the machine. All information (whether a number, character, instruction, etc.) is stored internally using a certain number of words, and each item can ultimately be considered as a string of 0s and 1s. Different machines use different word lengths and have different conventions for coding this information. Also, on any particular machine the same bit pattern may stand for different quantities depending on what type of information is being represented.

	Computer Type			
	IBM 370	Cray 1	DEC VAX 11/780	IEEE Standard
Bits/Word	32	64	32	–
Bits/Integer	32	62	32	–
Floating-point base	16	2	2	2
Precision (single)	6	48	24	24
(double)	14	96	56	53
Exponent (single)	$[-64, 63]$	$[-8192, 8192]$	$[-127, 127]$	$[-125, 128]$
range (double)	$[-64, 63]$	$[-8192, 8192]$	$[-127, 127]$	$[-1021, 1024]$

Figure 1.1   Typical machine values.

Most high level computer programming languages distinguish between at least two types of number, *integers* and *reals*. Positive integers are normally stored in a positional, base 2 representation, for example,

$$8741_{10} = 10001000100101_2$$
$$= 1 \times 2^{13} + 0 \times 2^{12} + 0 \times 2^{11} + 0 \times 2^{10} + 1 \times 2^9 + 0 \times 2^8$$
$$+ 0 \times 2^7 + 0 \times 2^6 + 1 \times 2^5 + 0 \times 2^4$$
$$+ 0 \times 2^3 + 1 \times 2^2 + 0 \times 2^1 + 1 \times 2^0.$$

Each integer value is stored in a fixed number of words and is thus represented by a fixed number of bits; some typical sizes are given in Figure 1.1. Typically a Fortran system running on a 16-bit word length machine uses just one word to store an integer, the number being right justified within the word.

If we use the leading bit to represent the sign of the number, 0 for a positive and 1 for a negative number, and use the remaining bits as a positional base 2 representation we obtain *sign and magnitude notation*. Thus, for example, using a 16-bit word -8741_{10} would be stored as

1	0	1	0	0	0	1	0	0	0	1	0	0	1	0	1

Using this convention the largest representable positive and negative integers are

$$32767_{10} = 011\ldots11_2$$
$$-32767_{10} = 111\ldots11_2,$$

where $32767 = 2^{15} - 1$.

Another common way of representing negative integers is in *two's complement* form. Using this convention the numbers may be thought of as being circular, as on a milometer, i.e.,

$$00\ldots10 = 2$$

$$00\ldots01 \; = \; 1$$
$$00\ldots00 \; = \; 0$$
$$11\ldots11 \; = \; -1$$
$$11\ldots10 \; = \; -2.$$

As with sign and magnitude, we treat all numbers with leading bit 1 as negative, so that the range of representable negative integers is

$$[10\ldots0_2, 11\ldots1_2] = [-32768_{10}, -1_{10}].$$

Similarly we treat all those with leading bit 0 as positive giving the range

$$[00\ldots0_2, 01\ldots1_2] = [0_{10}, 32767_{10}].$$

The name two's complement comes from the way we negate a given number; the rule is complement (switch all 1s to 0s and vice versa) and add 1. Thus -8741_{10} would be represented as

1	1	0	1	1	1	0	1	1	1	0	1	1	0	1	1

Addition is particularly simple in two's complement; the two operands are added and any carry off the left-hand end of the result is ignored.

Unless the result of an operation *overflows* (lies outside the range of values representable), the addition, subtraction and multiplication of integers will be exact. Many systems will abort the execution of a program if an integer overflow is detected. We remark that Fortran recognizes a division operator, /, as being an 'integer' division if both operands are integers, the rule is to perform the division and throw away any resulting fractional part.

To introduce real number representation we consider the base 10 number 8741.625. This is said to be written in *fixed-point format*. By shifting the decimal point to the left 4 places and multiplying the result by 10^4 we obtain a *floating-point representation*, 0.8741625×10^4. In this notation we may identify the following fields:

- An optional sign.

- The *mantissa* or *fractional part* which contains the *significant digits* of the number (8741625 in our example).

- The *base* of the number (10 in our example).

- The *exponent* — an integer value which determines how many places right the point needs to be shifted to obtain the fixed-point version (4 in the example). Note that a negative exponent is a shift left and padding zeros need to be inserted.

The floating-point representation of a number is not unique; for example, the above value could also be written as 8.741625×10^3 or 0.008741625×10^6 or 8741.625×10^0. However it is common practice to *normalize* floating-point numbers so that (in base ten) the mantissa lies in the range $[\frac{1}{10}, 1)$.

When using floating-point format to store real numbers, each may occupy one or more words. Most digital computers use a base, β, which is a power of 2, *binary* (2), *octal* (8) and *hexadecimal* (16) being the most popular. We may now write a general floating-point number in the form

$$\pm \left(\frac{d_1}{\beta} + \frac{d_2}{\beta^2} + \frac{d_3}{\beta^3} + \cdots + \frac{d_t}{\beta^t} \right) . \beta^e$$

where $0 \le d_i < \beta$ and e is an integer in some range $[L, U]$. If d_1 is constrained to be non-zero then the representation is normalized and t is known as the *precision* of the floating-point representation; see Figure 1.1 for some typical values. Because we do not usually require an enormous exponent range we allow the mantissa and exponent to be packed together. For a 32-bit word machine a single-precision number might typically be stored as

±	exponent	mantissa

whilst a double-precision number might occupy two words as

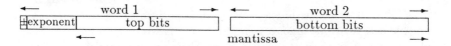

In our model the leading bit is used for the sign of the number, 0 for positive and 1 for negative. The exponent is stored as an integer; an eight-bit exponent stored in two's complement form would give a range of $[-128, 127]$ which, for base 2, allows numbers whose magnitude ranges between approximately 10^{-38} and 10^{38}, plus zero. The remaining bits are used to store the unsigned normalized mantissa which lies in the range $[\frac{1}{\beta}, 1)$. Our single-precision number allows for a 23-bit mantissa which gives approximately 7 decimal digits accuracy. Returning to our example, we note that we may write

$$8741.625 = 0.10001000100101101 \times 2^{14},$$

which would be stored, using the 32-bit format described above, as

0	0	0	0	0	1	1	1	0	1	0	0	0	1	0	0	0	1	0	0	1	0	1	1	0	1	0	0	0	0	0	0

$\pm \longleftarrow$ exponent $\longrightarrow \longleftarrow$ mantissa $\longrightarrow$

with the mantissa left justified.

In converting 8741.625_{10} into binary, we were fortunate in that the mantissa could be represented exactly using a finite number of bits, and that all significant bits could be accommodated in the space available. This is not always the case when we transform from one number base to another. For example,

$$0.1_{10} = (0.000110011001\ldots)_2$$
$$= (0.1999\ldots)_{16}.$$

Indeed 0.1_{10} cannot be stored exactly using a floating-point number base which is any power of 2. The solution is either to *round* or to *truncate* the number after t significant digits, which on its own may appear trivial, but could have disastrous consequences later on.

Because working with base 2 is so unnatural for most humans, we study the effects of rounding errors by considering a hypothetical base 10 computer which has 6-digit precision. We will assume that this machine rounds the result of every arithmetic operation, so that, for example, the result of the real division $22/7 = 3.142857142857\ldots$ would be stored as 0.314286×10^1. The *absolute error*, the magnitude of the difference between the true value and its approximation, is less than 0.000003 and, the *relative error* (absolute error divided by true value) is 0.9×10^{-6}. Similarly, representing $22000/7$ by 0.314286×10^4 involves an absolute error of 0.03 but a relative error of 0.9×10^{-6} again. In both cases we say that the number stored represents the true value correct to six significant figures.

Every time an arithmetic operation is performed on real numbers in our hypothetical machine, a rounding error is likely to occur; for example, the multiplication of two 6-digit numbers results in an 11- or 12-digit product. In general the resultant product cannot be represented without rounding error using only a 6-digit mantissa. Precisely how computers perform arithmetic operations is beyond the scope of this book. The interested reader is recommended to start with Knuth [49]. We remark that most machines which round compute the results of the basic arithmetic operations in slightly greater than basic precision before rounding.

At first sight it may appear that rounding errors are of little consequence. However, the two following examples should serve to demonstrate the danger of this assumption

EXAMPLE 1.1   (Cited by Fox and Wilkinson in the foreword to the NAG library manual.) Consider the evaluation of the definite integral

$$I_n = \frac{1}{e} \int_0^1 e^x x^n \, dx.$$

A little manipulation (integration by parts) yields the *recurrence relation*

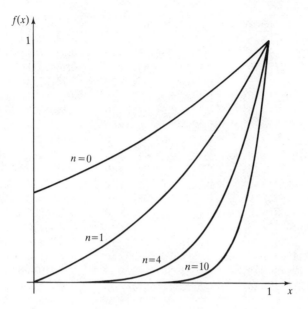

Figure 1.2   Plots of integrands.

$$
\begin{aligned}
I_0 &= 1 - e^{-1}, \\
I_n &= 1 - nI_{n-1}, \qquad n = 1, 2, \ldots.
\end{aligned}
\tag{1.1}
$$

Figure 1.2 shows the plot of the integrand for increasing values of n. We note that the integrand is always positive within the range of integration and that the area under the curve, and hence the value of I_n, decreases monotonically with n. Thus for all n we may deduce that $I_n > 0$ and $I_n < I_{n-1}$.

The results of running a program implementing (1.1) in single- and double-precision arithmetic on a VAX 11/780 are given in Figure 1.3 (the correct digits in the computed values are underlined). We see that for large n the values do not exhibit the behaviour forecast. In fact for $n > N_p$, where N_p is dependent on the precision of the arithmetic, successive values of I_n increase in magnitude and alternate in sign. Increasing the precision merely delays the point at which problems are noticed. For this example it is relatively easy to explain the difficulty.

The computation of I_0 is not exact; no matter how good the intrinsic functions are there will be an error, ϵ_0, in the value we obtain for I_0. We therefore start our recurrence with a value $\tilde{I}_0$ where $\tilde{I}_0 = I_0 + \epsilon_0$. Even if we incur no further rounding errors, the effect of this initial error is such that we compute a sequence $\{\, \tilde{I}_n \mid n = 1, 2, \ldots \,\}$ using the recurrence relation $\tilde{I}_n = 1 - n\tilde{I}_{n-1}$ rather than the values $\{\, I_n \mid n = 1, 2, \ldots \,\}$ defined by (1.1). If we let $\epsilon_n = \tilde{I}_n - I_n$ be the error at step n then, from the two recurrence

n	$\tilde{I}_n$ (single)	$\tilde{I}_n$ (double)	I_n (true)
0	0.632121×10^0	$0.6321205588286 \times 10^0$	$0.6321205588286 \times 10^0$
1	0.367879×10^0	$0.3678794411714 \times 10^0$	$0.3678794411714 \times 10^0$
2	0.264241×10^0	$0.2642411176571 \times 10^0$	$0.2642411176571 \times 10^0$
3	0.207277×10^0	$0.2072766470287 \times 10^0$	$0.2072766470287 \times 10^0$
4	0.170893×10^0	$0.1708934118854 \times 10^0$	$0.1708934118854 \times 10^0$
5	0.145534×10^0	$0.1455329405731 \times 10^0$	$0.1455329405731 \times 10^0$
6	0.126796×10^0	$0.1268023565615 \times 10^0$	$0.1268023565615 \times 10^0$
7	0.112430×10^0	$0.1123835040694 \times 10^0$	$0.1123835040693 \times 10^0$
8	0.100563×10^0	$0.1009319674451 \times 10^0$	$0.1009319674456 \times 10^0$
9	0.949326×10^1	$0.9161229299417 \times 10^{-1}$	$0.9161229298966 \times 10^{-1}$
10	0.506744×10^1	$0.8387707005829 \times 10^{-1}$	$0.8387707010339 \times 10^{-1}$
11	0.442581×10^0	$0.7735222935878 \times 10^{-1}$	$0.7735222886266 \times 10^{-1}$
12	-0.431097×10^1	$0.7177324769464 \times 10^{-1}$	$0.7177325364803 \times 10^{-1}$
13	0.570427×10^2	$0.6694777996972 \times 10^{-1}$	$0.6694770257562 \times 10^{-1}$
14	-0.797597×10^3	$0.6273108042387 \times 10^{-1}$	$0.6273216394138 \times 10^{-1}$
15	0.119650×10^5	$0.5903379364190 \times 10^{-1}$	$0.5901754087930 \times 10^{-1}$
16	-0.191438×10^6	$0.5545930172957 \times 10^{-1}$	$0.5571934593124 \times 10^{-1}$
17	0.325445×10^7	$0.5719187059731 \times 10^{-1}$	$0.5277111916900 \times 10^{-1}$
18	-0.585801×10^8	$-0.2945367075154 \times 10^{-1}$	$0.5011985495809 \times 10^{-1}$
19	0.111302×10^{10}	$0.1559619744279 \times 10^1$	$0.4772275579621 \times 10^{-1}$
20	-0.222605×10^{11}	$-0.3019239488558 \times 10^2$	$0.4554488407582 \times 10^{-1}$

Figure 1.3 I_n vs. n using single- and double-precision.

relations, we obtain $\epsilon_n = -n\epsilon_{n-1}$ and hence $\epsilon_n = (-1)^n n! \epsilon_0$. Using single-precision arithmetic ϵ_0 is at best $O\left(10^{-7}\right)$ and the rapid growth of the factorial is enough to destroy any accuracy we might hope to achieve. This error growth is clearly seen in Figure 1.4. We say that the recurrence relation (1.1) is *unstable*, and it is important that we recognize such situations, as well as those which are inherently stable.

EXAMPLE 1.2 Here we consider the problem of computing the roots of the quadratic polynomial equation $ax^2 + bx + c = 0$ with $a = c = 1$ and $b = 100$. The formulae for computing the roots are well known,

$$x_1 = \frac{-b + \sqrt{b^2 - 4ac}}{2a},$$

$$x_2 = \frac{-b - \sqrt{b^2 - 4ac}}{2a}.$$

Using our decimal computer to evaluate x_1 we look in detail at what happens when the expression under the square root is evaluated. We have

$$b^2 - 4ac = (0.100000 \times 10^3)^2$$
$$- 0.400000 \times 10^1 \times 0.100000 \times 10^1 \times 0.100000 \times 10^1.$$

n	$I_n - \tilde{I}_n$ (single)	$I_n - \tilde{I}_n$ (double)
0	-0.9150×10^{-08}	0.5551×10^{-16}
1	0.9150×10^{-08}	-0.2776×10^{-16}
2	-0.1830×10^{-07}	0.4163×10^{-16}
3	0.5490×10^{-07}	-0.1110×10^{-15}
4	-0.2196×10^{-06}	0.3608×10^{-15}
5	0.1098×10^{-05}	-0.1849×10^{-14}
6	-0.6588×10^{-05}	0.1103×10^{-13}
7	0.4611×10^{-04}	-0.7725×10^{-13}
8	-0.3689×10^{-03}	0.6180×10^{-12}
9	0.3320×10^{-02}	-0.5562×10^{-11}
10	-0.3320×10^{-01}	0.5562×10^{-10}
11	$0.3652 \times 10^{+00}$	-0.6118×10^{-09}
12	$-0.4383 \times 10^{+01}$	0.7342×10^{-08}
13	$0.5698 \times 10^{+02}$	-0.9544×10^{-07}
14	$-0.7977 \times 10^{+03}$	0.1336×10^{-05}
15	$0.1196 \times 10^{+05}$	-0.2004×10^{-04}
16	$-0.1914 \times 10^{+06}$	0.3207×10^{-03}
17	$0.3254 \times 10^{+07}$	-0.5452×10^{-02}
18	$-0.5858 \times 10^{+08}$	0.9813×10^{-01}
19	$0.1113 \times 10^{+10}$	$-0.1864 \times 10^{+01}$
20	$-0.2226 \times 10^{+11}$	$0.3729 \times 10^{+02}$

Figure 1.4   Errors in the computed integrals using single- and double-precision arithmetic.

(Each mantissa has been padded with zeros to emphasize the six-digit accuracy being used). When evaluated this expression yields 0.999600×10^4, and we obtain $\sqrt{0.999600 \times 10^4} = 0.999800 \times 10^2$, giving

$$x_1 = (-0.100000 \times 10^3 + 0.999800 \times 10^2) \times 0.500000 \times 10^0$$

after rounding. The expression within the brackets involves the sum of two quantities that are almost equal in magnitude but opposite in sign. Expressing the smaller number as 0.099980×10^3 and adding we obtain -0.000020×10^3 which, when normalized, gives -0.200000×10^{-1}. Unfortunately this result is only correct to 2 significant digits, the final 4 zeros appearing solely as a result of the normalization process. Continuing, we obtain $x_1 = -0.100000 \times 10^{-1}$ but it is, strictly, incorrect to express the result in this form since the implication is that we have six figure accuracy. We should really write the result as $x_1 = -0.10 \times 10^{-1}$ to emphasize the loss of four significant digits.

On the other hand we may determine $x_2 = -0.999900 \times 10^2$ with no loss of significance and use this value to compute x_1; since the product of the two roots is c we have $x_1 = c/x_2 = -0.100010 \times 10^{-1}$ which is also correct to all 6 digits.

Suppose we now attempt to check our results by substituting the computed six-digit roots back into the original equation. For reasons which will become apparent in Chapter 2 we choose to express the quadratic as $(ax + b)x + c$. Substituting $x = x_2$ our computer gives the value 0.100000×10^{-3}. We might initially conclude that x_2 is unacceptably inaccurate. However, compared with the size of the root (whose magnitude is approximately 100) the result is rather small and certainly on a six figure machine we cannot hope to do much better than this. To all intents and purposes x_2 does give 'zero' when substituted into the equation. Note that once again four significant digits have been lost and the final result should be quoted as 0.00×10^0. It is important to realize that when working with finite-precision arithmetic we will not, in general, obtain exact results. We must be prepared to accept results which contain errors; but it is important that we should be able to estimate the size of the errors involved. Such estimates are dependent both on the accuracy of the data used to define the problem, and on the method used to generate the approximate solution to the problem. Many of the routines in the NAG library allow the user to define the accuracy to which the approximation should be determined, and/or return an error estimate.

Finally, if the result of an arithmetic operation on two reals is larger than the maximum permitted real value, i.e., the exponent becomes too large, *floating-point overflow* will occur and the effect is generally catastrophic. If the result is smaller than the smallest representable real than a *floating-point underflow* occurs. Many systems set the result to zero in this case and do not bother to inform the user (see Exercise 1.20). Only in very exceptional circumstances is it likely to be important that underflow is detected, and it is usually possible for a program to continue execution if it occurs.

EXERCISE 1.4

(a) Write a short program to find out whether or not integer overflow is detected on your system. Does changing compiler and/or optimization level have any effect?

(b) How may overflow be detected in the result of the addition or subtraction of 2 two's complement integers?

(c) Using two's complement integers what operands would cause an overflow as the result of a division?

EXERCISE 1.5   IEEE single-precision floating-point format is defined by $\pm(1.m) \times 2^{e-127}$ where m is a 23-bit mantissa and e is an eight-bit positive integer. Because all real numbers are represented in normalized form, the first bit is always one and is not therefore stored; this is sometimes known as the *hidden bit*. The form of the exponent, $e - 127$, is termed a *biased exponent*.

(a) What are the largest and smallest floating-point numbers which may be represented? Give your answer in base 2 and base 10.

(b) What is the *machine epsilon*, i.e., the smallest positive floating-point number, ϵ, such that $1 + \epsilon > 1$?

(c) Show that taking the reciprocal of a floating-point number in this format cannot cause an overflow.

EXERCISE 1.6 Implement the recurrence relation for computing (1.1) by

(a) choosing a value $M > N$,

(b) setting $I_M = 0$,

(c) computing $I_{n-1} = (1 - I_n)/n$, for $n = M, M - 1, \ldots, 1$.

Compare I_0 obtained using this *backward recurrence relation* with the true solution. Rework the 'error analysis' given in the text for this case. For a given value of N, how should M be chosen to ensure that the set of values $\{ I_n \mid n = 0, 1, \ldots, N \}$ are obtained to a predefined tolerance?

EXERCISE 1.7 What is the effect of running the following programs on a base 2 machine?

(a)
```
        PROGRAM TEST
        DOUBLE PRECISION H,ONE,PTONE,X,ZERO
        PARAMETER (ONE = 1.0D0,ZERO = 0.0D0,PTONE = 0.1D0)
        X = ZERO
        H = PTONE
10      IF (X.EQ.ONE) THEN
           WRITE(*,100)
100        FORMAT(1X,'FINISHED')
        ELSE
           X = X+H
           GOTO 10
        ENDIF
        END
```

(b)
```
        PROGRAM TEST2
        DOUBLE PRECISION ONE,PTONE,TEN,X
        PARAMETER (TEN = 10.0D0,PTONE = 0.1D0,ONE = 1.0D0)
        X = PTONE
        WRITE(*,101)X,TEN*X
101     FORMAT(1X, 'X= ', D32.24, ' 10*X = ',D32.24)
        WRITE(*,102)ONE-TEN*X
102     FORMAT(1X,'1.0-10.0*X = ',D32.24)
        END
```

1.4 Fortran 77

We assume that the reader is conversant with the rudiments of *Fortran 77*, the basic control structure, the use of arrays in at least one dimension and the mechanism for calling subroutines and functions.

Fortran is not a purist's programming language. Its roots lie in the late 1950's and much has changed in computing since then. However there are several reasons why it is still an appropriate language in which to implement a scientific subroutine library:

- An international standard ANSI [4] was defined for Fortran 66. This was superseded by a standard for Fortran 77, ANSI [5]. All compilers which claim to implement Fortran 77 should contain the standard definition as a subset. (The Fortran 77 standard also defines a subset which is very similar to Fortran 66.) Unlike Ada, the definition of Fortran 77 does not forbid enhancements to the language, although the standard does advise that compiler writers should provide some means of identifying non-standard syntax. Provided that code adheres to the standard it should compile and have a similar effect on any Fortran supporting system (subject to limits imposed by the use of different precisions). This portability is an important ingredient in the success of a large subroutine library such as NAG; to make the library widely available it must be possible to implement it reasonably easily on a wide variety of different hardware configurations. The use of a standard conforming language is a stride in this direction.

- Fortran, which stands for FORmula TRANslation, was specifically designed for scientific (numerical) computation. Unlike many languages (e.g., Pascal, Basic, etc.) it has a rich set of built-in mathematical functions and a good range of numerical types (`INTEGER`, `REAL`, `DOUBLE PRECISION`, `COMPLEX`).

- There is an enormous investment in Fortran software which it is not financially viable to recode in another language. Upgrades to Fortran attempt to keep backwards compatibility. Plans are well advanced for the replacement of Fortran 77 by a new language, Fortran 8X (see Schonfelder [70] for a foretaste), which although extending the current language considerably, will contain almost all features of Fortran 77. Thus while new programs may be developed, and old ones modified using enhancements to the language, old style code will still compile, and can be freely intermixed with the new.

All code presented in this book is written in Fortran 77. Although prior to Mark 12 the NAG library was implemented entirely in Fortran 66, from Mark 12 onwards new routines and upgrades to existing codes will assume the availability of a Fortran 77 compiler. Also at Mark 12 all example programs in the main library manuals are presented in Fortran 77 (for more details see Section 1.8).

One of the major criticisms levelled at Fortran is that the language is a breeding ground for bad programming habits. It is true that Fortran allows a lot of laxity; however a good programmer can produce readable and efficient code by proper use of the control structures and type statements available, and by a refusal to use the more dangerous features. (For a good guide on how best to use Fortran 77 see Balfour and Marwick [7].) The question of Fortran style is a subject which invariably provokes long and vociferous arguments among programmers. We certainly want to avoid being dogmatic; it is to be hoped that the majority of points raised in the following sections are now included in the basic tenets of good programming practice.

1.5   Variable typing in Fortran

Like its predecessor, Fortran 77 allows default typing for integer and real variable names; unless otherwise indicated, names whose first character is I–N are taken to be integers, and those whose first letter is A–H or O–Z are taken to be reals. Any variable of type DOUBLE PRECISION, COMPLEX, LOGICAL or CHARACTER must be *explicitly* typed via a type declaration statement. The absence of the enforced explicit typing of all variable names leads to a number of very common programming errors:

- A name that should have been declared as, say DOUBLE PRECISION, but does not appear in a type declaration gets a default typing of either real or integer. At best this may cause loss of precision somewhere in the program; at worst it may result in totally incorrect results being generated.

- A misspelt variable name assumes a default type and, very often, a default initial value. Hence statements like

      X1 = XI + 1

  where XI is a misspelt X1 will compile and, using many compilers, generate executable code and incorrect results on running.

Some compilers can be persuaded to check that all variable names have been explicitly typed, and to refuse to compile source which uses untyped names. An example is the −u flag to the *Bell f77 compiler* under UNIX. Other systems will check, at run time, for the use of unset variables on the right-hand sides of assignment statements. This latter checking is very useful, and is often successful at picking up typographical errors like the example above. As a final resort such errors may be detected by looking at the cross reference tables optionally generated by most compilers. It is good programming practice explicitly to type all variables used within a program unit; there can then be no doubt as to what was intended.

Some NAG routines require the user to pass a function or subroutine name as a parameter, for example, to define a function whose root is sought. Since Fortran allows each program unit (subroutine, function or main program) to be compiled separately, it is necessary to inform the compiler that a particular name is a subprogram name whose definition will be provided later, and not a simple variable of the default type. This declaration of function names is performed using an EXTERNAL statement. For an external function name we also explicitly declare the type of the function name (e.g., REAL, DOUBLE PRECISION, etc.).

An example of a skeleton for a numerical integration routine is given in Code 1.1. There is no need to declare F in QUAD via an EXTERNAL statement since it is not defined to be an array, and the compiler will assume it is an external function reference. (Does your compiler complain, at compile time, of an unknown function reference when you forget to declare an array?) However, in line with our previous discussion of explicit typing, we recommend that all external subprogram (subroutine or function) names be 'typed' using an EXTERNAL statement. Thus, we should replace the EXTERNAL statement in the main program in Code 1.1 by

```
EXTERNAL QUAD,INTERN
```

and include

```
EXTERNAL F
```

in the subroutine QUAD.

```
      PROGRAM SKELTN
      DOUBLE PRECISION A,ANS,B,INTERN
      EXTERNAL INTERN
*
*.. INTERN IS A USER SUPPLIED FUNCTION DEFINING THE
*.. INTEGRAND OF THE FORM
*..      DOUBLE PRECISION FUNCTION INTERN(X)
*..      DOUBLE PRECISION X
*..      INTERN =
*..      END
*.. GIVEN A VALUE X IN THE RANGE OF INTEGRATION THIS
*.. FUNCTION RETURNS THE VALUE OF THE INTEGRAND
          .
          .
          .
      CALL QUAD(INTERN,A,B,ANS)
          .
          .
          .
```

```
      END
      SUBROUTINE QUAD(F,START,FINISH,RESULT)
      DOUBLE PRECISION F,FINISH,FVAL,RESULT,START,X
*
*.. F NEEDS TO BE TYPED EXPLICITLY OTHERWISE IT WOULD
*.. DEFAULT TO REAL.  IT DOES NOT NEED TO APPEAR
*.. IN AN EXTERNAL STATEMENT HERE UNLESS IT IS PASSED
*.. AS A PARAMETER TO ANOTHER SUBPROGRAM WITHIN QUAD
        .
        .
        .
*
*.. EVALUATE THE INTEGRAND AT X
      FVAL = F(X)
        .
        .
        .
      END
```

Code 1.1   Example of use of EXTERNAL statement.

Variable names which are typed but not used within a program unit do not have any effect, except possibly to confuse a reader of the source code; some helpful compilers may flag such variables. On the other hand routines whose names appear in EXTERNAL statements but are otherwise unreferenced in the program may cause problems. In addition, some systems consider a program to be incomplete if a non-existent subprogram is referenced, either implicitly or explicitly, even when the missing routine is not actually called during execution of the program.

EXERCISE 1.8   Assuming that all variables take on default types and $N \neq 0$, what is the result of the following Fortran 77 code fragment?

```
      H = 1/N
      DO 10 I = 1,N
        A(I) = I*H
10    CONTINUE
```

EXERCISE 1.9   In the following program Mr Quick is experimenting with how to compute one third of H as accurately as possible. For which of the assignments does the ANSI Fortran 77 standard [5] guarantee a result to full double-precision accuracy? For which does your compiler generate full double-precision accuracy?

```
        PROGRAM QUICK
        DOUBLE PRECISION H,S,T,THIRD,U,V,W,X,Y,Z
        THIRD = 0.1/0.3
        ATHIRD = 0.1D0/0.3D0
        READ(*,100)H
100     FORMAT(D14.6)
        S = 0.1/0.3*H
        T = THIRD*H
        U = DBLE(0.1/0.3)*H
        V = 0.1D0/0.3*H
        W = 0.1/0.3D0*H
        X = DBLE(THIRD)*H
        Y = 0.1D0/0.3D0*H
        Z = ATHIRD*H
        END
```

1.6   Arrays, workspace and paging

It is unfortunate that in order to understand precisely how arrays are passed as parameters, and how the elements of arrays may be accessed as efficiently as possible, the programmer has to be aware of how information is stored and accessed at a lower level.

The Fortran standard defines exactly how the elements of an array are ordered. Basically, this is a mapping between the user defined, possibly multidimensional, object and Fortran's one-dimensional internal memory view. We confine our discussion to two-dimensional arrays although the generalization to higher dimensions is straightforward. Two-dimensional arrays may be thought of as being stored by columns in the linear algebra sense. Thus an array declared as $A(6,2)$, possessing 6 rows and 2 columns, has its elements stored in consecutive locations in memory in the order

$$
\begin{array}{ccccc}
1 & 2 & 6 & 7 & 12 \\
A(1,1) & A(2,1) & \cdots \quad A(6,1) & A(1,2) & \cdots \quad A(6,2)
\end{array}
$$

Thus element $A(I,J)$ is in position $(J-1) \times 6 + I$ in the one-dimensional internal vector. When passing a two-dimensional array as a parameter to a subprogram it is essential to pass the leading dimension, as declared in the calling routine. This extra information is needed to compute the offsets in a manner consistent with that used in the calling routine. For example, in Code 1.2 the array AM is defined to have 10 rows and columns in the main program. If we wish to zero the top left 3×3 portion of this array in MATING we must pass AM itself, its leading dimension (NDIM), and the size of the portion to be zeroed (N). The corresponding array in MATING is defined to have NDIM rows and only N columns, but this is perfectly acceptable. However, if we tried to slice off the first three rows and columns of AM by declaring A as A(N,N), we would get erroneous results (see Exercise 1.11).

```
      PROGRAM MAIN
      INTEGER NDIM,N
      PARAMETER (NDIM = 10)
      EXTERNAL MATING
      DOUBLE PRECISION AM(NDIM,NDIM)
*
*.. ZERO ELEMENTS OF 3x3 SUBMATRIX
*.. NOTE AM (1,1), AM (1,2) AND AM (1,3) ARE IN ELEMENTS
*.. 1, NDIM+1 AND 2*NDIM+1 OF THE ONE-DIMENSIONAL VECTOR
      N = 3
      CALL MATING(AM,NDIM,N)
      .
      .
      .
      END

      SUBROUTINE MATING(A,IA,N)
      INTEGER I,IA,J,N
      DOUBLE PRECISION A(IA,N),ZERO
      PARAMETER (ZERO = 0.0D0)
      DO 20 I = 1,N
        DO 10 J = 1,N
          A(I,J) = ZERO
10       CONTINUE
20      CONTINUE
      END
```

Code 1.2   Passing a two-dimensional array as a parameter.

It should be recognized that in a Fortran function or subroutine call, each parameter is passed in a *call by name* fashion. Any assignment to a parameter within the subprogram automatically changes its value in the calling segment. When an array is passed, we effectively simply indicate where, in memory, the array starts (that is, we pass the address of the first element). In consequence, the 'shape' of the array can be changed within the subprogram. Passing across the leading dimension as a parameter, and using it in the internal declaration of the array, ensures compatibility with the calling sequence.

Standard Fortran makes no provision for dynamic storage allocation. All type statements must appear before any executable statements, and hence, for example, we cannot write

```
PROGRAM MAIN
READ(5,'(I3)')N
DOUBLE PRECISION A(N,N)
```

and thereby reserve just the amount of array space required by the problem in hand. Fortran uses static storage allocation, that is, the total amount of storage required for all the program's variables must be known at compile time. To make a program 'size independent' the usual trick is to use the main program principally for the declaration of suitably large arrays, and then pass these, along with the problem size, to subprograms where the arrays are 'reduced' as required. This applies not only to the arrays which are necessary to define the problem and its solution, but also to any auxiliary arrays which are needed for temporary storage by the algorithm. For example, the numerical solution of a general system of linear equations may require, amongst other things, an integer array of length n, where n is the size of the system (see Exercise 4.8). We could declare the array in the main program using

```
INTEGER MAXEQS
PARAMETER (MAXEQS=100)
INTEGER IPVT(MAXEQS)
```

This would restrict the maximum size of problem solvable to 100, and would require the main program to be edited and recompiled for this to be changed. The subprogram would then look like

```
SUBROUTINE LINSYS(IPVT,N,...)
INTEGER N,IPVT(N)
```

and this would be adequate for *any* size problem, requiring compilation once only. The alternative, to define IPVT locally only in LINSYS using

```
SUBROUTINE LINSYS(<problem definition parameters only>)
INTEGER MAXEQS
PARAMETER (MAXEQS=100)
INTEGER IPVT(MAXEQS)
```

would require the subroutine itself to be recompiled, should a problem of size greater than 100 be encountered. We pay for generality by an increase in the number of parameters in the argument list, and some routines in the NAG library appear quite menacing until it is realized just how many of the parameters are associated with providing workspace arrays.

Finally we consider the efficient access of array elements. If a particular computer uses fixed memory allocation then all array accesses take the same time. The address of the required element is computed and the value is retrieved by accessing that address in main memory. With a machine which supports *virtual memory management*, the data and code of an executing program are divided into *pages* and, generally, only a subset of these pages is resident in memory at any one time, the rest being stored on disc. When any attempt is made to access code or data not in memory, a resident page is written out to disc, and the required page is transferred from disc to memory. This process is known as *swapping*. The transfer of pages of information from disc to memory is very expensive compared to the cost of accessing information directly in memory. Programs which perform an abnormally large number of transfers (*page turns*) are said to *thrash*. Thus, for efficiency reasons on virtual machines we would like to minimize the number of page turns.

To illustrate both the problem and its solution, we consider a very naive paging system, in which each page can hold 100 real numbers and only three pages of data may be resident in memory at any one time. Assume that we wish to implement a matrix/vector multiply for a matrix of order 100, that is, we wish to compute $b = Ax$ where A and x are given. If the vectors b and x each occupy a page and are not swapped out of memory, then the last page of data will contain the array element which is currently being accessed. Being stored by columns, the first column of A will be on one page, the second on another, and so on.

The code fragment

```
      DO 20 I = 1,100
        SUM = 0.0D0
        DO 10 J = 1,100
          SUM = SUM+A(I,J)*X(J)
10        CONTINUE
        B(I) = SUM
20      CONTINUE
```

would thus cause a page turn for each access of an element of the array. This causes 10000 page turns. A simple solution is to reverse the order of the two loops to give

```
      DO 10 J = 1,100
        B(I) = 0.0D0
10      CONTINUE
      DO 30 J = 1,100
        DO 20 I = 1,100
          B(I) = B(I)+A(I,J)*X(J)
20        CONTINUE
30      CONTINUE
```

which, although unnatural, requires only 100 page turns. The rule is to try and arrange the computation so that elements of a two-dimensional array are accessed by columns rather than by rows.

The figures given in the above example are *not* realistic; they serve only to illustrate the underlying principle. Where possible, NAG routines which work with two-dimensional arrays use algorithms which have been especially designed to reduce page turns. Optimizations of this type are very important when using certain types of vectorizing (pipelined) machines like Crays. For more details see Metcalf [58].

EXERCISE 1.10 What is the output from the following code? See if you can work out what happens without running it.

```
      PROGRAM TEST
      INTEGER I,J
      INTEGER A(6,2)
      EXTERNAL JUNK
      DO 20 J = 1,2
        DO 10 I = 1,6
          A(I,J) = (J-1)*6+I
10      CONTINUE
20    CONTINUE
      DO 30 I = 1,6
        WRITE(6,100) (A(I,J),J = 1,2)
30    CONTINUE
100   FORMAT(1X,2I3)
      CALL JUNK(A)
      END

      SUBROUTINE JUNK(A)
      INTEGER I,J
      INTEGER A(4,3)
      DO 10 I = 1,4
        WRITE(6,100) (A(I,J),J = 1,3)
10    CONTINUE
100   FORMAT(1X,3I3)
      END
```

Run the program to check your answer. Although legal Fortran 77, such programming tricks should be avoided.

EXERCISE 1.11 What is wrong with the following code?

```
      PROGRAM TRASH
      INTEGER I,J,N,NDIM
      PARAMETER (NDIM = 10)
      INTEGER A(NDIM,NDIM)
      EXTERNAL JUNK
*
*.. FILL IN TOP 3x3 CORNER
      N = 3
      DO 20 I = 1,N
        DO 10 J = 1,N
```

```
         A(I,J) = (J-1)*NDIM+I
10       CONTINUE
20     CONTINUE
       CALL JUNK(A,N,N)
       END

       SUBROUTINE JUNK(A,IA,N)
       INTEGER I,IA,J,N
       INTEGER A(IA,N)
       DO 10 J = 1,N
          WRITE(6,100)(A(I,J),I = 1,N)
10       CONTINUE
100    FORMAT (1X,3I3)
       END
```

Does your compiler allow this sort of rubbish to run? If so complain to somebody!

EXERCISE 1.12 Compare the number of array accesses needed by the code fragments used to compute the matrix/vector multiply above. How can this number be reduced in the second case?

EXERCISE 1.13 If you have access to a machine which uses virtual memory try to quantify the savings of the loop reversal in the matrix/vector multiply given above. Plot graphs of N vs. execution time for each implementation. Can you find out what the page size is on your machine? Hint: Look in the X02 section of the NAG machine dependent documentation.

EXERCISE 1.14 What happens when you attempt to execute the following code?

```
       PROGRAM SILLY
       INTEGER N
       DOUBLE PRECISION A(10)
       N = 0
       CALL TRIPE(A,N)
       END

       SUBROUTINE TRIPE(A,N)
       INTEGER N
       DOUBLE PRECISION A(N)
       END
```

1.7   Introduction to the NAG library

The NAG library is a collection of user callable functions and subroutines, each of which implements a specific numerical task, for example, computing the approximate solution to a system of linear equations, finding the roots of a polynomial, etc. It exists in a number of different forms:

- The main Fortran library — the most widely distributed version of the NAG library. This consists of some 688 user callable routines. It has been implemented on a wide range of machines from scientific workstations to large mainframes.

- The *Workstation* library consists of just over 150 of the most commonly used routines from the main library. It is specifically designed for users of machines at the top end of the personal computer range, and scientific workstations.

- The *Fortran PC50* library contains the 50 most frequently used routines and is aimed at improving the scientific problem solving capabilities of the personal computer.

- Versions of the library in programming languages other than Fortran

  - *Algol 60* — kept pace with the main Fortran library during the early stages of the NAG project. It has remained largely unchanged since 1980,

  - *Algol 68* — has many of the capabilities of the main Fortran library. It also contains a number of extra facilities which exploit some of the 'nice' facilities of the language, e.g., a multiple-precision arithmetic package,

  - *Pascal* — another subset library which like the Fortran PC50 library above is targetted at the personal computer user.

Although this book is primarily aimed at users who have access to Fortran versions of the library, the underlying numerical methods and their practical implementation should be of interest to all NAG users. Note that it is frequently possible to call a NAG Fortran routine from a program written in another language. However, almost every other language stores arrays by rows, not by columns, and this can give rise to severe difficulties.

The main Fortran library and its subsets are available on a wide range of different machine/compiler combinations using either single- or double-precision arithmetic, depending on the word length of the particular machine. Each version of the library, known as an *implementation*, is tested separately to ensure that the routines are performing correctly in that particular environment. Use of NAG routines is an advantage when programs are to be transported from one machine to another, since the tailoring of the library to each individual environment reduces the amount of code which requires attention.

The care taken over individual implementations reinforces the NAG philosophy of reliable and robust code at the expense of optimal efficiency. This means that the software will

- correctly solve all problems within the solution domain of the algorithm;

- reject, by signalling an error condition (see Section 1.9), all problems that are not within its solution domain;

- return, where possible, either a good upper bound of the error in the final result, or guarantee a solution to within a user supplied tolerance, or both.

The extra time spent in checking the input data, computing error bounds, etc., is repaid by the increased confidence that the user is able to place in the final results.

General purpose algorithms are used to reduce the number of routines in the library. This is important since the library already contains over 200,000 lines of Fortran. This also leads to an overhead; specific problems may invariably be solved more efficiently by using purpose built code rather than a general purpose routine. However, only in very exceptional circumstances will the gains in computational efficiency justify the effort required to design, code, and debug such specialized programs. When recognition of a large, commonly occurring, subclass of a problem domain (e.g., polynomials with real, as opposed to complex, coefficients) leads to a marked gain in efficiency by reducing storage and/or execution time, specialist routines have been included in the library. It is up to users to identify their problems as belonging to such a subclass; the more general routine would also solve the problem — it would just cost more.

The library is dynamic; as improved algorithms appear old routines are replaced. A replacement routine may

- solve a wider class of problems,

- provide a more efficient solution than the existing routine,

- provide a more accurate error estimate to the computed solution, etc.

New releases of the NAG library are known as *marks*; these releases allow

- improvements to be made to existing routines, such as bug fixes,

- new routines to be introduced into the library,

- obsolete routines to be removed from the library.

A new routine may either extend the facilities (problem solving capabilities) of the library in existing, or additional, problem areas, or it may be destined to replace an existing routine. In the latter case both the routine to be replaced and its replacement will coexist for at least one mark of the library to enable users to change their programs. Adequate warning is always given of the intention to remove library routines.

1.8   Routine names and documentation

The NAG routines are arranged in a three-level tree structure:

- Chapter level — the topmost level, used to divide the library into general areas of numerical analysis or statistics. Each chapter has a three letter name and a title, for example,

    H01 — operations research,

    F04 — simultaneous linear equations.

- User callable routine level — within each chapter. This contains a number of routines each of which performs a specific numerical, or statistical task. Full details of how to call these routines may be found in the documentation. Examples are

    F04AHF — solves a general linear system with a real coefficient matrix,

    F04AKF — solves a general linear system with a complex coefficient matrix.

- Implementation routine level — the lowest level. Routines at this level, which are not designed to be called by the user of the library, arise either from a desire to split user callable routines into manageable pieces, or because several user callable routines can be made to share the same code.

The NAG routine naming convention may appear a little daunting at first sight. (It is not mnemonic, but is based on the Modified SHARE Classification; a list of the main classifications can be found at the head of the Index to Algorithms in [3].) All NAG routine names consist of six characters, the maximum identifier length allowed by Fortran. The first three characters determine the chapter to which the routine belongs; the leading character is a letter and the next two are *always* digits. Note, therefore, that the round symbol in the second and occasionally the third, position is always a zero and is *never* the letter 'Oh'. The next two characters, both letters, are used to label the routines within each chapter. The final character is used to specify the language the routine is implemented in according to the scheme

    F — Fortran

    A — Algol 60

    B — Algol 68

    C — Pascal.

Some installations may support both the single- and double-precision versions of the library on the same system, and the two versions are normally distinguished by changing the last letter of each routine name. Users at such sites will need to ascertain what the letter for their preferred precision is. Throughout this book we will assume that it is F.

The available documentation for the full subroutine library falls into a number of categories:

- Encyclopaedic — the main library manuals. These consist of 6 large volumes which contain all the information required to use each user callable routine in the library. This includes a full description of all the arguments, a description of the error conditions that may arise, references for more information on the underlying algorithms, and an example program along with data and results. Chapter introductions give a general overview of the particular problem area, references for further reading, a synopsis of the routines available, and recommendations on which routine should be used in specific circumstances. In many cases a flow diagram is provided to assist the user to find the routine best suited to his problem. If you do not know the answers to questions in the flow diagram answer 'no', and never admit to being anything but an inexperienced user! This is the bible for the NAG library; all other documentation, including this book, are meant to complement, and not replace, these manuals.

- Condensed edition — the *mini manual*, meant as an *aide-mémoire*, it contains the chapter introductions and the chapter contents from the main library manuals.

- Detective — the on-line interactive *help system*. This contains a subset of the routine documentation, including a full description of the parameters and meanings of the error flags, for all user callable routines. It also contains a description of the subject areas covered by the library, and gives advice on choice of routine. Not all installations have this system available.

- Specialized — the implementation dependent documentation. This contains all information which is dependent on the particular machine/compiler/precision combination used to implement the library. For example, it lists the values of machine dependent arithmetic constants whose values are returned by functions from the X02 chapter (see Exercise 1.20), and valid ranges for the arguments of the special functions (chapter S) (q.v. Section 3.8). Also included are

  - the interpretation of a number of terms which appear in italics in routine descriptions,
  - how the published example programs need to be modified to run under the user's system. The modified programs may even be

available to users in machine readable form, and these can often make a good starting point for one-off programs,

- the differences, if any, which may be expected when running the published example programs,

- in exceptional circumstances, the names of routines which are documented but not available.

This information is provided in a site dependent way and is often included in the on-line supplement; try the keyword LOCAL.

NAG users who do not possess a reasonably strong mathematical background often experience difficulties both in the selection and efficient use of library routines. This may be due either to unfamiliarity with the mathematical problem descriptions given in the library manuals, which often appear to such users to be written in a foreign language, or to the inability of users to identify the particular parts of their problem which may be solved using library routines. Even when the problem is well-defined mathematically, it may still be difficult to choose the most effective routine for the job in hand; this is where the chapter introduction and, we hope, this volume may prove helpful.

1.9   The error handler

Only in very exceptional circumstances will a NAG routine fail to terminate in a 'tidy' manner. (For example, a user defined function, provided as an argument to a NAG routine, may attempt to compute the square root of a negative number.) Nevertheless, this does not necessarily mean that a routine will invariably determine a solution.

A NAG routine may 'fail' for a number of reasons:

- A programming or data error produces an invalid input parameter, e.g., a negative error tolerance, an incorrect array dimension, etc.

- The problem to be solved is outside the problem domain of the routine. For example, an input vector whose elements are expected to be in ascending order is found not to be in this form.

- The routine cannot solve the problem to the requested/required accuracy.

A cursory glance at the NAG routine documents reveals that almost all argument lists contain a parameter IFAIL. When present, IFAIL is usually the last parameter, although there are curious exceptions, e.g., E01ACF. This argument has a dual role:

- it allows the calling routine to decide what action is to be taken in the event of a failure within a NAG routine, and

- it allows the called sequence to communicate any failure to the calling routine.

An error is ultimately communicated to the user as a positive integer value; an explanation of the failure values for each routine may be found in Section 6 of the routine documentation. Only the first error detected will be flagged; correction of this will not necessarily result in a subsequent successful run.

The user may choose which of two courses of action is to be taken if an error is detected:

- The *hard error option*, selected by setting the value of IFAIL to zero on entry to the routine. Any error detected within the routine causes the program to terminate, and an error message of the form

```
** ABNORMAL EXIT from NAG Library routine C05NAF : IFAIL = 1

** NAG hard failure:   execution terminated
```

  is sent to the currently selected error message channel (see below). This is the more common mode of operation since it is not possible to recover from many of the error conditions.

- The *soft error option*, selected by giving IFAIL the value 1 or −1 on entry. On detection of an error this causes IFAIL to be set to the relevant error number and control is returned to the calling unit. If the initial setting of IFAIL is −1 (*noisy exit*) an error message of the form

```
** ABNORMAL EXIT from NAG Library routine C05NAF : IFAIL = 1

** NAG soft failure:   control returned
```

  is output. With an initial value of 1 (*silent exit*) no error message is generated. On successful exit IFAIL has a return value of zero. Note that if more than one call is made to a NAG routine, selection of the soft error option means resetting IFAIL before each call. This mode of error handling is useful for interactive applications, and in some special circumstances where it may still be possible to proceed, for example, if a numerical integration routine has returned a slightly less accurate result than requested. When using the soft error option it is imperative to test the value of IFAIL on return from a routine and ensure that the correct action is taken should a failure occur.

Note that since IFAIL will be assigned a value inside the routine, this parameter should always be a variable and not a constant.

By default all error messages are sent to an installation dependent channel number, usually connected to a terminal for interactive work or to a printer in batch mode. The routine X04AAF may be used both to find the current value of this channel number and to reset it. A call takes the form

    CALL X04AAF(IFLAG,NERR)

If IFLAG is set to zero on entry then on exit NERR contains the current setting of the channel number; if IFLAG is set to one on entry then the error channel number is reset to NERR.

Some routines (for example, F04QAF) produce advisory information on the progress of the algorithm. The amount of output generated may be controlled by other parameters to such routines. As with error messages, this output is produced on a site dependent default advisory channel. The value of this channel number may be manipulated by X04ABF, which is used in exactly the same way as X04AAF above.

The default channel numbers for the advisory and error messages may also be found in the implementation dependent documentation (see Section 1.8).

EXERCISE 1.15   Use X04AAF and X04ABF to discover the default error and advisory channel numbers. Are they the same as those given in the implementation dependent documentation? Use X04AAF to divert the error messages from a NAG routine into a user defined file.

EXERCISE 1.16   Design a user interface to a library routine to perform a matrix–matrix product. What error conditions would you check for?

1.10   Calling a NAG routine

The first thing a potential user should do is to read the appropriate NAG chapter introduction. Using this he should attempt to classify his problem and identify which routine (if any) most meets his requirements. He should then thoroughly read the routine documentation to ascertain how the input data required by the routine is related to the problem definition. A study of the example problem and its associated program (Section 13 of the routine documentation) may help to clarify any difficulties with the meaning and/or use of the parameters.

Certain parameters may not form part of the original problem definition, for example, error tolerances, starting values for iterative methods, etc., and their setting may require some additional analysis. Poor choices for these may lead to extremely inefficient programs. Many users grossly overestimate the accuracy they require in their final results. It is almost invariably a waste of time to compute results to a greater precision than that inherent in the data defining the problem. When debugging, it is advisable often to set larger error tolerances than required on the production runs in order to reduce execution times. For programs which need to be run many times on different data sets, some experimentation may be necessary to obtain a balance between efficiency and accuracy.

It is important to realize that library routines are not infallible; the results of all numerical computations should be checked carefully. Several relatively simple checking procedures may be used to increase confidence in results:

- All error bounds returned by routines should be checked to ensure that they are compatible with those requested/expected.

- Routines which allow the user to set error tolerances should be run with at least two different values for these to ensure that the results reflect the change in the requested accuracy.

- The effects on the final results of *perturbations* (small relative changes) to the input data should be investigated. This may be done by randomly perturbing the input data at the level of known errors in the data, and rerunning the program. Substantial changes in the output data may indicate an ill-conditioned problem, see Section 1.15.

- For iterative procedures the program should be rerun with different starting values. This may cause the algorithm to converge to other 'solutions'. This type of experimentation is especially useful with, for example, function minimization routines since they may converge to local rather than global minima.

Finally we turn to the question of how the precompiled NAG routines are actually combined with a user's program to form an executable module. This procedure will differ from one system to another. Some systems automatically search for missing routines in a standard library; others need to be told where to look. For example, under bsd UNIX using the Bell Laboratories *f77* compiler on a VAX, we need to specify a particular object code library. Even this specification may change from one installation to another. At Kent,

```
f77 myprog.f -lnag
```

will produce an executable file containing all the NAG library routines called by myprog.f and any dependencies (NAG routines called by NAG routines). You should consult your local system documentation to discover how to access the library.

1.11   Passing parameters to NAG routines

All implementations of the NAG library are derived from a small number of base versions, depending on the underlying floating-point arithmetic supported by the target hardware (q.v. Section 1.3). The choice of base version determines whether the basic floating-point type used by the library is real or double-precision. Generally, if the real arithmetic has a mantissa length of less than about 30 bits, then a double-precision version of the library is produced. In fact the double-precision (`REAL*8`) version is by far the most common, since architectures which mimic IBM floating-point hardware and the new IEEE floating-point standard all define the mantissa of real variables to be around 24 bits. This accuracy is not sufficient for reliable numerical computation. Single-precision versions of the NAG library are thus reserved for machines with their own individual floating-point formats (e.g., Cray and CDC).

It is important to ascertain the precision of your implementation of the library since this will dictate whether you write your calling programs using real or double-precision variables. This information is contained in the implementation dependent documentation (q.v. Section 1.8). If you are a 'good' programmer the library should use the same precision as you normally work with!

Some NAG routines destroy their input data during execution, for example, the polynomial root finders destroy the original polynomial coefficients whilst computing the zeros. It may therefore be necessary to make a copy of this information prior to the routine call. The routine documentation indicates whether or not input parameters are changed by a routine and, if so, what values they have on exit. As a general rule it is not advisable to use constants as parameters. For example,

```
CALL NAGSUB(A,B,10,12,IFAIL)
```

would be better written as

```
M = 10
N = 12
CALL NAGSUB(A,B,M,N,IFAIL)
```

Indeed it is illegal to use a constant or an expression as an actual argument if the associated dummy parameter is assigned to during the execution of the subroutine. (What would you expect to happen if you did?) Since it is virtually impossible for the run-time system to detect such errors the subroutine may appear to execute correctly; but the program may fail at some later stage or produce results which are completely erroneous. In such cases the error is often extremely difficult to detect.

It is sometimes possible to save storage by allowing the same actual argument to be used as an input (read only) parameter, and as an output (write only) parameter. For example, when solving a linear system of the form $Ax = b$, some NAG routines allow the solution vector, x, to be written into the same array as was used to store the vector b, the given right-hand side. Although this is sometimes useful it is strictly against the

Fortran 77 standard (ANSI [5], Section 15.9.3.6) and this practice should
be avoided. In some circumstances such a doubling up of parameter use
may have disastrous results.

In what many people would describe as Fortran's usual anomalous
fashion, the standard only defines one complex data type which is equivalent to a pair of real numbers. Obviously, if real precision is not accurate
enough for real computations, complex precision will suffer from the same
inadequacies. To obtain double-precision accuracy it is necessary either to
use non-standard and therefore machine dependent extensions to the language, or to implement complex arithmetic using the double-precision type
provided.

In the first case a number of different extensions exist for declaring
double-precision complex variables, for example,

- COMPLEX*16 — IBM and IBM look-alike compilers,

- DOUBLE COMPLEX — UNIX Bell *f77* compiler.

There is also some diversity in the naming convention used for the associated extensions to the intrinsic functions (e.g., CMPLX, CABS, etc.), and the
definition of user defined external functions. This solution decreases the
portability of the library.

The alternative is to declare all complex variables as double-precision
arrays of length 2; the first element storing the real part, and the second
the imaginary part. This means that all simple variables of complex type
become arrays and all complex arrays have their dimensionality increased
by one. It also means that the programmer loses the convenience of complex arithmetic which has to be performed either explicitly or by using
subroutine calls.

As an illustration, assume C and D are double-precision complex variables, TEMP1 and TEMP2 are double-precision variables and CA and DA are
double-precision arrays of length two, then we have

- using double-precision complex arithmetic

```
C = C+D
C = C*D
```

- using double-precision arrays and explicit complex arithmetic

```
*
*.. COMPLEX ADD
    CA(1) = CA(1)+DA(1)
    CA(2) = CA(2)+DA(2)
*
*.. COMPLEX MULTIPLY
```

```
TEMP1 = CA(1)*DA(1)-CA(2)*DA(2)
CA(2) = CA(2)*DA(1)+CA(1)*DA(2)
CA(1) = TEMP1
```

- using double-precision arrays and subroutine calls

```
CALL CADD(CA(1),CA(2),DA(1),DA(2),TEMP1,TEMP2)
CA(1) = TEMP1
CA(2) = TEMP2

CALL CMULT(CA(1),CA(2),DA(1),DA(2),TEMP1,TEMP2)
CA(1) = TEMP1
CA(2) = TEMP2
```

The use of double-precision arrays is obviously likely to be error prone and, along with the use of subroutines as above, makes the code far less readable, especially if complicated expressions are involved.

The use of complex data within the NAG library is restricted to a relatively small number of routines primarily in the linear algebra chapter. The, possibly implementation dependent, method for passing complex parameters to these routines is described in the machine dependent documentation (see Section 1.8).

1.12   Example calls

We give here examples of user programs which call NAG routines. Readers unfamiliar with calling library routines are advised to study this section in detail, and to input and run at least one of the examples on their computer system. This exercise will require knowledge of how to

- input a program on your system; this usually involves the use of a screen or context editor,

- compile a Fortran 77 program,

- link your calling program with the precompiled subroutines in the NAG library,

- run the complete executable module.

The examples have been chosen to illustrate the use of

- a NAG routine called as a function,

- a NAG routine called as a subroutine,

- the failure mechanism,

- two-dimensional and workspace arrays,

- the interaction of NAG routines.

Several of the examples have been chosen from chapters of the library which are not covered in detail elsewhere in this book, but which contain routines a user may find useful for day-to-day computational problems. Note that throughout the book we assume that the user has access to the double-precision version of the NAG library. All code presented in the text uses generic function names. The changes required to run these programs with a single-precision library are

- replace all DOUBLE PRECISION type statements by REAL type statements,

- replace all double-precision constants (e.g., 0.0D0) by single-precision constants (e.g., 0.0E0).

The first two examples are from the complex arithmetic chapter, A02, and illustrate the use of NAG routines both as functions and subroutines. The function A02ABF returns the modulus of a complex number, that is, given a complex number of the form $x + iy$ the routine returns $\sqrt{x^2 + y^2}$. Although this appears to be trivial the NAG implementation attempts to guard against the possibilities of unnecessary overflow and loss of accuracy; for more details see Exercise 1.17. The routine has two double-precision parameters, X and Y, which define the real and imaginary part of the complex number whose modulus is required. The double-precision result is returned through the function name which must appear in a DOUBLE PRECISION type statement in the calling program unit. Code 1.3 gives a simple program which computes the modulus of $3.0 + 4.0i$.

```
      PROGRAM TEST1
*
*.. DEFINE THE OUTPUT CHANNEL
      INTEGER NOUT
      PARAMETER (NOUT = 6)
*
*.. WE NEED TO TYPE THE NAG FUNCTION
      DOUBLE PRECISION A02ABF
      EXTERNAL A02ABF
*
*.. THE REAL AND IMAGINARY PARTS OF THE COMPLEX NUMBER
*.. ARE TO BE STORED IN THE DOUBLE PRECISION NUMBERS X AND Y
*.. RES WILL CONTAIN THE MODULUS
      DOUBLE PRECISION RES,X,Y
```

```
*
*.. THE DOUBLE PRECISION COMPLEX DATA TYPE IS NOT
*.. PART OF THE ANSI 1977 STANDARD
      X = 3.0D0
      Y = 4.0D0
      RES = A02ABF(X,Y)
      WRITE(NOUT,100)X,Y,RES
100   FORMAT(1X,'MODULUS OF ',F5.2,' +I ',F5.2,
     +          ' = ', F5.2)
      END
```

Code 1.3   Example call of A02ABF.

Our second example shows the use of the subroutine A02ACF to compute the result of a complex division. Like A02ABF above, a pair of double-precision values is used to define each complex number. The six parameters to A02ACF consist of three pairs of variables defining the input dividend, the divisor, and the resulting quotient, in that order. The program given in Code 1.4 computes the result of $(4 + 3i)/(3 + 2i)$.

```
      PROGRAM TEST2
*
*.. DEFINE THE OUTPUT CHANNEL
      INTEGER NOUT
      PARAMETER (NOUT = 6)
      EXTERNAL A02ACF
*
*.. THE REAL AND IMAGINARY PARTS OF THE COMPLEX NUMBERS
*.. ARE TO BE STORED IN THE DOUBLE PRECISION VARIABLES
*.. V, W, X AND Y. THE RESULT GOES INTO A AND B.
      DOUBLE PRECISION A,B,V,W,X,Y
      V = 4.0D0
      W = 3.0D0
      X = 3.0D0
      Y = 2.0D0
      CALL A02ACF(V,W,X,Y,A,B)
      WRITE(NOUT,100)V,W,X,Y,A,B
100   FORMAT(1X,F5.2,' +I ',F5.2, '/',F5.2,' +I ',F5.2,
     +          ' = ', F5.2, ' +I ',F5.2)
      END
```

Code 1.4   Example call to A02ACF.

We next consider a slightly more advanced example which illustrates the use of the NAG error mechanism. M01CBF sorts a portion of an array of integers into either ascending (non-decreasing) or descending order. A call takes the form

```
CALL M01CBF(IV,M1,M2,ORDER,IFAIL)
```

IV is an integer array of length at least M2 whose elements M1 to M2 contain the integer values to be sorted. On successful exit these values are rearranged into sorted order. The parameter ORDER is of type CHARACTER*1 and is used to indicate whether ascending ('A' or 'a') or descending ('D' or 'd') order is required. The routine will fail only if one or more of the defining parameters M1, M2 or ORDER contains an illegal value, i.e., M1 < 1 or M2 < 1; M1 > M2; or ORDER not set to 'A', 'a', 'D' or 'd'. In Code 1.5 we illustrate the use of both the hard and soft error options. We reiterate that if the soft error option is used (IFAIL set to 1 or −1 on entry) it is imperative that the value of IFAIL is tested on exit to ensure that the routine has executed successfully.

```
      PROGRAM TEST3
      INTEGER MMAX,NOUT
      PARAMETER (MMAX = 100,NOUT = 6)
      INTEGER I,IFAIL,IV(MMAX),M1,M2
      CHARACTER*1 ORDER
      EXTERNAL M01CBF
*
*.. SET UP VALUES IN ASCENDING ORDER
      M2 = 10
      DO 10 I = 1,M2
        IV(I) = I
10    CONTINUE
*
*.. SORT THE FIRST M2 ELEMENTS INTO DESCENDING ORDER
*.. USE THE HARD ERROR OPTION
      M1 = 1
      ORDER = 'D'
      IFAIL = 0
      CALL M01CBF(IV,M1,M2,ORDER,IFAIL)
*
*.. PRINT RESULTS
      WRITE(NOUT,102)
      WRITE(NOUT,103)(IV(I),I=1,M2)
*
*.. ATTEMPT TO SORT ELEMENTS 5 TO 10 INTO ASCENDING
```

```
*.. ORDER BUT GET M1 AND M2 INTERCHANGED
*.. USE THE SOFT ERROR OPTION
      M1 = 10
      M2 = 5
      ORDER = 'A'
      IFAIL = 1
      CALL M01CBF(IV,M1,M2,ORDER,IFAIL)
*
*.. TEST IFAIL ON EXIT
      IF(IFAIL.EQ.1)THEN
         WRITE(NOUT,100)IFAIL,M1,M2
      ELSE IF(IFAIL.EQ.2)THEN
         WRITE(NOUT,101)IFAIL,ORDER
      ELSE
*
*.. IF WE GET HERE THE ROUTINE HAS TERMINATED SUCCESSFULLY
         WRITE(NOUT,102)
         WRITE(NOUT,103)(IV(I),I = 1,M2)
      ENDIF
*
100   FORMAT(1X,' ERROR ',I5,' IN CALL TO M01CBF : M1 = ',I4,
     +          ' M2 =',I4)
101   FORMAT(1X,' ERROR ',I5,' IN CALL TO M01CBF : ORDER = ',
     +          A1)
102   FORMAT(1X,' FINAL SORTED VALUES ARE')
103   FORMAT(1X,12I5)
      END
```

Code 1.5   Example call to `M01CBF`.

To illustrate the use of both two-dimensional and workspace arrays we turn to `F01AAF` which computes the inverse of a real matrix. That is, given a square matrix A with all elements real, the routine computes another matrix, A^{-1}, such that $A^{-1}A = I$ where I is the unit matrix. For example, if we consider the *Hilbert* matrix whose $(i,j)^{th}$ element is given by $1/(i+j-1)$, then for the third-order matrix we have

$$A = \begin{pmatrix} 1 & \frac{1}{2} & \frac{1}{3} \\ \frac{1}{2} & \frac{1}{3} & \frac{1}{4} \\ \frac{1}{3} & \frac{1}{4} & \frac{1}{5} \end{pmatrix} \quad \text{and} \quad A^{-1} = \begin{pmatrix} 9 & -36 & 30 \\ -36 & 192 & -180 \\ 30 & -180 & 180 \end{pmatrix}. \quad (1.2)$$

(Note that `F01AAF` should NEVER be used to solve systems of linear equations; see Chapter 4 for details of how such problems should be tackled.) A call to `F01AAF` takes the form

CALL F01AAF(A,IA,N,AINV,IAINV,WKSPCE,IFAIL)

The integer N defines the order of the matrix to be inverted. The input matrix is stored in the two-dimensional, double-precision array A whose leading dimension (the number of rows A was declared as having) is IA, which must be at least N. As an example, suppose A were declared to be of size (5,5). Then the matrix A in (1.2) would be set up as

$$
A = \begin{pmatrix}
1 & \frac{1}{2} & \frac{1}{3} & * & * \\
\frac{1}{2} & \frac{1}{3} & \frac{1}{4} & * & * \\
\frac{1}{3} & \frac{1}{4} & \frac{1}{5} & * & * \\
* & * & * & * & * \\
* & * & * & * & *
\end{pmatrix}
$$

where the elements marked as * are not assigned, and the values of 5 and 3 are given to IA and N respectively. On successful exit the two-dimensional array AINV will contain the inverse matrix. We need to provide the declared leading dimension of this array in IAINV, which must also be at least N. The vector WKSPCE, which must be declared with length N or more, is used as workspace. The user does not need to initialize the elements of this array and, on exit, the array contains no useful information. An example program showing how to invert the matrix A in (1.2) is given by Code 1.6.

```
      PROGRAM TEST4
      INTEGER MAXSZ,NOUT
*
*.. NOUT FOR THE OUTPUT CHANNEL
*.. MAXSZ IS THE MAXIMUM SIZE OF MATRIX INVERTIBLE
      DOUBLE PRECISION ONE
      PARAMETER (MAXSZ = 10,NOUT = 6,ONE = 1.0D0)
      DOUBLE PRECISION A(MAXSZ,MAXSZ),AINV(MAXSZ,MAXSZ),
     +                 WKSPCE(MAXSZ)
      INTEGER I,IFAIL,J,N
      EXTERNAL F01AAF
*
*.. SET UP THE N TH ORDER HILBERT MATRIX
      N = 3
      DO 20 I = 1,N
         DO 10 J = 1,N
            A(I,J) = ONE/(I+J-ONE)
10       CONTINUE
20    CONTINUE
```

```
*
*.. INVERT THE MATRIX A
    IFAIL = 0
    CALL F01AAF(A,MAXSZ,N,AINV,MAXSZ,WKSPCE,IFAIL)
    DO 30 I = 1,N
      WRITE(NOUT,'(1X,3E12.4)')(AINV(I,J),J = 1,N)
30  CONTINUE
    END
```

Code 1.6   Example call to `F01AAF`.

For the final example we illustrate how a pair of NAG routines may work together by considering the generation of *pseudo-random numbers*. When writing simulation programs it is often convenient, for debugging purposes, to be able to use the same sequence of random numbers with each run of the program. However, for production runs we obviously want to be able to generate different sequences for each run. The NAG random number generator provides both these facilities by using a combination of two routines. The first sets the random number generator into either a repeatable or non-repeatable state, and must be called before the second which actually returns a random number each time it is called. The double-precision function `G05CAF` returns a random number from a uniform distribution between zero and one. A call takes the form

```
    RNDNO = G05CAF(DUMMY)
```

where `DUMMY` is a double-precision variable whose input value is ignored. This parameter was required by the Fortran 66 standard which insisted that external functions had at least one argument; in Fortran 77 it is possible to have a function with no arguments although the brackets are still required. The generator may be set to a repeatable state by preceding the first call of `G05CAF` with a call to `G05CBF` which takes a single integer parameter allowing different sequences to be generated. For a given value of this parameter the generated sequences are not guaranteed to be the same across different implementations of the library. For a sequence which differs on each program run, we precede our first call of `G05CAF` with a call to the parameterless subroutine `G05CCF`, which initializes the sequence of random numbers using the real time clock. It is therefore highly unlikely that two runs of the program will result in the same sequence of random numbers being generated. Code 1.7 illustrates the use of `G05CAF` and `G05CBF`.

We shall present a number of further examples of calls to NAG routines in subsequent sections of this book. Where this is given, as here, in terms of a main driving unit the types of all identifiers will be apparent from the code itself. Elsewhere we shall be content to give just an example of the call itself. Unless specified explicitly variables will conform to the convention

that they are of type integer if the name begins with a letter in the range
I–N, and real/double precision (as appropriate) otherwise.

```
      PROGRAM TEST5
*
*.. NOUT IS THE OUTPUT CHANNEL
      INTEGER NOUT
      PARAMETER (NOUT = 6)
      DOUBLE PRECISION G05CAF,DUMMY
      INTEGER I,J
      EXTERNAL G05CAF
*
*.. FOR A NON-REPEATABLE STATE REPLACE THE NEXT THREE
*.. STATEMENTS BY
*      EXTERNAL G05CCF
*      CALL G05CCF
      EXTERNAL G05CBF
      I = 42
      CALL G05CBF(I)
*
*.. GENERATE AND PRINT 10 RANDOM NUMBERS
      DO 10 J = 1,10
         WRITE(NOUT,'(1X,E16.8)') G05CAF(DUMMY)
10    CONTINUE
      END
```
Code 1.7   Example call to G05CAF and G05CBF.

EXERCISE 1.17 Show that if the modulus

$$|z| = \sqrt{x^2 + y^2}$$

of the complex number $z = x+iy$ is less than the maximum floating-point number
it may be computed without overflow as

$$|z| = a\sqrt{1 + \left(\frac{b}{a}\right)^2}$$

where a is the larger in magnitude of x and y, and b is the smaller.

EXERCISE 1.18 Discover, from a manual, how fast the basic floating-point
operations are on your machine. How would you attempt to validate these timings
practically? For example, why would timing the code fragment

```
      DO 10 I = 1,1000000
         A = B+C
10    CONTINUE
```

not give a good estimate of the time taken to perform a floating-point addition? What other problems might you need to take into consideration if you wanted to obtain a more accurate estimate?

How does the time taken to perform some of the common mathematical functions (for example, SQRT, COS, etc.) compare with simple arithmetic operations on your machine?

EXERCISE 1.19

(a) If you are using a double-precision version of the library, discover whether your Fortran compiler has an extension to allow double-precision complex variables to be declared and manipulated. Find out how to declare such variables, and the names of the associated intrinsic functions.

(b) (Double-precision implementations only.) What happens on your machine if you forget to type A02ABF as double precision? Remember if it returns erroneous results, you are bound to make this type of mistake sooner or later!

(c) Compare the performance of A02ABF with the complex modulus routine provided by the compiler. You could attempt to compare

  (i)   the average time taken by each of the routines,

  (ii) their robustness when operating on large values of x and/or y. Try values of x and y which may cause intermediate values to overflow and which will generate an overflow in the final results (see Section 1.3).

(d) The complex division routine A02ACF is designed to avoid unnecessary underflows and overflows in the intermediate stages of the computation. Compare its performance with the division operation provided by your Fortran compiler.

EXERCISE 1.20   The X02 chapter of the NAG library contains a number of functions which return various machine dependent constants, e.g., the value of the largest storable integer, the base of the floating-point arithmetic, etc. These assist in the movement of the library from one machine to another and provide users of the library with the same portability aids for their own programs.

Several algorithms are available (see, for example, Malcolm [55] and Cody and Waite [18], pp. 258–264) which try to discover such machine characteristics automatically. Although these algorithms are logically correct their implementations may produce incorrect results on some machine/compiler combinations due to idiosyncrasies of the hardware or compiled code. We present codes to compute a number of the machine parameters and invite the reader to run them on as many machine/compiler combinations as possible (try switching on the code optimizer on the compiler if there is one), and to compare the results with those returned by the equivalent X02 routine. If the results differ, attempt to discover why — be warned, though, the reason for the difference may be far from obvious.

• The base (β) and the precision (t) of the floating-point arithmetic.
  To obtain the base we note that the following integer values can be represented as floating-point numbers

$$1, 2, \ldots, \beta^t - 1, \beta^t, \beta^t + \beta, \beta^t + 2\beta, \ldots, \beta^{t+1}, \beta^{t+1} + \beta^2, \ldots$$

i.e., consecutive values in the range $[1, \beta^t]$ differ by 1 whilst those in the range $[\beta^t, \beta^{t+1} - \beta]$ differ by β. Thus we may compute β by subtracting two of these consecutive values in the range $[\beta^t, \beta^{t+1} - \beta]$. The precision can then be calculated by finding the smallest power of β for which the distance between successive integer floating-point values exceeds one. Code 1.8 attempts to compute β and t, when β is a power of 2; there is no guarantee that it will always succeed.

```
      PROGRAM MCPAR1
      DOUBLE PRECISION A,B,ONE,TWO
      INTEGER BETA,T,NOUT
      PARAMETER (ONE = 1.0D0,TWO = 2.0D0,NOUT = 6)
*
*.. STARTING WITH TWO KEEP DOUBLING UNTIL CONSECUTIVE
*.. INTEGER VALUES (STORED AS FLOATING-POINT NUMBERS)
*.. ARE MORE THAN ONE APART
      A = TWO
10    IF((A+ONE)-A.EQ.ONE) THEN
         A = TWO*A
         GOTO 10
      ENDIF
*
*.. NOW FIND THE NEXT INTEGRAL NUMBER IN THE RANGE
      B = TWO
20    IF((A+B).EQ.A) THEN
*
*.. NEED TO ADD ON MORE TO GET TO NEXT VALUE
         B = TWO*B
         GOTO 20
      ENDIF
*
*.. CAN NOW COMPUTE THE BASE - NOTE THAT IT ISN'T
*.. NECESSARILY B AS ROUNDING AND TRUNCATION CAN
*.. TAKE PLACE
      BETA = INT((A+B)-A)
*
*.. WE CAN NOW COMPUTE T, THE NUMBER OF DIGITS
*.. PRECISION IN THE FLOATING-POINT REPRESENTATION
*.. BY FINDING THE SMALLEST POWER OF BETA FOR WHICH
*.. THE DISTANCE TO THE NEXT FLOATING-POINT INTEGER
*.. EXCEEDS ONE
      T = 1
      A = TWO
30    IF((A+ONE)-A.EQ.ONE) THEN
         A = A*BETA
         T = T+1
         GOTO 30
```

```
      ENDIF
      WRITE(NOUT,100)T,BETA
100   FORMAT(1X,'THERE ARE ',I4,' BASE ',I4,
     +    ' DIGITS IN THE FLOATING-POINT REPRESENTATION')
      END
```

Code 1.8   Program to discover the base and precision of the floating-point arithmetic.

- The machine epsilon (ϵ).
  This is defined as β^{1-t} and can be computed directly from the parameters found above. Alternatively we may note that it is the difference between 1.0 and the next larger representable floating-point number. This is the method we use in Code 1.9.

```
      PROGRAM MCPAR2
      DOUBLE PRECISION EPS,EPSP1,HALF,NEPS,ONE
      PARAMETER (ONE = 1.0D0,HALF = 0.5D0,NOUT = 6)
*
      NEPS = ONE
*
*.. KEEP HALVING UNTIL THE ADDITION TO ONE HAS NO EFFECT
*.. KEEP THE PREVIOUS VALUES IN EPS
10    EPS = NEPS
      NEPS = HALF*NEPS
      IF((NEPS+ONE).GT.ONE)GOTO 10
*
*.. NOW COMPUTE EPS BY ADDING TO ONE TO TAKE ACCOUNT
*.. OF ROUNDING AND SUBTRACT ONE AGAIN TO OBTAIN THE
*.. REQUIRED VALUE
      EPSP1 = EPS+ONE
      EPS = EPSP1-ONE
      WRITE(NOUT,100)EPS
100   FORMAT(1X,'COMPUTED VALUE FOR THE MACHINE',
     +          ' EPSILON =',E16.8)
      END
```

Code 1.9   Program to compute the machine epsilon.

Use X02DAF to test whether your system sets all underflowing quantities to zero.

EXERCISE 1.21

(a) Change the example program given in Code 1.3 so that values for X and Y are read in, rather than assigned within the program. Why is it a good idea to change the FORMAT statement?

(b) Extend the changes to allow for the repeated input of values. Can you find out how to stop the program cleanly? You should try and do this using a

```
READ(NIN,100,END = <label>)X,Y
```

statement rather than 'special' values of X and Y.

(c) How would you test the accuracy of A02ABF?

EXERCISE 1.22   The function A02ABF does not use the error parameter IFAIL. Discuss the problems of using an error indicator with a function.

EXERCISE 1.23

(a) What are the advantages of using a PARAMETER statement to define array bounds and other program constants?

(b) Change the example program in Code 1.6 to allow matrices of order up to 15 to be inverted.

(c) Inverting A and then inverting the result should bring us back to A. Amend your program to keep a copy of the original matrix and measure the error (see Section 1.17) caused by using the inverse routine twice. Do this for Hilbert matrices of order $3, 5, \ldots, 15$. For more details of the computational difficulties arising in this problem see Section 4.7.

EXERCISE 1.24

(a) Run the example program given in Code 1.7 several times to convince yourself that the sequence repeats! What is the effect of changing the value of the parameter to G05CBF? If possible run the program on two different machines. Do you get the same sequence of numbers? Amend the program to generate a non-repeating sequence. Convince yourself it does not repeat on separate runs. What happens if the call to G05CCF is accidentally placed inside the loop generating the random numbers? Account for the behaviour on your system.

(b) What do you think will happen if the dummy parameter to G05CAF is either declared real or omitted on your system? What actually does happen?

(c) What is the probability of G05CCF producing the same sequence on two runs of a program?

(d) Plot 200 random coordinate pairs. Do they look random? How could you reduce the amount of pen movement whilst plotting such a picture?

(e) Consider the random number generator

$$X_0 \text{ odd},$$
$$X_{n+1} = (65539X_n) \bmod 2^{31}, \qquad n = 0, 1, \ldots .$$

Values in the range $(0,1)$ are then formed as $X_n/2^{31}$. Implement this generator, using floating-point arithmetic if necessary, and repeat the plot of random coordinate pairs. Compute $(9X_n - 6X_{n+1} + X_{n+2}) \bmod 2^{31}$. Is this what you expect from a random number generator?

EXERCISE 1.25 Accurate values are available for several common mathematical constants via calls to functions provided in the X01 chapter. Note that use of these functions is an aid to making programs portable, since if code is moved to a system with different floating-point parameters (see Section 1.3), the constant values will remain correct without changes to the source. Find the name of the function in X01 which returns the value of π. Print out its value to as many decimal places as possible; how many correct decimal digits are there? Compare its value with other common 'machine independent' methods of computing π, e.g.,

- use the Fortran intrinsic function, ATAN, to compute $4\tan^{-1}(1)$,
- using the recurrence

$$a_n = (a_{n-1} + b_{n-1})/2,$$
$$b_n = \sqrt{a_{n-1}b_{n-1}},$$

where $a_0 = 1$ and $b_0 = 1/\sqrt{2}$, compute

$$c_n = \frac{4a_n b_n}{1 - \sum_{i=0}^{n-1} 2^i(a_i - b_i)^2},$$

for $n = 1, 2, \ldots,$ until $|c_n - c_{n-1}| < \epsilon$. This recurrence relation converges astonishingly quickly (c_6 is correct to 100 digits provided you have enough precision in your arithmetic). For more details see Borwein and Borwein [10].

1.13   Common problems

Although it is possible for a NAG routine to return incorrect or misleading results the most likely source of error remains the user. This section highlights a number of common pitfalls; the list is by no means exhaustive. Most mistakes occur either through a misunderstanding of the documentation resulting in an inappropriate routine being selected, or an error in setting up the input parameters for the particular problem being solved, or in the interpretation of the output parameters. We concentrate here on common programming mistakes. In the following the error messages are purely for illustration purposes — real system errors are very often far more cryptic!

- 'Unsatisfied external reference *name*' or 'missing routine *name*' — an error of this type signifies that a call has been made to a function, or subroutine, but no routine of that name has been provided. Depending on the system such a failure may occur before the program starts executing, or when the call is actually made. If the missing name looks like a NAG routine name, check that it is correct. Two common mistakes are

  - using the letter 'Oh' instead of the digit zero as the second character,

- reversing the two routine name characters, e.g., using G05ACF instead of G05CAF.

Another possibility is forgetting to link in the NAG library and attempting to run just the user provided part of the program.

- 'Wrong number or type of parameters' — on many systems programming errors of this type go unrecognized, despite the fact that the Fortran 77 standard insists that formal and actual parameters should agree in number and type (ANSI [5], Section 15.9); any results produced are likely to be wrong. Possible problems are

  - wrong number of parameters in the actual call to the NAG routine — it is very easy, especially with long argument lists, to miss out parameters,

  - wrong type of parameters in actual call
    * a parameter has been omitted from a type statement. This error is most common when using a double-precision implementation of the library, although occasionally using a single-precision version a variable may accidentally take on an incorrect type (e.g., a variable named MULT would default to type integer unless explicitly declared real),
    * an array parameter has had its dimensioning information omitted. Remember that it is illegal to use a simple variable as an actual argument if the dummy argument is an array, even if the actual argument is unused by the routine, or if the actual array would be of length 1,
    * an actual parameter which is a function or subroutine name has not been included in an EXTERNAL statement. Note that a function name should also appear in a type statement and, in the case of a double-precision implementation, that this is mandatory for functions returning a double-precision result,
    * a constant or expression used as an actual argument is of the wrong type. Note that it is illegal to use an integer constant as an actual argument when the dummy parameter is a real or double-precision variable (i.e., real zero must be one of 0.0, 0.0E0 or 0.0D0, but never 0),
    * a constant or expression has been used as an actual argument when the dummy parameter is an output parameter (i.e., it is assigned to in the body of the routine). See also Section 1.11,
    * a parameter has been misspelt,

  - passing wrong dimensioning information. When the dummy argument is a two-dimensional array the value of the argument defining the leading dimension must be set to the value used in the dimension statement in the calling routine. See also Section 1.6.

- Logical or data errors when setting up input parameters. Even if your compiler/run time system checks for unassigned variables (variables which are used in a program before they have been assigned a value), it is quite probable that these checks will have been turned off when compiling the library. Many systems initialize all variables to a default value like zero. Check carefully all the code used to initialize input parameters; as a final resort input values may be printed out just prior to the NAG routine call.

- Knock on effects caused by

  - data, required for further computation, being destroyed by a call to a NAG routine,

  - using the soft error option and failing to trap an error. This may result in incorrect values being used in the ensuing computation. See also Section 1.9.

Some of these problems can be caused by continuing a statement past column 72. Remember, a statement can only occupy positions 7–72 of a line; anything in columns > 72 is ignored. Long statements must be continued on a new line.

If you have thoroughly checked all the above points and are quite convinced that

- the input parameters have been initialized correctly,

- the NAG routine has been called correctly,

- your original analysis of the problem was correct,

- you have chosen the correct routine from the library,

it is probably time to consult an expert. Remember that very few bugs are found in NAG routines, and so the most likely cause of failure is simple error or misuse.

EXERCISE 1.26 Try running some incorrect programs which call NAG routines with, for example,

- incorrect routine names,
- the wrong number of parameters,
- the wrong type of parameters,
- simple variables instead of arrays, etc.

Make a note of any resultant system error messages that are reported. It is always easier to find errors when you know what you are looking for!

1.14   Iterative methods

In order to make sensible use of a NAG routine it is often useful, and some-times necessary, to have some basic idea of the underlying methods used to compute a solution. One aim of this volume is to study the methods on which certain routines in the library are based. To do this we need to discuss a number of fundamental concepts in numerical analysis. We start with the difficulties of implementing iterative methods. We are especially concerned with the problem of terminating such methods at a point which ensures the user's accuracy requirements have been attained, with the min-imum amount of computation. Note that we are endeavouring to obtain a final value to within a predefined distance of the true solution whilst working solely with approximate values.

We begin with a simple illustrative example of the use of an iterative method to find an approximate square root of a positive number, a. The method is derived by applying *Newton's method* (see Section 2.6 for more details) to the problem of finding the *root* (zero) of the equation $x^2 - a = 0$. We write the method as

$$x_{n+1} = \frac{1}{2}\left(x_n + \frac{a}{x_n}\right), \qquad n = 0, 1, \ldots, \tag{1.3}$$

where the subscripts indicate the iteration number. Starting with x_0, an initial guess at $\sqrt{a}$ (say $a/2$), we form what we hope will be an improved estimate using $x_1 = (x_0 + a/x_0)/2$, and then refine this result by forming $x_2 = (x_1 + a/x_1)/2$, and so on. In principle this iteration is an infinite process which will, we hope, converge to the required result. That is, we anticipate that each successive iterate will get closer to $\sqrt{a}$. We express this desired property as

$$\lim_{n \to \infty} x_n = \sqrt{a}, \tag{1.4}$$

that is, the limit as n 'tends to infinity' (increases in value) of the sequence of iteration values $x_0, x_1, x_2, \ldots$ is $\sqrt{a}$. If the sequence does not converge we say that it diverges. An alternative to (1.4) is

$$\lim_{n \to \infty} |x_n - \sqrt{a}| = 0,$$

which says that the process converges if the absolute difference between x_n and $\sqrt{a}$ gets smaller as we increase the number of iterations performed. In fact, for this particular problem, we can prove that the Newton iteration (1.3) will converge for all starting values x_0.

To implement (1.3) we will need to halt the process after a finite number of steps and accept the latest iterate as an approximation to $\sqrt{a}$. Ideally we would like to be able to halt the process when, for some value of n, x_{n+1} is considered to be close enough to $\sqrt{a}$ for further computation to be unnecessary. There are several ways in which such a stopping criterion may be interpreted depending both on the problem and the reasons for solving it. For example, we might require the iteration to continue until the relative error in x_{n+1} is less than some user specified accuracy tolerance ϵ (say 5×10^{-7}). That is, we would like to halt when

$$\left| \frac{\sqrt{a} - x_{n+1}}{\sqrt{a}} \right| < \epsilon, \tag{1.5}$$

which would provide a value with six significant digits accuracy. If we had used an absolute accuracy criterion,

$$\left| \sqrt{a} - x_{n+1} \right| < \epsilon,$$

we would obtain six decimal place accuracy, that is, when expressed in fixed-point format, six figures after the decimal point may be quoted as correct. We note that some care needs to be exercised when providing such accuracy requirements to ensure that the demanded tolerances are actually attainable given the arithmetic capabilities of the equipment being used. This is especially true when using an absolute error criterion as then the attainable accuracy is dependent on the magnitude of the value being computed.

In our example $\sqrt{a}$ is the unknown value we are attempting to determine, and so we cannot impose condition (1.5) directly. Instead we continue the iteration until, say,

$$\left| \frac{x_{n+1} - x_n}{x_{n+1}} \right| < \epsilon, \qquad x_{n+1} \neq 0, \tag{1.6}$$

that is, until the relative distance between successive iterates is considered to be small enough. Our condition for convergence has become

$$\lim_{n \to \infty} |x_{n+1} - x_n| = 0.$$

In the absence of further information, there is no guarantee that this will provide us with an approximation to $\sqrt{a}$ within the desired tolerance.

In deriving an iterative method the numerical analyst must attempt to determine

- under what conditions the iteration is guaranteed to converge to the required value, α, say,

- a relationship between $|x_{n+1} - x_n|$ and $|x_{n+1} - \alpha|$.

Unfortunately, in many circumstances it is not possible to produce solutions to these problems which are of practical use, and when implementing an iterative scheme as a computer program it is often necessary to impose an upper limit ITMAX, say, on the number of cycles made to guard against possible divergence. Precisely what value should be chosen for ITMAX depends on many things, but in particular it depends on the convergence rate of the process. This is defined to be the value of β for which

$$|x_{n+1} - \alpha| \sim |x_n - \alpha|^\beta.$$

In many practical instances β is an integral value; if $\beta = 1$ we say that convergence is *linear*, if $\beta = 2$ *quadratic*, and so on. For high values of β we would expect convergence to be evident after only a relatively small number of cycles and hence ITMAX would be assigned a low value to avoid unnecessary computation. For smaller values of β ITMAX would need to be increased. Other factors which may influence the value chosen for ITMAX are the distance the initial value, x_0, is from α, and the accuracy tolerance ϵ. Indeed, for some higher-order schemes, such as Newton's method, the fast rate of convergence only occurs near the solution; for iterates a long way from α the convergence rate may be substantially slower (or the process may even diverge).

EXERCISE 1.27   Implement the Newton iteration, (1.3), for computing the square root of a positive number using the relative error criterion (1.5). Set the tolerance to generate results as close to the full machine precision as possible. How does the choice of starting value affect the number of iterations required? What is the rate of convergence when the starting value is not close to $\sqrt{a}$? Try to compute $\sqrt{0}$ starting with $x_0 = 2$; why are there problems?

EXERCISE 1.28   Using (1.3) show that

- for any starting value, x_0, in the range $(0, \infty)$, $x_1 \geq \sqrt{a}$,
- $x_{n+1} - \sqrt{a} = (x_n - \sqrt{a})^2/(2x_n)$.

Hence show that $x_1 \geq x_2 \geq \cdots \geq \sqrt{a}$ and $\lim_{n\to\infty} x_n = \sqrt{a}$. What does this result say about the convergence properties of the Newton iteration for the equation $x^2 - a = 0$?

EXERCISE 1.29   Newton's method may be used to compute $\sqrt[3]{a}$ for $a > 0$ using the iteration $x_{n+1} = x_n - (x_n^3 - a)/(3x_n^2)$. Use a similar analysis to that presented in Exercise 1.28 to investigate whether this iteration could be used as a general purpose function for computing real, positive cube roots.

EXERCISE 1.30   The Pythagorean sum is defined as $a \oplus b = \sqrt{a^2 + b^2}$. Show that if $|a| \geq |b|$ then $|b| \leq |a| \leq a \oplus b$ and that $a \oplus b$ is in the range $[\,|a|, \sqrt{2}|a|\,]$. Hence construct and implement an algorithm, based on Newton's method, which computes $a \oplus b$ to a given relative precision ϵ. Extend this algorithm to compute the Euclidean norm of a vector

$$\|\mathbf{x}\|_2 = \sqrt{\sum_{i=1}^{n} x_i^2},$$

where $\mathbf{x}^T = (x_1, x_2, \ldots, x_n)$.

EXERCISE 1.31   A second iteration for computing the Pythagorean sum, $a \oplus b$ (see Exercise 1.30 above) is given by the following pseudocode:

```
double precision function pythag(a, b)
double precision a,b,p,q,r,s
p = max(|a|, |b|)
q = min(|a|, |b|)
while (q is numerically significant) do
   r = (q/p)²
   s = r/(4 + r)
   p = p + 2 × s × p
   q = s × q
endwhile
pythag = p
end
```

Implement this iteration using a relative stopping criterion. Empirically find the rate of convergence.

Consider the following methods for computing the Pythagorean sum

(a)   form $a^2 + b^2$ and use the Fortran intrinsic square root routine,

(b)   use the method described in Exercise 1.30,

(c)   use the method given in this exercise,

(d)   use the NAG routine for computing the Pythagorean sum, F06BNF.

For both (b) and (c) use as a stopping criterion the fact that two successive iteration values are equal. Compare the methods from the point of view of

- speed,
- accuracy,
- robustness, i.e., if $a \oplus b$ is representable, there should be no overflow at an intermediate step.

If any of your implementations fail the robustness requirement what steps can be taken to improve the algorithm's performance?

1.15   Ill-conditioning and instability

The two important concepts of *ill-conditioning* and *instability* are both concerned with the effects that errors can have on a computed solution. They are both qualitative rather than quantitative terms, and we introduce them by means of examples.

First we consider the system of linear equations

$$
\begin{aligned}
x & + & 2y & = & -1 \\
2x & + & (4 + \epsilon)y & = & -2 - \epsilon.
\end{aligned}
\tag{1.7}
$$

Each equation may be interpreted as a straight line in the x-y plane; the solution (a pair of values for x and y which satisfies both equations simultaneously) is the point of intersection of the two lines. It is simple to verify that the solution of (1.7) is given by $x = 1$ and $y = -1$, and that this solution is independent of the value of ϵ provided it is non-zero. Suppose that ϵ is very small, say the same order of magnitude as the rounding errors on your computer, and that the values appearing on the right-hand side, which are subject to calculation, or observational, error are recorded as -1 and -2. We then obtain the pair of equations

$$\begin{aligned} x \;+\; & 2y = -1 \\ 2x \;+\; & (4+\epsilon)y = -2, \end{aligned} \tag{1.8}$$

which has solution $x = -1$ and $y = 0$. Whenever a relatively small change to the parameters of a problem results in a disproportionately large change to the solution we say that the problem is *ill-conditioned* and therefore its numerical implementation will automatically be unstable.

We have already seen (Example 1.1) that a recurrence relation can exhibit unstable characteristics. As a further manifestation of instability caused by the introduction of rounding errors, we consider the evaluation of the exponential function, e^x, for some given value x. If we first look at

$$\frac{e^{x+\delta} - e^x}{e^x} = e^\delta - 1$$

we see that a small change in the argument, from x to $x + \delta$, will not alter the value of the exponential by a disproportionately large amount. The problem is, therefore, reasonably well-conditioned.

Now, if $f(x)$ is some function, then under suitable conditions (in particular provided that a sufficient number of derivatives exist) the *Taylor series expansion*

$$\begin{aligned} f(x) \;=\; & f(a) + (x-a)f'(a) + \frac{(x-a)^2}{2!}f''(a) + \cdots \\ & + \frac{(x-a)^p}{p!}f^{(p)}(a) + \frac{(x-a)^{p+1}}{(p+1)!}f^{(p+1)}(\zeta) \end{aligned} \tag{1.9}$$

exists where a is the chosen point of expansion, ζ is some, unknown, value between a and x, and p is an integer (see Conte and de Boor [19], p. 27 for details). Choosing $f(x) = e^x$ and $a = 0$, and noting that the derivative of e^x is e^x, leads to the expansion

$$e^x = 1 + x + \frac{x^2}{2} + \frac{x^3}{6} + \cdots + \frac{x^p}{p!} + \frac{x^{p+1}}{(p+1)!}e^\zeta, \tag{1.10}$$

and this result holds for all $p > 0$. We write (1.10) as

$$e^x = \sum_{i=0}^{p} T_i + E$$

where $T_i = x^i/i!$ and $E = x^{p+1}e^\varsigma/(p+1)!$.

For $-1 < x < 1$ the values of T_i decrease in magnitude as i increases, but for large values of $|x|$ the terms will initially grow in magnitude until the factorial in the denominator dominates when, once again, the T_is will decrease in size. This suggests that if p is chosen large enough the first $p+1$ terms in the Taylor series expansion may be summed to provide an approximation to e^x.

If we consider the case $x = -10$, we have, for example

$$T_6 = \frac{1000000}{6!} = 1388.88\ldots.$$

But on, say, a 6 decimal digit machine the rounding error incurred in representing this value is of greater magnitude than the quantity we are trying to compute, namely $e^{-10} \approx 0.45 \times 10^{-4}$. Rounding errors introduced into the computation will therefore totally dominate the computed solution. Numerical methods which exhibit this behaviour are said to be *unstable*.

To summarize, ill-conditioning is a property of the problem itself and is independent of any chosen method of solution. If a problem is ill-conditioned any small errors introduced into the parameters will tend to have a dramatic effect on any computed solution, no matter how that solution is computed. In contrast, instability is a property of a proposed method of solution. Given the use of exact arithmetic, an unstable algorithm will perform as well as any other. However, the introduction of rounding errors can have such a disastrous effect on the solution as to render it completely worthless.

EXERCISE 1.32   Explain geometrically ill-conditioning in a system of two linear equations. What happens in (1.7) when $\epsilon = 0$?

EXERCISE 1.33   What is the solution of the system (1.7) if the right-hand side is

(a) $(-1, -2 + \epsilon)^T$,
(b) $(-1 - \epsilon, -2)^T$.

Would the system be considered ill-conditioned if $\epsilon = 2$?

EXERCISE 1.34   Suggest ways of rewriting e^{-10} so that the Taylor series expansion (1.10) can be used to provide an accurate approximation to this value on a six decimal digit machine.

1.16   Orders

For a given class of mathematical problem there will often be a number of numerical methods of solution available to us, and we need some measure by

Operation count	Max. size problem per cpu minute	Time to solve problem of size $n = 100$	
n	60000	0.1	second
$n \ln n$	6799	0.5	second
n^2	244	10	seconds
n^3	39	16.6	minutes
$n!$	8	$> 10^{100}$	years

Figure 1.5   Operation count v. feasible problem size.

which we can compare them. One criterion, discussed briefly in Section 1.2, is efficiency. We may measure the efficiency of a method by its *operation count*, the total number of arithmetic operations required to compute a solution. For a direct method this may be determined *a priori*; for example, we know precisely how many operations are required to compute the roots of a quadratic polynomial using the formulae given in Section 1.3. However, for an iterative method we are unlikely to know in advance how many iterations will be required for convergence and, hence, we may have to be content to measure the cost per iteration. In addition, we may use *a posteriori* counts of the number of iterations required by a given method to solve sets of certain 'model' problems.

The operation count is often quoted as a function of the problem size, n. For example, the size of the linear system $A\mathbf{x} = \mathbf{b}$, where A and $\mathbf{b}$ are known and $\mathbf{x}$ is to be determined, is the number of equations in the system. We say that the method is $O(n)$ (order n) if the operation count is directly proportional to n, that is, equals $Cn + D$ for some constants C and D. For large n, D will be insignificant and hence doubling n will result in approximately doubling of the amount of work performed by the algorithm. For an $O(n^2)$ method the operation count is of the form $Cn^2 + Dn + E$. The first term dominates for large n and so doubling n will tend to quadruple the number of arithmetic operations required. In Figure 1.5 we compare the effect of various orders on the size of problem solvable. For each order we assume that the constant C is unity and that each 'operation' takes a millisecond of cpu time. We tabulate the maximum size of problem solvable using one minute of cpu time and the time needed to solve a problem of size 100. We see that algorithms possessing low-order operation count are likely to be considerably more efficient. Of course, we must not forget that the constant C will have some effect; if it is very large for a low-order method then the use of a higher-order method with a relatively small constant may be preferable for small values of n. However, as n increases a point will eventually be reached beyond which the low-order method should be used, see Figure 1.6. Such cross-overs are often difficult to locate; indeed values for the coefficients are seldom available and, hence, the use of low-order methods is normally recommended in all circumstances.

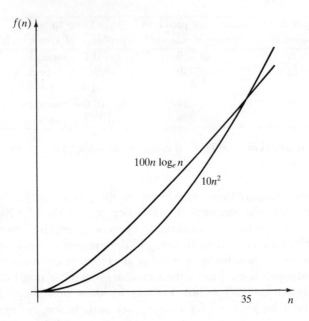

Figure 1.6   Cross-over of operation counts.

A second way of measuring the performance of a numerical method is to consider its *accuracy*. Often this can be determined as a function of a *discretization parameter*, h. For example, suppose we wish to estimate the value of the definite integral

$$I = \int_a^b f(x)\,dx.$$

We choose to discretize the interval $[a, b]$ by introducing a set of points $\{x_i \mid i = 1, 2, \ldots, n\}$ with $x_1 = a$ and $x_n = b$, and the points a constant distance h apart. The parameter h is the inverse of the 'size' of the discretized problem. We now approximate the integral over each subinterval $[x_i, x_{i+1}]$ by the area under the straight line joining $f(x_i)$ and $f(x_{i+1})$. Whence, summing over $i = 1, 2, \ldots, n-1$, we obtain an approximation, R, to I (see Figure 1.7). It can be shown that the error in this approximation is $O(h^2)$ (order h^2), that is, $|I - R| = Ch^2$ for some constant C. The effect of halving h on this numerical method, known as the *trapezium rule*, will, in principle, reduce the error by a factor of four. From this point of view higher-order methods are likely to be more efficient; that is, for a given value of $h < 1$ they are likely to produce more accurate results. Unfortunately, high-order accuracy methods often require more work and/or information than low-order methods. For example, to obtain quadratic convergence Newton's method (Section 2.6) requires information not only about the function whose root is required, but also its derivative. We also

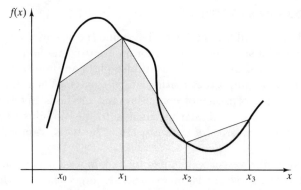

Figure 1.7   Trapezium rule.

find that, because of the use of finite-precision arithmetic, high-order methods may not perform as well as expected. An example of this behaviour
may be found in Section 3.2.

EXERCISE 1.35   On a certain machine a floating-point multiplication takes
300 nanoseconds and an addition 200 nanoseconds. Ignoring loop overheads,
what is the largest order of square matrices which may be added and multiplied
together in 5 minutes processor time? What other considerations need to be
taken into account in order to obtain a practical estimate of these maximum
sizes?

EXERCISE 1.36   We may compare the performances of sorting algorithms
by considering the expected number of comparisons, $E(C_n)$, and interchanges,
$E(I_n)$, required to sort a random set of n data items. For the simple bubble sort
we have

$$
\begin{aligned}
E(C_n) &= \tfrac{1}{2}\left(n^2 - n\ln(n) - (\gamma + \ln(2) - 1)n\right) + O\left(\sqrt{n}\right), \\
E(I_n) &= \tfrac{1}{4}n(n - 1),
\end{aligned}
$$

and for quicksort

$$
\begin{aligned}
E(C_n) &= 2(n + 1)S_n - 4n, \\
E(I_n) &= (n + 1)\left(S_n - \tfrac{2}{3}\right),
\end{aligned}
$$

where γ is Euler's constant, $0.5772156649\ldots$ (see XO1ABF) and S_n is the nth
Harmonic number defined by

$$
S_n = \gamma + \ln(n) + \frac{1}{2n} - \frac{1}{12n^2} + \frac{1}{120n^4} - \frac{1}{252n^6} + \frac{1}{240n^8} + \cdots
$$

(see Gonet [38] or Knuth [50] for more details).

Prepare cross-over graphs to compare the expected number of comparisons
and interchanges for the two methods. Attempt to quantify the time taken
to compare and interchange a pair of floating-point numbers on your machine.
Hence attempt to compare the expected run times of the two algorithms. For
what values of n is bubble sort likely to be the faster method?

1.17   Measures of size

We have already used a modulus to determine the size of a real number in
the formation of a stopping criterion for an iterative method. Its use is not
restricted to algorithm development; it also plays an important role in the
derivation of many important mathematical results. When describing and
evaluating the performance of computational methods involving vectors
and matrices, we need a similar mechanism to the modulus to allow us to
compare the relative sizes of such objects. The tools we use for this are
known as *norms*.

The norm of a vector $\mathbf{x}$, denoted by $\|\mathbf{x}\|$, is used as a means of deter-
mining whether one vector is, in some sense, larger than another. In Euclidean two-space, the length of the vector $(x_1, x_2)^T$ is equal to $\sqrt{x_1^2 + x_2^2}$.
This idea can readily be extended to n-space, and we say that the size of
the vector $\mathbf{x} = (x_1, x_2, \ldots, x_n)^T$ is $\sqrt{x_1^2 + x_2^2 + \cdots + x_n^2}$. This is just one
means of calculating the size of $\mathbf{x}$; it is known as the *2-norm* (or *Euclidean
norm*) of $\mathbf{x}$ and is denoted by $\|\mathbf{x}\|_2$ (see Exercise 1.30). Other measures
may be derived, but each must satisfy the following conditions for it to be
classified as a norm:

(1)   the norm of a vector is always non-negative;

(2)   the norm of a vector is zero if, and only if, each component of the
      vector is exactly zero;

(3)   the norm of a scalar multiple of a vector is equal to the modulus of
      that scalar times the norm of the vector;

(4)   the norm of the sum of two vectors is less than, or equal to, the sum
      of the norms of the vectors.

Whilst in theory many norms may be defined, the *p-norms* given by

$$\|\mathbf{x}\|_p = \left(\sum_{i=1}^n |x_i|^p \right)^{1/p},$$

with $p = 1, 2$ and ∞ are the ones generally used in practice. It can be shown
that the ∞-*norm*, defined as $\lim_{p \to \infty} \|\mathbf{x}\|_p$, is equal to the component of
maximum modulus.

For a measure on matrices to be classed as a norm it must satisfy
axioms equivalent to (1)–(4) for vector norms. In addition, we require the
following:

(5)   The norm of the product of two matrices is less than, or equal to, the
      product of the norms of the matrices.

As with vector norms, it is theoretically possible to define a number of
matrix norms, but the ones most commonly used are those based on the

vector p-norms (with $p = 1, 2$ or ∞), referred to as subordinate matrix norms . The formal definition of such a norm is

$$\|A\|_p = \sup_{\mathbf{x} \neq 0} \frac{\|A\mathbf{x}\|_p}{\|\mathbf{x}\|_p},$$

that is, the p-norm of an $n \times n$ matrix A is equal to the supremum (lowest upper bound) over all non-zero n-vectors $\mathbf{x}$ of the ratio of the p-norms of the vector $A\mathbf{x}$, and $\mathbf{x}$ itself. The implication that, in order to determine the p-norm of A, we have to consider all non-zero vectors (and there are an infinite number of them, of course), calculate the required p-norms and ratio and determine an upper bound on these quantities is somewhat frightening! Fortunately for the cases $p = 1$ and $p = \infty$ a simplification is possible. We have

$$\|A\|_1 = \max_{1 \leq j \leq n} \sum_{i=1}^{n} |a_{ij}|,$$

$$\|A\|_\infty = \max_{1 \leq i \leq n} \sum_{j=1}^{n} |a_{ij}|,$$

that is, the *1-norm* of a matrix is equal to the maximum absolute column sum, whilst the ∞-*norm* of a matrix is equal to the maximum absolute row sum, and both these quantities are easily computed.

When deriving and analysing numerical methods it does not, in general, matter which norm we choose to use. (An important exception to this will be encountered in Chapter 5 where we make reference to the vector 2-norm when describing the least squares method for curve fitting.) Indeed, it may be shown that for any two vector p-norms, $\|\cdot\|_a$ and $\|\cdot\|_b$, there exist constants, α and β such that $\alpha\|\mathbf{x}\|_a \leq \|\mathbf{x}\|_b \leq \beta\|\mathbf{x}\|_a$ for all vectors $\mathbf{x}$, and we say that vector p-norms are *equivalent*. We may, therefore, work with an unspecified norm and leave the choice, if one needs to be made, until the program development stage.

The F06 chapter contains three functions, based on the BLAS routines of Lawson et al. [54], to compute vector norms. For the 1- and 2-norms the functions FO6EKF and FO6EJF respectively are available and for the ∞-norm the integer function FO6JLF returns the index of the element of largest magnitude. An example call is of the form

```
ENORM = FO6EJF(N,X,INCX)
```

where the N-vector $\mathbf{x}$ whose norm is required is represented by the two arguments X and INCX. INCX is known as the increment argument and x_i is taken to be $X(1 + (i - 1) \times INCX)$. If $INCX = 1$ then $X(1), X(2), \ldots, X(N)$ define the vector.

EXERCISE 1.37 Show that if $\|\cdot\|$ is a subordinate matrix norm then $\|A\mathbf{x}\| \leq \|A\| \|\mathbf{x}\|$.

EXERCISE 1.38 Verify that the vector Euclidean norm, $\|\mathbf{x}\|_2$, and the vector maximum norm, $\|\mathbf{x}\|_\infty$, both satisfy the four properties of a vector norm for a real vector $\mathbf{x}$. Compare the amount of work required to compute each norm; which is faster to compute on your machine? Hint: for the 2-norm case use the *Cauchy–Schwarz inequality* (see Stewart [78], p. 165) which, in n-space, takes the form

$$\left(\sum_{i=1}^{n} x_i y_i \right)^2 \leq \left(\sum_{i=1}^{n} x_i^2 \right) \left(\sum_{i=1}^{n} y_i^2 \right).$$

EXERCISE 1.39 Prove that the vector 1- and ∞- norms are equivalent.

EXERCISE 1.40 Show that the *Frobenius norm* (also referred to as the *Schur* or *Euclidean* norm) of a real matrix, defined by

$$\|A\|_F = \sqrt{\sum_{i=1}^{n} \sum_{j=1}^{n} a_{ij}^2},$$

satisfies the five conditions for a matrix norm. [Note: the Frobenius norm is actually the ordinary Euclidean vector norm of a vector of length n^2 whose elements are the matrix elements a_{ij}. It is *not* the subordinate norm to the Euclidean vector norm. In fact $\|A\|_2^2$ is equal to the largest eigenvalue (*spectral radius*) of $A^T A$ which, although possessing a number of useful theoretical properties, is not as computationally convenient as the Frobenius norm.]

EXERCISE 1.41 What is the value of the 1-norm of the nth-order Hilbert matrix, H_n? What is the value of the 1-norm of H_3^{-1} (Section 1.12)? A measure of the ill-conditioning of a system of equations is the size of the *condition number* $\|A\| \, \|A^{-1}\|$, where A is the coefficient matrix. Use the program of Code 1.6 to find the inverse of H_n for $n = 3, 4, \ldots, 10$. Modify the code so that it computes the 1-norm of H_n and H_n^{-1}, and the product $\|H_n\| \, \|H_n^{-1}\|$. How good are the inverse approximations?

Now perturb each element of H_n slightly, by 0.01%, say. What effect does this have on the elements of H_n^{-1}? How good are the inverse approximations this time? Use both the original and perturbed form of H_n to check this.

EXERCISE 1.42 Write a function which uses F06EKF to determine the 1-norm of a matrix.

Chapter 2
Roots of Functions

In this chapter we consider the problem of locating a root of a function, with polynomial equations being given special attention. We

- describe some simple methods for locating a real root of a real-valued function of one real variable;

- apply Newton's method to the solution of polynomial equations and consider both real and complex roots;

- extend Newton's method to the location of a real root of a vector-valued function of severable variables.

In each case attention is concentrated on those methods which have been implemented in the NAG library, and we discuss calls to the appropriate routines.

2.1 Introduction

One of the more common problems the reader is likely to encounter is that of locating the roots (zeros) of an equation. In this chapter we concentrate on the solution of

$$f(x) = 0, \qquad (2.1)$$

where f is a given real-valued function of the real independent variable x. It may be possible to express f in some simple algebraic form. More generally f will be some expression that can be evaluated for any value of x; this may involve the determination of an integral, the solution of a differential equation, etc. In the next two sections we shall consider two methods for locating the real roots of such a general equation, and then discuss the routines within the NAG library which implement these schemes. If f is a polynomial an alternative approach is preferable and we give attention to this special case. We also consider the extension of the method to the location of complex roots of polynomial equations. Finally we outline methods for nonlinear systems of equations.

Before attempting to make use of software for locating a root of an equation some preliminary analysis is advisable. In some cases it will be possible to determine the roots with ease. For example, the roots of a quadratic polynomial may be determined using a well-known formula (see Section 1.3). For more complicated equations it is possible that a careful inspection will yield one or more of the roots immediately. It is easily seen that the equation $xe^x + x = 0$ has a root at $x = 0$. If we rewrite the equation as $x(e^x + 1) = 0$ we may conclude that there are no further (real) roots since $e^x + 1$ is everywhere positive.

Roots which cannot be located directly will need to be determined numerically. For a general function we cannot hope to devise an algorithm which will find all roots since, with the exception of polynomial equations, we have no automatic means of determining how many there might be. Some functions have an infinite number of roots (e.g., $\sin(x) = 0$). We start with one or more initial guesses in the vicinity of one of the roots and attempt to iterate towards it.

We classify a root, α, as being *simple* if $f(\alpha) = 0$ but $f'(\alpha) \neq 0$, and *isolated* if it is well separated in some sense from all other roots. The numerical methods we describe in this chapter will, in general, be able to locate simple, isolated roots with ease. The only constraint we impose on f, apart from the ability to evaluate it at any chosen point, is that it be *continuous*, at least in the vicinity of the root. Simple, non-isolated roots may give problems as the iterative scheme may become confused as to which root it is attempting to locate. Non-simple, or multiple, roots occur when $f(\alpha) = f'(\alpha) = 0$; these can also be difficult to locate (and in certain circumstances the first numerical method we consider cannot even be used to determine such a point). Such values correspond to local *maxima*, or *minima*, of f, or *points of inflexion*. If we suspect the existence of a multiple root then it may be necessary to resort to the use of *minimization* techniques which are outside the scope of this chapter.

The performance of any iterative method for locating a root is influenced by the accuracy of the original guess. If we start too far away from the root the number of iterations, and hence function evaluations made, may be considerable. For certain methods we may not even be able to locate a root if the initial guess is poor, or the method may converge to some root we are not interested in. The situation is more critical if the roots are not isolated. A preliminary to any attempt to locate a root is, therefore, the production of a rough sketch of the function, or of its factors. If the roots are not isolated then a very rough sketch may fail to reveal the true form of f near the x-axis. In contrast, due to rounding error, too fine a sketch may suggest a large number of false roots.

Figure 2.1 illustrates the first problem using the function

$$f(x) = 10.1e^{-3x} - \frac{0.1}{x^2 + 0.01}.$$

The left-hand sketch is obtained by taking steps of 0.1 from $x = 0$ to $x = 2$. Only when the step size is reduced to 0.05 does the pair of roots, shown

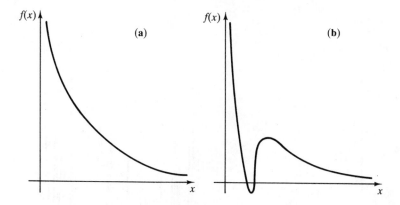

Figure 2.1   Rough sketches showing how roots may be missed.

in the second sketch, become visible. Figure 2.2 shows how accurate root finding may be affected by computer arithmetic. The graph shows the result of plotting the polynomial $p(x) = x^5 - 5x^4 + 10x^3 - 10x^2 + 5x - 1$ (which has a root of multiplicity 5 at $x = 1$) at a number of equally spaced points in the range $[0.999, 1.001]$ and joining successive points with straight lines. This suggests the presence of a large number of spurious roots around the multiple root at $x = 1$ and illustrates why numerical methods will experience problems with this sort of equation.

Consider now the equation $xe^x \tan(x) - e^x = 0$. We observe that this may be factored as $e^x (x \tan(x) - 1) = 0$, and hence we may concentrate on locating the zeros of $x \tan(x) - 1 = 0$ since e^x is everywhere positive. We may rewrite this as

$$\tan(x) - \frac{1}{x} = 0, \qquad (2.2)$$

and we note that this has roots at the intersections of the graphs $y = 1/x$ and $y = \tan(x)$, see Figure 2.3. We conclude that

- there are an infinite number of roots of the equation, and as x increases in magnitude these approach the zeros of $\tan(x)$, namely $n\pi$ for $n = \pm 1, \pm 2, \ldots$;

- despite the rather innocuous appearance of the equation, its value is unbounded at $x = 0$ (because of the singularity in $1/x$), and at the asymptotes of $\tan(x)$. In addition, the equation is discontinuous at each of these points.

We must bear these properties in mind when choosing starting values for an iterative method, and for analysing the subsequent performance of the method.

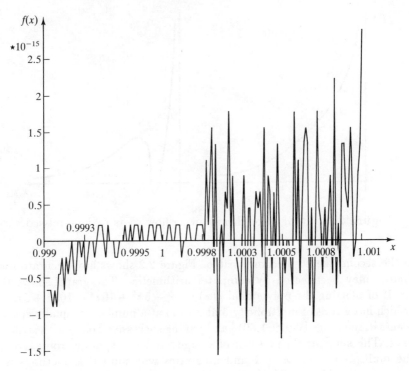

Figure 2.2   Polynomial plot showing the effect of floating-point arithmetic.

EXERCISE 2.1   Use a rough sketch to locate the approximate positions of the roots of the equation $200(4x - 25) - 500(1.2)^x = 0$.

EXERCISE 2.2   If (2.1) is recast as $x = g(x)$ the *fixed-point iteration* $x_{k+1} = g(x_k)$ converges to a root of f provided that $|g'| < 1$ in an interval containing the root and the initial guess x_0. Recast $x^3 + 1.5x - 1.5 = 0$ in ways which ensure that this condition is satisfied. Attempt to locate the real root near 0.7. What happens when you choose starting values some distance from the root? Why?

EXERCISE 2.3   Newton's method for locating a root of $f(x) = 0$ (for details see Section 2.6) requires f (and its first derivative) to be evaluated at the iterates x_k. Discuss the relative merits of applying it to the following equations: (a) $x \tan(x) - 1 = 0$; (b) $\tan(x) - 1/x = 0$; (c) $x - 1/\tan(x) = 0$ and (d) $x - \arctan(1/x) = 0$.

EXERCISE 2.4   The roots of the cubic polynomial equation $a_0 x^3 + a_1 x^2 + a_2 x + a_3 = 0$ may be determined from the formulae

$$x_1 = (s_1 + s_2) - \frac{a_1}{3a_0},$$

$$x_2 = -\frac{1}{2}(s_1 + s_2) - \frac{a_1}{3a_0} + \frac{i\sqrt{3}}{2}(s_1 - s_2),$$

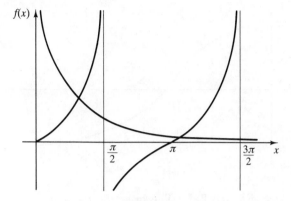

Figure 2.3   Roots of simple example.

$$x_3 = -\frac{1}{2}(s_1 + s_2) - \frac{a_1}{3a_0} - \frac{i\sqrt{3}}{2}(s_1 - s_2),$$

where $i = \sqrt{-1}$,

$$s_1 = \left(r + (q^3 + r^2)^{\frac{1}{2}}\right)^{\frac{1}{3}},$$

$$s_2 = \left(r - (q^3 + r^2)^{\frac{1}{2}}\right)^{\frac{1}{3}},$$

and

$$q = \frac{1}{3a_0^2}\left(a_2 a_0 - \frac{a_1}{3}\right),$$

$$r = \frac{1}{a_0^2}\left(\frac{a_1 a_2}{6} - \frac{a_0 a_3}{2} - \frac{a_1^3}{27}\right).$$

Under what circumstances will these formulae give rise to loss of significance? Show how the iteration of Exercise 2.2 can be used to determine the square and cube roots in the definitions of s_1 and s_2.

2.2   The method of bisection

Suppose we wish to find a root of the equation $f(x) = 0$ and that we have, by some process as yet to be determined, been able to find values $x_a^{[0]}$ and $x_b^{[0]}$ (with $x_a^{[0]} < x_b^{[0]}$) such that

$$f(x_a^{[0]}).f(x_b^{[0]}) < 0.$$

Provided f is continuous in the interval $\left[x_a^{[0]}, x_b^{[0]}\right]$, then since $f(x_a^{[0]})$ and $f(x_b^{[0]})$ are of opposite sign, f must be zero at at least one point in the open interval $\left(x_a^{[0]}, x_b^{[0]}\right)$. For simplicity we assume the existence of a single root

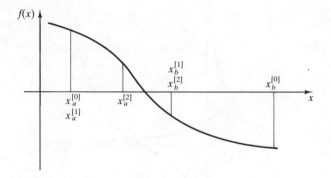

Figure 2.4   The bisection method.

only. Suppose now we bisect $\left(x_a^{[0]}, x_b^{[0]}\right)$ by introducing the midpoint $x_{mid} = \left(x_a^{[0]} + x_b^{[0]}\right)/2$. If we are fortunate this will be the required root. If not by examining the sign of the product $f\left(x_a^{[0]}\right).f\left(x_{mid}\right)$ we can determine within which half of $\left(x_a^{[0]}, x_b^{[0]}\right)$ the root lies; a negative product means that the root must be in $\left(x_a^{[0]}, x_{mid}\right)$, otherwise it is in $\left(x_{mid}, x_b^{[0]}\right)$. Let $\left(x_a^{[1]}, x_b^{[1]}\right)$ denote the new bracket for the root. We can continue this search by repeatedly applying the *bisection* process, generating $\left(x_a^{[2]}, x_b^{[2]}\right)$, $\left(x_a^{[3]}, x_b^{[3]}\right)$, and so on (see Figure 2.4). The code fragment Code 2.1 describes one iteration. At each step of the process we are able to compute both an approximation to the root, and a bound on the absolute distance of this approximation from the true root, $|x_a^{[k]} - x_b^{[k]}|/2$. In the limit, the interval $\left(x_a^{[k]}, x_b^{[k]}\right)$ will be of zero length, and hence convergence to the root is guaranteed, albeit at a slow rate.

```
FA = F(XA)
XMID = XA+0.5*(XB-XA)
FMID = F(XMID)
IF (FA*FMID.LT.0.0D0) THEN
   XB = XMID
ELSE
   XA = XMID
ENDIF
```
Code 2.1   Bisection iteration.

When developing code which implements the method of bisection care must be taken with the termination criterion. The iteration may be stopped if one of the following situations is reached:

- The value of $|f(x_{mid})|$ is small enough for x_{mid} to be taken as the root. We must bear in mind that the computer executing the program only works to finite precision, and that the evaluation of f will involve rounding errors. We should be careful not to impose a condition which is not achievable.

- The size of the interval $\left(x_a^{[k]}, x_b^{[k]}\right)$ is so small that there is little to be gained from further bisection.

If f has a shallow slope near the root, a small value of f is encountered some distance from it. In contrast, if the slope is very steep it may not be possible to reach a point at which f is regarded as being small enough.

The subroutine BISECT in Code 2.3 implements the method of bisection. The function f whose zero is sought is defined by the parameter F. An example of this user supplied function is given in Code 2.2. It should be noted that the actual name used for the function must appear in an EXTERNAL statement in any program unit which calls BISECT, and also that the name should be typed explicitly. The function used in this example is very simple. Rather more complex functions are, of course, possible, and their evaluation may involve prior calls to one or more NAG routines.

```
DOUBLE PRECISION FUNCTION F(X)
DOUBLE PRECISION X
INTRINSIC SIN,EXP
F = SIN(X)-EXP(-X)
END
```

Code 2.2   Example of a user defined function.

In Code 2.3 ETA and EPS are used to control the iteration using the magnitude of f and the length of the bracketing interval respectively. Since

$$x_b^{[k]} - x_a^{[k]} = \left(x_b^{[k-1]} - x_a^{[k-1]}\right)/2 = \cdots = \left(x_b^{[0]} - x_a^{[0]}\right)/2^k,$$

an integer limit on the number of cycles made can be derived by deducing that

$$n = \left\lceil \ln\left(\left(x_b^{[0]} - x_a^{[0]}\right)/eps\right)\Big/\ln(2)\right\rceil,$$

where ln denotes the natural logarithm, and $\lceil x \rceil$ is the smallest integer greater than x. Control using EPS alone is allowed by setting ETA to zero. On successful exit from the routine IFAIL is set to 0, A and B define the end-points of the final bracketing interval, ROOT is an approximation to the root, and FROOT returns the value of f at ROOT. The failure parameter IFAIL returns a value of 1 if EPS is non-positive, A = B, or f is not of opposite sign at A and B.

```
      SUBROUTINE BISECT(A,B,EPS,ETA,F,ROOT,FROOT,IFAIL)
*
*.. USES THE METHOD OF BISECTION TO LOCATE A ROOT OF A
*.. CONTINUOUS FUNCTION, F.
*.. F SHOULD BE OF OPPOSITE SIGN AT A AND B.
*.. EPS IS USED TO TERMINATE THE PROCESS WHEN THE CURRENT
*.. INTERVAL IN WHICH A ROOT IS KNOWN TO LIE IS CONSIDERED
*.. TO BE SMALL ENOUGH. ETA TERMINATES THE ITERATION WHEN A
*.. SMALL F-VALUE IS ENCOUNTERED. ON EXIT ROOT CONTAINS AN
*.. APPROXIMATION TO THE ROOT AND FROOT THE VALUE OF
*.. F AT THIS POINT.
      DOUBLE PRECISION A,B,EPS,ETA,F,FATA,FROOT,HALF,PROD,
     +    ROOT,TWO,ZERO
      INTEGER IFAIL,K,KMAX
      INTRINSIC ABS,INT,LOG
      PARAMETER (HALF = 0.5D0,TWO = 2.0D0,ZERO = 0.0D0)
      EXTERNAL F
      IFAIL = 0
      FATA = F(A)
      IF(EPS.LE.ZERO .OR. A.EQ.B .OR. FATA*F(B).GT.ZERO)THEN
        IFAIL = 1
        RETURN
      ENDIF
*
*.. DETERMINE HOW MANY INTERVAL BISECTIONS ARE PERMITTED.
*.. THE HALF IS ADDED TO ENSURE THAT WE ROUND
*.. TO THE NEAREST INTEGER, RATHER THAN JUST TRUNCATE.
      KMAX = INT(LOG((B-A)/EPS)/LOG(TWO)+HALF)
      DO 10 K = 1,KMAX
        ROOT = (A+B)/TWO
        FROOT = F(ROOT)
*
*.. TEST FOR A SMALL VALUE OF F.
        IF(ABS(FROOT).LT.ETA)RETURN
        PROD = FATA*FROOT
*
*.. TEST FOR WHICH HALF OF THE CURRENT
*.. INTERVAL THE ROOT IS KNOWN TO EXIST.
        IF(PROD.LT.ZERO)THEN
          B = ROOT
        ELSE
          A = ROOT
          FATA = FROOT
```

```
         ENDIF
10       CONTINUE
*
*..  COMPUTE BEST APPROXIMATION TO ROOT
*..  AND FUNCTION VALUE AT EXIT POINT.
         ROOT = (A+B)/TWO
         FROOT = F(ROOT)
         RETURN
         END
```

Code 2.3   Subroutine BISECT.

Before discussing other root finding methods we look at how a root bracketing interval may be found automatically. One way of proceeding is as follows. Starting at some point x_0 which is an estimate of the root, we determine the direction of decreasing $|f|$ by choosing some positive step length, h, and computing $f(x_0)$ and $f(x_0 + h)$. If the second function value is the smaller of the two in magnitude we conclude we are going in the right direction; if not we negate h. We could then evaluate $f(x_0 + 2h), f(x_0 + 3h), \ldots,$ until a point is reached at which f changes sign. However, if x_0 is some distance from the root or h is small we may need to make a large number of function calls before obtaining an appropriate bracket. To overcome this we might consider a *binary search* in which the sequence $f\left(x_0 + (2^i + 1)h\right)$ for $i = 0, 1, \ldots$ is taken. The code fragment Code 2.4 illustrates this scheme.

```
         F0 = F(X0)
         X0 = X0+H
         F1 = F(X0)
         IF(ABS(F1).GT.ABS(F0))THEN
           X0 = X0-2.0D0*H
           H = -H
           F1 = F(X0)
         ENDIF
*
10       IF(F0*F1.GE.0.0D0)THEN
           H = 2.0D0*H
           X0 = X0+H
           F0 = F1
           F1 = F(X0)
           GOTO 10
         ENDIF
```

Code 2.4   Interval search.

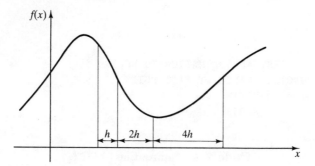

Figure 2.5   Failure to locate a bracket because h is too small.

Unfortunately, this method will fail if f has a local minimum but no root in the search direction. In Figure 2.5 a search starts in the direction of decreasing f but fails to locate a sign change. We thus need to impose a limit on the number of steps taken. An unsuccessful search would then be followed by a search with a different value of h. In Figure 2.5 we would need to increase h in order to make the search go in a negative direction, and hence locate the root shown. In other cases, the initial value of h may be so large that it misses the root (see Figure 2.6) and a reduction in the size of h is called for.

EXERCISE 2.5   Why can the bisection method not be used directly to locate a root α of $f(x) = 0$ of multiplicity r when r is even? Verify that the method can be used to locate α as a single root of $f(x)/f'(x) = 0$.

2.3   Methods of linear interpolation

The method of bisection uses only the sign of function values to locate a root. Its slow convergence rate leads us to enquire whether the situation could be improved by making use of the magnitude of f as well. In the method of *linear interpolation* we again begin with two values x_0 and x_1 which bracket the root. Assume $x_1 > x_0$. We join the coordinates $(x_0, f(x_0))$, and $(x_1, f(x_1))$ with a straight line, and note where this crosses the x-axis, to give x_2. This point, the *linear interpolant*, is defined by

$$x_2 = \frac{f_1}{f_1 - f_0}x_0 + \frac{f_0}{f_0 - f_1}x_1,$$

where $f_i = f(x_i)$, and this provides a new estimate of the root lying within the interval (x_0, x_1). We then proceed as for the bisection method; if $f_0 . f_2 < 0$ we discard x_1, otherwise we discard x_0. Linear interpolation is then used on the remaining two points to give x_3, and so on.

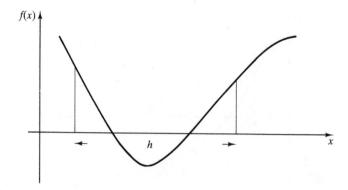

Figure 2.6 Failure to locate a bracket because h is too big.

Unfortunately the convergence rate of the method (known as *regula falsi* or the *method of false position*) is only linear (see Ralston and Rabinowitz [67], p. 339). By this we mean that if α is the root, $|\alpha - x_{k+1}| \approx L\,|\alpha - x_k|$ for some constant value L. Further, near the root one of the iteration points becomes fixed (see Exercise 2.6). An alternative approach is to drop the requirement that f be of opposite sign at the two interpolation points. The iterates are then generated using the formula

$$x_{k+1} = \frac{f_k}{f_k - f_{k-1}} x_{k-1} + \frac{f_{k-1}}{f_{k-1} - f_k} x_k. \qquad (2.3)$$

If f_k and f_{k-1} happen to be of opposite sign then x_{k+1} will simply be the linear interpolant, otherwise it will be the *linear extrapolant* (see Figure 2.7). The convergence rate of this, the *secant method*, is *superlinear*, that is, between 1 and 2, which is better than false position. It can be shown (Ralston and Rabinowitz [67], p. 341) that the rate is 1.62, that is $|\alpha - x_{k+1}| \approx L\,|\alpha - x_k|^{1.62}$. Unfortunately, convergence can no longer be guaranteed. Indeed, if $f_k = f_{k-1}$ the interpolating chord will be parallel to the x-axis!

The secant iteration (2.3) may also be written

$$x_{k+1} = x_k - \frac{x_k - x_{k-1}}{f_k - f_{k-1}} f_k. \qquad (2.4)$$

Our discussion of finite differences, to be given in Section 2.10, suggests that this is

$$x_{k+1} = x_k - f_k/f'_k \qquad (2.5)$$

with f'_k approximated by $(f_k - f_{k-1})/(x_k - x_{k-1})$. (2.5) is known as *Newton's method*; for a general function, f, it requires knowledge of f', which may be inconvenient. For polynomial equations derivatives are relatively easily determined, and we will return to the method in Section 2.6.

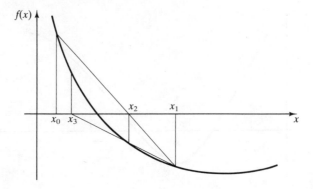

Figure 2.7   Secant iteration.

The subroutine SECANT uses the secant method to locate a zero of a function (see Code 2.5). The first two parameters X and H are used to define the two starting values as X and X + H. The routine is similar to BISECT in Section 2.2 in that termination is achieved either by a small f-value, or by a small value of $|x_{k+1} - x_k|$. Note that unlike the bisection method, there is no simple relation linking the distance between successive iterates and the size of $|\alpha - x_{k+1}|$. Since divergence is possible additional checks must be included in the code. In the example routine we limit the number of iterations, and hence the number of function evaluations, used to locate a root. The routine may return a failure indication, IFAIL = 3, even for a convergent process, if the accuracy tolerances are very strict and NFMAX has not been assigned a large enough value for the iteration to get sufficiently close to the root. This situation may arise if the starting values, X and X + H, are some distance from the root. Other failure exits are IFAIL = 1 if either NFMAX or the error tolerance EPS is found to be non-positive, and IFAIL = 2 if the method has encountered a chord parallel to the x-axis. If the program does fail to find a root, f should be analysed further with a view to determining more accurate starting values.

```
      SUBROUTINE SECANT(X,H,EPS,ETA,NFMAX,F,ROOT,FROOT,IFAIL)
*
*.. USES THE SECANT METHOD TO FIND A ROOT OF THE FUNCTION F.
*.. THE FIRST TWO ITERATES ARE X AND X+H.
*.. EPS IS USED TO TERMINATE THE ITERATION WHEN TWO
*.. SUCCESSIVE ITERATES ARE  CONSIDERED TO BE CLOSE.
*.. ETA TERMINATES THE PROCESS WHEN A SMALL VALUE OF F IS
*.. ENCOUNTERED. SETTING ETA TO ZERO ALLOWS EPS TO BE
*.. USED ONLY FOR CONTROLLING CONVERGENCE.
*.. NFMAX IS A LIMIT ON THE NUMBER OF SECANT ITERATIONS MADE.
```

```
*.. IF IFAIL = 0 ON EXIT
*..    ROOT CONTAINS THE APPROXIMATION TO THE ROOT
*..    AND FROOT THE VALUE OF F AT THAT POINT.
*.. IF IFAIL = 1 ON EXIT
*..    EPS .LE. 0 .OR. NFMAX .LE. 0
*.. IF IFAIL = 2 ON EXIT
*..    CHORD PARALLEL TO X AXIS DETECTED
*.. IF IFAIL = 3 ON EXIT
*..    SECANT ITERATION FAILED TO CONVERGE TO THE SPECIFIED
*..    TOLERANCES. ROOT CONTAINS THE LATEST ESTIMATE OF THE
*..    ROOT AND FROOT THE VALUE OF F AT THAT POINT.
      DOUBLE PRECISION EPS,ETA,F,FK,FKMIN1,FKPLU1,FROOT,H,
     +   ROOT,X,XK,XKMIN1,XKPLU1,ZERO
      INTEGER IFAIL,K,NFMAX
      PARAMETER (ZERO = 0.0D0)
      EXTERNAL F
      INTRINSIC ABS
      IFAIL = 0
      IF(EPS.LE.ZERO .OR. NFMAX.LE.0)THEN
        IFAIL = 1
        RETURN
      ENDIF
      XKMIN1 = X
      XK = XKMIN1 + H
      FKMIN1 = F(XKMIN1)
      FK = F(XK)
      DO 10 K = 1,NFMAX
*
*.. TEST FOR ZERO DERIVATIVE APPROXIMATION
        IF(FK-FKMIN1.EQ.ZERO)THEN
          IFAIL = 2
          RETURN
        ENDIF
        XKPLU1 = (FK*XKMIN1-FKMIN1*XK)/(FK-FKMIN1)
        FKPLU1 = F(XKPLU1)
*
*.. TEST FOR A SMALL VALUE OF F.
        IF(ABS(FKPLU1).LT.ETA)GOTO 20
*
*.. TEST FOR CONVERGENCE OF THE ITERATES.
        IF(ABS(XKPLU1-XK).LT.EPS)GOTO 20
*
*.. UPDATE THE ITERATION.
        XKMIN1 = XK
        FKMIN1 = FK
```

```
        XK = XKPLU1
        FK = FKPLU1
10      CONTINUE
*
*.. ITERATION FAILED TO CONVERGE
        IFAIL = 3
20      ROOT = XKPLU1
        FROOT = FKPLU1
        END
```

Code 2.5   Subroutine SECANT.

The secant scheme is a reasonably fast and efficient method for locating a root. However, the fact that convergence cannot be guaranteed is a serious drawback. The bisection method guarantees convergence, but at a slower rate. It would appear that some compromise between the two might offer an attractive alternative. Such a hybrid bisection/secant method has been proposed by Bus and Dekker [13] (see also Dekker [25] and Brent [11]). Starting with an initial bracketing interval a secant iteration is performed at each step. If this yields a point outside the bracketing interval the value is rejected and a bisection iteration is performed instead.

An alternative approach to improving the performance of the secant method is the *continuation method*, proposed by Swift and Lindfield [81]. Consider the sequence of problems

$$g\left(x, \theta_r\right) = f(x) - \theta_r f\left(x_0\right), \qquad 1 > \theta_1 > \theta_2 > \cdots > \theta_m = 0,$$

and observe that $g\left(x_0, 1\right) = 0$. Using x_0 as the initial guess, we use the secant method to solve $g\left(x, \theta_1\right) = 0$ to yield x_1. Repeating the process, we determine x_2, the root of $g\left(x, \theta_2\right) = 0$, using x_1 as the starting value, and so on. Eventually we solve $g\left(x, \theta_m\right) = 0$, whose root we label x_m. But $g\left(x, \theta_m\right) = g(x, 0) = f(x)$, and so x_m is a root of $f(x) = 0$. At each stage the next θ_i is chosen in a way which attempts to ensure that only a small number of iterations are required to solve the next subproblem. For further details and a performance analysis, see Swift and Lindfield [81].

EXERCISE 2.6   Near a root a function is likely to exhibit monotonic behaviour. Using a rough sketch, indicate what effect this has on the iterates in the method of false position.

EXERCISE 2.7   Write a subroutine which implements the rule of false position. By selecting a problem for which you know the root, monitor the convergence of the iterates. Using the values of $|\alpha - x_{n+1}|$ and $|x_{n+1} - x_n|$ determine experimentally the convergence rate of the method, and of secant iteration.

EXERCISE 2.8   The secant method may be written as (2.3), (2.4) or $x_{k+1} = \left(f_k x_{k-1} - f_{k-1} x_k\right)/\left(f_k - f_{k-1}\right)$. Discuss the relative merits of these formulae when x_k is close to a root.

2.4   The NAG C05 routines

Routines for determining a root of a general nonlinear function are to be found in the C05 chapter of the NAG library. (For locating all roots of a polynomial the routines in the C02 chapter should be used; these are discussed in Section 2.9.) The routines are divided into two groups, those for solving single equations (all classified under C05A) and those for solving systems of equations (named C05N, C05P and C05Z). The latter we discuss in Section 2.10.

We first consider C05ADF, C05AGF and C05AJF. These operate in what is known as *direct communication*. Essentially this means that they are true black-box routines.

C05AJF attempts to locate a simple zero of a function using an algorithm based on the continuation method described in the previous section. A call, which is similar to that for SECANT (Code 2.5), takes the form

    CALL C05AJF(X,EPS,ETA,F,NFMAX,IFAIL)

On entry X should contain an estimate of the position of the required root; on exit it contains the computed solution. EPS, ETA and NFMAX are again used to control the iteration. We remark that F, the routine which defines the function whose root we seek, must be declared in an EXTERNAL statement in the calling program unit.

We use this routine to solve equation (2.2), choosing initial values for X,EPS,ETA, and NFMAX of 0.5, 0.5×10^{-6}, 0.0 and 100 respectively. The number of function calls made can be monitored by declaring, in a COMMON block, an integer counter which is initialized to 0 before the call of C05AJF, and incremented by 1 inside the definition of F. A value for X of 0.860333435 is returned after 12 function evaluations, the value of f at this point being -0.569×10^{-6}.

We next consider the equation $e^{-x} - \delta = 0$ which has the root $\alpha = -\ln(\delta)$ if $\delta > 0$. In particular, if $\delta = 0.1 \times 10^{-8}$, $\alpha = 0.207232658 \times 10^2$. Suppose that we choose to use C05AJF to locate this root and select values for EPS, ETA and NFMAX as before, and X = 0.0. The routine fails with IFAIL = 3 (no zero near X or accuracy criteria too stringent). This is to be expected since our initial guess is some considerable distance from the actual root. Setting X = 21.0, the routine locates an approximate root at 0.207233331×10^2 after 12 function evaluations, the value of F at this point being -0.673×10^{-13}. Finally, if we choose X = 50.0 as a starting value C05AJF fails with IFAIL = 2 (internally calculated scale factor has the wrong order of magnitude for the problem). At this point the user is advised to change to the routine C05AXF where direct control over this factor is permitted. We return to this later but observe here that e^{-50} is very small indeed. If we evaluate $e^{-50} - 0.1 \times 10^{-8}$ within the program we see that the value returned is -0.1×10^{-8}; that is, e^{-50} is of no significance whatever. Near our initial value, therefore, F behaves like the constant function $F = -0.1 \times 10^{-8}$ (which, of course, has no root). It is not surprising that CO5AJF experiences some difficulty with this problem.

The hybrid bisection/secant algorithm of Bus and Dekker is implemented as NAG routine CO5ADF, a call of which takes the form

        CALL CO5ADF(A,B,EPS,ETA,F,X,IFAIL)

Using this routine to solve (2.2) with an initial bracketing interval, $[A,B]$, of $[0.5, 1.0]$, and EPS and ETA as before, we obtain a root estimate at $X = 0.860333572$ ($f(X) = -0.645 \times 10^{-7}$) after just 6 function evaluations. It is tempting to infer from this one example that CO5ADF might be more efficient than CO5AJF. However, the comparison made is unfair since the determination of an appropriate bracket, required by CO5ADF, is itself likely to involve function calls. Instead, a comparison should be made between CO5AJF and CO5AGF, for which a typical call takes the form

        CALL CO5AGF(X,H,EPS,ETA,F,A,B,IFAIL)

This routine also implements the Bus and Dekker algorithm, but its initial task is to locate a bracket for a root. Starting with an initial guess at the position of a root (supplied via X), the binary search method is used to locate a bracket, the step used in this search being the user supplied value H. If the routine fails to locate a bracket with this value of H it halts with IFAIL = 2. No attempt at increasing/reducing H is made; this must be done by the calling sequence. (Refer back to Section 2.2 for a discussion of the problems which are likely to be encountered in determining a bracket.) Note that H must be large enough so that $X + H \neq X$. On successful exit from the routine A and B define the bracket determined by the search, unless a point is subsequently found at which $f(x) < \text{ETA}$, in which case $X = A = B$ on exit.

Using CO5AGF to solve (2.2) again, with parameter values as in the call to CO5AJF, and $H = 0.1$, a root of 0.860333589 is located after 9 function calls, the value of f at this point being -0.183×10^{-9}. The bracket determined by the search process is $[0.8, 1.2]$. The number of function calls made is less than that for CO5AJF, but we could easily make CO5AGF appear less efficient by choosing a poor value for H. There can be no hard and fast rule as to which of CO5AGF and CO5AJF will prove superior for a given type of problem. We can say, however, that if a bracket for a root is available, CO5ADF is likely to make the least number of function calls.

We now look at the difference between using direct and reverse communication by considering CO5AZF, which is the reverse communication counterpart of CO5ADF. Its call sequence is

        CALL CO5AZF(X,Y,FX,TOLX,IR,C,IND,IFAIL)

This routine only performs a single step of the Bus and Dekker algorithm before returning to the calling unit for a function evaluation at X. A further call to CO5AZF must then be made, with FX equal to this value. Note that this is the only way that CO5AZF may obtain function information. However, not all the responsibility for controlling the computation falls on the user; for, while it is his job to construct the iteration loop, and to ensure that the calculation does not use too many function calls, the routine still looks after the termination criteria and error handling. If on exit the indicator variable IND has value 0 then the iteration should be terminated. The value of IFAIL returned will indicate whether a root has been found to the required accuracy, or an error has been detected. TOLX is a convergence tolerance and IR indicates whether a mixed (IR = 0), absolute (IR = 1) or relative (IR = 2) condition should be used. The workspace array C must be of length at least 17.

The two example drivers, Code 2.6 and Code 2.7, illustrate the difference in programming effort and style required to use CO5ADF and CO5AZF. The advantage of reverse communication is that the user may exercise more control over the iteration procedure. For example, if certain constraints on the required root are broken the user may choose either to abandon the computation, or to restart it with different initial values. In particular, the user may monitor the function values being used and observe whether a singularity or discontinuity exists. (The error flag IFAIL = 4 often indicates the presence of a pole in f, but the routine may halt with this error for other reasons.) CO5AZF is also to be preferred when the function F does not have a natural calling sequence as defined by CO5ADF.

```
      PROGRAM TSTADF
*
*.. EPS CONTROLS THE ACCURACY WITH WHICH THE
*..     ROOT IS COMPUTED
*.. ETA CONTROLS THE MAGNITUDE OF F(X). GENERALLY
*..     WE SET ETA = 0.
      INTEGER NOUT,IFAIL
      DOUBLE PRECISION A,B,EPS,ETA,F,ONE,TOL,X,ZERO
      PARAMETER (ZERO = 0.0D0,ONE = 1.0D0,TOL = 1.0D-5,
     +           NOUT = 6)
      EXTERNAL CO5ADF,F
*
      EPS = TOL
      ETA = ZERO
*
*.. A AND B BRACKET THE ROOT
      A = ZERO
      B = ONE
*
```

```
*.. C05ADF DOES THE WHOLE JOB - USE HARD ERROR OPTION
      IFAIL = 0
      CALL C05ADF(A,B,EPS,ETA,F,X,IFAIL)
*
*.. HARD FAIL OPTION - ROUTINE MUST HAVE SUCCEEDED
*.. TO HAVE GOT HERE - CHECK SIZE OF F
      WRITE(NOUT,101)X,X,F(X)
*
  101 FORMAT(' APPROXIMATION TO THE ZERO = ',E16.8
     +  /' F(',E12.4,' )=',E12.4)
      END
```

Code 2.6   Use of C05ADF.

Routine C05AVF also works in reverse communication. Its call sequence is

```
      CALL C05AVF(X,FX,H,BOUNDL,BOUNDU,Y,C,IND,IFAIL)
```

This routine just attempts to locate a bracket for a root using the binary search technique. Searches with step sizes equal to H, $0.1 \times$ H, $0.01 \times$ H and $0.001 \times$ H are used in an attempt to determine a bracket. The parameters BOUNDL and BOUNDU restrict the region in which a search is to be made. The workspace array C needs to be of length at least 11. On successful final exit the bracket for a root is defined by X and Y.

```
      PROGRAM TSTAZF
*
*.. SET THE ACCURACY REQUIREMENT - THIS USES BOTH
*.. TOL AND IR. FOR GENERAL PURPOSE USE IR=0 IS
*.. GENERALLY APPROPRIATE.
      INTEGER IFAIL,IND,IR,NOUT
      DOUBLE PRECISION BRACK,C(17),F,FX,ONE,TOL,X,ZERO
      PARAMETER (NOUT = 6,TOL = 1.0D-5,ZERO = 0.0D0,
     +           ONE = 1.0D0)
      EXTERNAL C05AZF,F
*
      IR = 0
*
*.. X CONTAINS AN APPROXIMATION TO THE ROOT, BRACK
*.. IS A SECOND APPROXIMATION SUCH THAT
*..        F(X)*F(BRACK) .LT. 0
      X = ZERO
```

```
      BRACK = ONE
*
*.. WE USE IND=1 TO START - THIS MEANS WE DON'T
*.. COMPUTE ANY INITIAL FUNCTION VALUES. THE SOFT ERROR
*.. OPTION IS SELECTED TO ILLUSTRATE THE ERROR HANDLING.
      IND = 1
      IFAIL = 1
*
*.. MAIN ITERATION LOOP
   10 CONTINUE
      CALL C05AZF(X,BRACK,FX,TOL,IR,C,IND,IFAIL)
*
*.. CHECK FOR FAILURE - IF HARD ERROR OPTION IS USED
*.. THE PROGRAM WILL HALT INSIDE C05AZF
      IF(IND.EQ.0 .AND. IFAIL.NE.0)THEN
         WRITE(NOUT,100)IFAIL
         WRITE(NOUT,101)X,BRACK
*
*.. C05AZF REQUIRES A FUNCTION VALUE
      ELSE IF(IND.NE.0)THEN
         FX = F(X)
*
*.. THE PREVIOUS STATEMENT COULD HAVE BEEN IN-LINE CODE
*.. E.G.  FX = EXP(-X)-SIN(X)
*.. CODE MAY ALSO BE INSERTED HERE TO COUNT FUNCTION
*.. EVALUATIONS, ETC.
         GOTO 10
*
*.. OTHERWISE WE HAVE IND=0 AND IFAIL=0 AND THE
*.. ROUTINE HAS SUCCESSFULLY TERMINATED
      ELSE
         WRITE(NOUT,102)
         WRITE(NOUT,101)X,BRACK
         WRITE(NOUT,103)X,F(X)
      ENDIF
*
  100 FORMAT(' C05AZF FAILS - IFAIL = ',I3)
  101 FORMAT(' BEST APPROXIMATION TO ROOT = ',
     +   E16.8/' BRACKETING VALUE          = ',E16.8)
  102 FORMAT(' C05AZF SUCCESSFUL')
  103 FORMAT(' F(',E12.4,' ) = ',E12.4)
      END
```

Code 2.7   Use of C05AZF.

C05AXF is the reverse communication version of C05AJF. Its call sequence takes the form

CALL C05AXF(X,FX,TOL,IR,SCALE,C,IND,IFAIL)

where C is a workspace array of length at least 26. If the gradient of f is close to zero at, or near, the root (e.g., α is a point of inflexion) the secant method may experience difficulties. Users of C05AXF have some control over this via the parameter SCALE. It is used to decide whether the calculation of derivative estimates involves loss of significance due to cancellation errors.

We return to the problem of locating the root of $e^{-x} - \delta = 0$ where $\delta = 0.1 \times 10^{-8}$. Recall that if we look for a root near 50.0 and EPS, ETA and NFMAX are given the values 0.5×10^{-6}, 0.0 and 100 respectively, C05AJF fails with IFLAG $= 2$. We take the advice given in the routine documentation and attempt to solve the problem using C05AXF. We take the further advice to give SCALE the value $50.0 \times \sqrt{\text{X02AJF}(0.0)}$. (The function X02AJF returns the machine epsilon – see Exercise 1.20.) To TOL, IR and IND we assign the values 0.5×10^{-6}, 1 and 1, respectively. The routine fails with IFAIL $= 3$ (SCALE too small or significant derivatives cannot be computed). If we multiply the original value of SCALE by 10, 100, etc. the routine fails in the same manner and we conclude that it is derivative estimation which is proving to be a problem. We have already observed (Section 2.3) that secant iteration is a limiting case of Newton's method, with the derivative of f replaced by a finite difference approximation. However, near $x = 50$ the derivative of $e^{-x} - \delta = 0$ is, computationally, zero, and hence we can expect C05AJF and C05AXF to experience difficulties in making progress towards the root.

EXERCISE 2.9   Make a comparative study of the performance of C05AGF and C05AJF. Construct equations and starting values which trip the error mechanisms. Such testing of the error exits forms part of the stringent test programs which are used to test thoroughly each implementation of the NAG library. These programs attempt to exercise all the code in each library routine and to ensure that each routine performs to specification even at the extremities of the floating-point number system (see also Exercise 1.19).

2.5   Polynomial equations

Due to their simple form, mathematically speaking, and the ease with which they may be evaluated, differentiated and integrated, polynomials occur in many areas of numerical analysis, e.g., quadrature, interpolation and the approximation of data and functions.

A polynomial of degree n is defined to be a function of the form

$$p_n(x) \equiv a_0 x^n + a_1 x^{n-1} + \cdots + a_n = \sum_{j=0}^{n} a_{n-j} x^j, \qquad (2.6)$$

where the coefficients $\{ a_j \mid j = 0, 1, \ldots, n \}$ may be real or complex. One nice property polynomials possess is that the number of roots may be determined by inspection; a polynomial of degree n has exactly n roots. These may be real or complex but, provided the coefficients are real, complex roots always occur in *conjugate pairs*, that is, if $a + ib$ is a root then so is $a - ib$, where $i = \sqrt{-1}$. Roots may not be simple, in which case we refer to a multiple root. For example the cubic polynomial, $p_3(x) \equiv x^3 - 4x^2 + 5x - 2 = (x - 1)^2(x - 2)$ has a simple root at $x = 2$, but a double root (root of multiplicity two) at $x = 1$. We recall that at a non-simple root the gradient of $p_n(x)$ will be zero, and we shall need to use special techniques to take account of this behaviour.

Finding the roots of polynomials is of practical importance. An example application area is in *critical phenomena* calculations where the required critical points are defined to be the poles (points of singularity) of certain rational functions of the form $p_n(x)/q_m(x)$, where $p_n(x)$ and $q_m(x)$ are polynomials of degree n and m respectively. The explicit determination of these critical points requires the computation of the roots of $q_m(x)$. For more details see, for example, Amit [2].

One important area of linear algebra, the calculation of the *eigenvalues* of an $n \times n$ matrix A (that is, values of λ for which $Ax = \lambda x$ for some *eigenvector* **x**), may be formulated in terms of the roots of a polynomial of degree n. While this reformulation is of interest theoretically, computationally it is not an effective approach to the problem. It is, however, not surprising that many of the problems associated with the efficient and accurate numerical calculation of eigenvalues have direct counterparts in numerical polynomial root finding.

Of course, any of the root-finding methods discussed earlier in this chapter may, in principle, be used to locate a real root of a real-valued polynomial. However, because of the ease with which polynomials and, more importantly, their derivatives may be evaluated, the use of Newton's method is potentially more attractive, and we now consider this algorithm in detail.

EXERCISE 2.10 Let $p_n(x)$ be a polynomial of degree n with real coefficients. *Descartes' rule of signs* states that the number, *pos*, of real positive roots of $p_n(x)$ is at most m, the number of sign changes in the sequence $a_0, a_1, \ldots, a_n$ (zero coefficients are ignored). Moreover, $m - pos$ is a nonnegative even integer. How may the number of negative real roots of $p_n(x)$ be determined? What can you say about the roots of $8x^5 - 6x^4 - 3x^2 + 5x - 4$?

2.6 Newton's method

Suppose f is a general real-valued function and consider Figure (2.8). If x_{k+1} is the point at which the gradient to the curve $y = f(x)$ at $x = x_k$

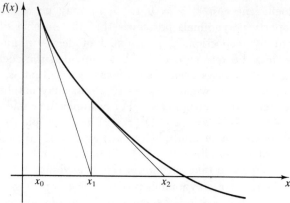

Figure 2.8   Newton iteration.

cuts the x-axis, we have $f'(x_k) = f(x_k)/(x_k - x_{k+1})$ or, rearranging,

$$x_{k+1} = x_k - \frac{f(x_k)}{f'(x_k)},$$

and this is *Newton's method* (often referred to as the *Newton–Raphson* method). In the vicinity of a simple root α, it can be shown that the iteration is quadratically convergent (see Nonweiler [60], p. 165), provided that certain conditions are satisfied. In essence quadratic convergence means that if x_k is 'sufficiently close' to the root then $|\alpha - x_{k+1}| \approx L|\alpha - x_k|^2$ for some L, and this is a much faster rate of convergence than that possessed by any of the root-finding methods of earlier sections. In computational terms it means that we would expect to double the number of correct significant digits at each iteration as long as $L \sim O(1)$. This promise of rapid convergence near the root makes Newton's method an extremely popular scheme, although it does have its drawbacks. For example, if x_k is a local maximum of f the method breaks down since $f'(x_k)$ is zero. Like the secant method, of which Newton's method may be considered a limiting case, there is no guarantee of convergence to a root from an arbitrary initial value, x_0. Even where the method does converge it may not find the root closest to x_0. Furthermore, it may not be efficient, or indeed possible, to compute the derivative of the function; the evaluation of f may, for example, involve the numerical solution of a system of ordinary differential equations. It is therefore necessary to consider carefully whether the advantages of the increased speed of convergence are outweighed by the additional and repeated cost of the derivative evaluations. However, in the particular case when f is a polynomial, both function and derivative values may be computed cheaply and easily as we show in Section 2.7.

For a polynomial equation, a single Newton iteration takes the form

$$x_{k+1} = x_k - \frac{p_n(x_k)}{p'_n(x_k)}. \tag{2.7}$$

Now, if α is a root of $p_n(x)$ of multiplicity r, then

$$p_n(x) \equiv (x - \alpha)^r q_{n-r}(x), \tag{2.8}$$

where $q_{n-r}(x)$ is a polynomial of degree $n - r$. Differentiating (2.8) we find

$$p'_n(x) = (x - \alpha)^r q'_{n-r}(x) + r(x - \alpha)^{r-1} q_{n-r}(x),$$

and hence, for $r > 1$, α is also a root of $p'_n(x)$, and we can expect trouble with (2.7) since both the numerator and the denominator of the *correction term* $p_n(x_k)/p'_n(x_k)$ will tend to zero as x_k approaches α. Any rounding errors in the computation of $p_n(x_k)$ will be magnified. Furthermore it may be shown (see Nonweiler [60], p. 175) that the rate of convergence of Newton's method close to the root is reduced from quadratic to linear. Assuming that we are in the, admittedly unlikely, position of knowing the multiplicity of the root, we can restore quadratic convergence by using the *modified Newton iteration*

$$x_{k+1} = x_k - r\frac{p(x_k)}{p'(x_k)},$$

(see Nonweiler [60], p. 177 and Exercise 2.11).

EXERCISE 2.11 Newton's method may be written as $x_{k+1} = g(x_k)$ where $g(x) = x - f(x)/f'(x)$. Show that $g'(x) = 0$ when f has a simple root. Connect this result with Exercise 2.2. Show also that $g(x) = x - rf(x)/f'(x)$ satisfies $g'(\alpha) = 0$ when f is a polynomial and has a root of multiplicity r at α.

EXERCISE 2.12 Write down the Taylor series (1.9) for $f(x + h)$, expanded about the point x. Show that if α is a root of $f(x) = 0$, Newton's method may be derived by setting $\alpha = x + h$ and ignoring second-order terms in h.

2.7 Polynomial evaluation and deflation

A naive approach to the evaluation of a polynomial of degree n would involve

- $O(n)$ exponentiations to form the x^k for $k = 0, 1, \ldots, n$,

- $O(n)$ multiplications to multiply the powers of x by their respective coefficients,

- $O(n)$ additions to sum the individual terms.

We may reduce the operation count by observing that (2.6) may be written in *nested multiplication* form as

$$p_n(x) \equiv \left(\cdots\left((a_0 x + a_1)x + a_2\right)x + \cdots + a_{n-1}\right)x + a_n.$$

If x_k is an iteration value for Newton's method, we may evaluate $p_n(x_k)$ using the recurrence relation

$$
\begin{aligned}
b_0 &= a_0, \\
b_j &= a_j + b_{j-1}x_k, \qquad j = 1, 2, \ldots, n, \qquad \text{(2.9)} \\
p_n(x_k) &= b_n.
\end{aligned}
$$

This may be implemented by the following code fragment

```
      PNXK = A(0)
      DO 10 J=1,N
      PNXK = A(J)+PNXK*XK
10    CONTINUE
```

where A is an array of length $N+1$ containing the polynomial coefficients $\{\, a_j \mid j = 0, 1, \ldots, n \,\}$. Using (2.9) to evaluate $p_n(x_k)$, we avoid the need to perform any exponentiations, and are left with $O(n)$ multiplications and additions. Note that it is not necessary to store the b_js in an array; we simply overwrite PNXK as the next term is added in. Later we shall show that the b_js are of some significance if we wish to compute more than one root of $p_n(x)$. For even faster ways of evaluating polynomials see Knuth [49], pp. 471–505.

Consider the polynomial of degree $n-1$

$$q_{n-1}(x) \equiv b_0 x^{n-1} + b_1 x^{n-2} + \cdots + b_{n-1},$$

where the b_js have been calculated using (2.9). Then

$$
\begin{aligned}
(x - x_k)\, q_{n-1}(x) &= b_0 x^n + (b_1 - b_0 x_k)x^{n-1} + \cdots + (b_{n-1} - b_{n-2}x_k)x \\
&\quad + (-b_{n-1}x_k).
\end{aligned}
$$

Now, from (2.9),

$$
\begin{aligned}
a_0 &= b_0, \\
a_j &= b_j - b_{j-1}x_k, \qquad j = 1, 2, \ldots, n-1,
\end{aligned}
$$

and we see that $(x - x_k)\, q_{n-1}(x) = p_n(x) - b_n$, giving

$$p_n(x) = (x - x_k)\, q_{n-1}(x) + p_n(x_k). \qquad \text{(2.10)}$$

Differentiating (2.10) we find $p_n'(x) = (x - x_k)\, q_{n-1}'(x) + q_{n-1}(x)$, so that $p_n'(x_k) = q_{n-1}(x_k)$, and the value of the derivative required by (2.7) may be determined using the nested multiplication algorithm on $q_{n-1}(x)$. The code fragment in Code 2.8 illustrates the method.

```
      PNXK = A(0)
      PDIFF = A(0)
      DO 10 J = 1,N-1
         PNXK = A(J)+PNXK*XK
         PDIFF = PNXK+PDIFF*XK
10    CONTINUE
      PNXK = A(N)+PNXK*XK
```

Code 2.8   Nested multiplication.

Note that by evaluating the polynomial and its derivative at the same time we again avoid the need to store the b_js explicitly.

Considering (2.10) further, we see that if α is a root of $p_n(x)$ then $p_n(x) = (x - \alpha)q_{n-1}(x)$, that is, the $\{\, b_j \mid j = 0,1,\ldots,n-1 \,\}$ are the coefficients of the polynomial obtained by dividing out the root from the original polynomial. The roots of $q_{n-1}(x)$ are the remaining $n - 1$ roots of $p_n(x)$. We call this process of factoring out known roots *deflation*, and the resulting polynomial, $q_{n-1}(x)$, is referred to as the *deflated polynomial*. The use of (2.9), sometimes known as *Horner's algorithm*, to deflate a polynomial is also referred to as *forward division*. It is possible to perform the deflation process so as to obtain the coefficients of the deflated polynomial in reverse order. Thus we may write

$$
\begin{aligned}
p_n(x) &= a_n + a_{n-1}x + \cdots + a_0 x^n \\
&= (-\alpha + x)\left(c_{n-1} + c_{n-2}x + \cdots + c_0 x^{n-1}\right) - \alpha c_{-1}x^n,
\end{aligned}
$$

which, when $\alpha \neq 0$, leads to the recurrence relation

$$
\begin{aligned}
c_{n-1} &= -a_n/\alpha, \\
c_j &= \left(c_{j+1} - a_{j+1}\right)/\alpha, \qquad j = n-2, n-1, \ldots, -1,
\end{aligned}
\tag{2.11}
$$

and this is known as *backward division*. Using an exact root and exact arithmetic throughout would give the same deflated polynomial using backward or forward division. In such a case we would find $b_n = 0$ in (2.9) and $c_{-1} = 0$ in (2.11). If an approximate root and finite-precision arithmetic are used this condition is unlikely to hold exactly.

To study the effect of backward and forward deflation we consider the cubic polynomial equation $x^3 + 9792.9001x^2 + 8813.6931x + 0.80372736 = 0$. The exact roots are at -9792.0000, -0.90000000 and $-0.91200000 \times 10^{-4}$, all correct to 8 significant digits. We look at how errors in the first computed root may affect the accuracy with which the remaining roots are calculated. Figure 2.9 shows the results obtained using the forward division

First root extracted	Roots of the deflated quadratic polynomial		b_3
-0.979201×10^4	Complex roots		-0.957903×10^{-6}
-0.900001×10^0	-0.901980×10^{-4}	-0.979200×10^4	0.882947×10^{-2}
-0.920000×10^{-4}	-0.899999×10^0	-0.979200×10^4	-0.704952×10^{-2}

Figure 2.9   Results using forward division algorithm.

algorithm which has been implemented as subroutine HORNER in Code 2.9. In each case we have perturbed the first root extracted by a small amount. We see that the order in which the roots are computed can have a dramatic effect on the accuracy of any remaining roots. In fact if the largest root is extracted first the deflated polynomial is found to have complex roots even though b_3 is small. The results obtained using the smallest root first are acceptable.

For the backward division algorithm we find (Figure 2.10) that the reverse is true. Removing the smallest root first leads to a large relative error in both roots of the deflated polynomial, while dividing out the largest root first gives acceptable results.

We must obviously exercise caution when implementing the deflation process and, in particular, we need to ensure that the roots obtained from the reduced polynomials do not diverge from those of the original. If the roots of a polynomial are very ill-conditioned with respect to their defining coefficients, it would be unreasonable to expect deflation to improve the situation.

The numerical problems illustrated in the example above are due principally to the accuracy with which the initial roots have been computed. Increasing the precision of the arithmetic used in calculating the deflated polynomial coefficients will make little difference. Thus we must ensure that the roots are found in a particular order and/or modify the deflation algorithm. In fact the algorithm is easily improved by combining the forward and backward division algorithms into a *composite deflation* scheme. This method generates a deflated polynomial of the form

$$b_0 x^{n-1} + b_1 x^{n-2} + \cdots + b_{r-1} x^{n-r} + \tfrac{1}{2} \left(b_r + c_r \right) x^{n-r-1}$$
$$+ c_{r-1} x^{n-r-2} + \cdots + c_{n-1},$$

where the b_j and c_j are computed using (2.9) and (2.11) respectively. The choice of r may be made in several ways; a simple method is to deter-

First root extracted	Roots of the deflated quadratic polynomial		c_{-1}
-0.979201×10^4	-0.912000×10^{-4}	-0.899999×10^0	0.104192×10^{-9}
-0.900001×10^0	-0.912000×10^{-4}	-0.979199×10^4	-0.134574×10^{-1}
-0.920000×10^{-4}	-0.103672×10^{-1}	-0.842675×10^6	0.984030×10^{14}

Figure 2.10   Results using backward division algorithm.

mine the value of r which minimizes $|b_r - c_r| / (|b_r| + |c_r|)$ taken over all non-zero $(|b_r| + |c_r|)$. For a full discussion, including other methods for choosing r, see Peters and Wilkinson [62].

```
      SUBROUTINE HORNER(A,N,XVAL,B,PNX,PDIFFX)
*
*.. GIVEN A POLYNOMIAL, P(X), DEFINED BY THE COEFFICIENTS
*.. A(0),...,A(N) SUCH THAT
*..     P(X) = A(0)X**N + A(1)X**N-1+ ... + A(0)
*.. AND A REAL VALUE XVAL, THIS ROUTINE RETURNS THE VALUES
*.. OF P(XVAL) AND P'(XVAL) IN PNX AND PDIFFX RESPECTIVELY.
*.. ON EXIT B(0),...,B(N-1) CONTAIN THE COEFFICIENTS
*.. OF THE POLYNOMIAL Q(X) DEFINED BY
*..     P(X) = (X - XVAL)Q(X) + P(XVAL)
*.. NOTE N IS ASSUMED TO BE AT LEAST ONE.
      INTEGER I,N
      DOUBLE PRECISION A(0:N),B(0:N-1),PDIFFX,PNX,XVAL
      B(0) = A(0)
      PDIFFX = A(0)
      DO 10 I = 1,N-1
        B(I) = XVAL*B(I-1)+A(I)
        PDIFFX = PDIFFX*XVAL+B(I)
10    CONTINUE
      PNX = XVAL*B(N-1)+A(N)
      END
```

Code 2.9   Subroutine HORNER.

Roots obtained using the deflation technique should always be *purified* by iteration, using Newton's method, on the original polynomial with the deflated roots as starting values. Generally this requires only a single iteration (plus one more to check for convergence). We could attempt to locate all the roots from the original polynomial without using deflation at all. However, this will only be achieved at an increased cost. Furthermore, we have no control over which root the iteration will find from a given starting value so that the process may well locate the same root more than once.

EXERCISE 2.13 Write a routine to replace HORNER which implements

- the backward division algorithm,
- the composite deflation technique.

2.8   Newton's method and complex roots

The discussion so far has assumed that both the coefficients of the polynomial and its roots are real. We now show how Newton's method may be modified to allow for the possibility of complex roots. Consider the polynomial

$$p_n(z) = a_0 z^n + a_1 z^{n-1} + \cdots + a_n, \qquad (2.12)$$

where $\{\, a_j \mid j = 0, 1, \ldots, n \,\}$ are real coefficients and z may be real or complex. We may write $p_n(z) = R(x,y) + iS(x,y)$, where R and S are real functions and $z = x + iy$. To locate a root of (2.12) we attempt to solve the pair of nonlinear simultaneous equations

$$
\begin{aligned}
R(x,y) &= 0 \\
S(x,y) &= 0.
\end{aligned}
\qquad (2.13)
$$

The extension of Newton's method to solve the system of m equations in m unknowns

$$\mathbf{f}(\mathbf{x}) = 0,$$

where $\mathbf{f}^T(\mathbf{x}) = \big(f_1(\mathbf{x}), f_2(\mathbf{x}), \ldots, f_m(\mathbf{x})\big)$ and $\mathbf{x}^T = (x_1, x_2, \ldots, x_m)$, is

$$\mathbf{x}_{k+1} = \mathbf{x}_k - J_k^{-1} \mathbf{f}(x_k), \qquad (2.14)$$

where J, the *Jacobian* matrix associated with $\mathbf{f}$, is given by

$$
J_k =
\begin{pmatrix}
\frac{\partial f_1}{\partial x_1} & \frac{\partial f_1}{\partial x_2} & \cdots & \frac{\partial f_1}{\partial x_m} \\[6pt]
\frac{\partial f_2}{\partial x_1} & \frac{\partial f_2}{\partial x_2} & \cdots & \frac{\partial f_2}{\partial x_m} \\[6pt]
\vdots & \vdots & & \vdots \\[6pt]
\frac{\partial f_m}{\partial x_1} & \frac{\partial f_m}{\partial x_2} & \cdots & \frac{\partial f_m}{\partial x_m}
\end{pmatrix}
\qquad (2.15)
$$

and the subscript k indicates that all partial derivatives are evaluated at $\mathbf{x}_k$. Note that when implementing (2.14), the inverse Jacobian matrix need not be calculated; rather, each iteration is recast as a system of linear equations of the form

$$J_k \Delta \mathbf{x}_k = -\mathbf{f}(\mathbf{x}_k),$$

where $\Delta \mathbf{x}_k = \mathbf{x}_{k+1} - \mathbf{x}_k$ is the correction term.

For the system (2.13), the iteration (2.14) gives

$$
\begin{pmatrix} x_{k+1} \\ y_{k+1} \end{pmatrix}
=
\begin{pmatrix} x_k \\ y_k \end{pmatrix}
-
\begin{pmatrix} (R_x)_k & (R_y)_k \\ (S_x)_k & (S_y)_k \end{pmatrix}^{-1}
\begin{pmatrix} R_k \\ S_k \end{pmatrix}
\qquad (2.16)
$$

where $R_x = \partial R / \partial x$, $R_y = \partial R / \partial y$, $S_x = \partial S / \partial x$, $S_y = \partial S / \partial y$, and the subscript k again denotes evaluation at $\mathbf{x}_k$. Using the Cauchy–Riemann equations (see Copson [20], pp. 40–41) we obtain $(R_x)_k = (S_y)_k$ and $(R_y)_k = -(S_x)_k$ so that the system (2.16) becomes

$$\begin{pmatrix} x_{k+1} \\ y_{k+1} \end{pmatrix} = \begin{pmatrix} x_k \\ y_k \end{pmatrix} - \frac{1}{(R_x)_k^2 + (S_x)_k^2} \begin{pmatrix} R_k(R_x)_k + S_k(S_x)_k \\ S_k(R_x)_k - R_k(S_x)_k \end{pmatrix}. \qquad (2.17)$$

Now, since all coefficients in (2.12) are real, we know that if $x + iy$ is a root then so is $x - iy$. We consider a factor of the form

$$\big(z - (x + iy)\big)\big(z - (x - iy)\big) = z^2 - uz - v,$$

where it is assumed that $y \neq 0$ and $u \equiv 2x$ and $v \equiv x^2 + y^2$ are real. Hence, for arbitrary z we may write

$$p_n(z) = (z^2 - uz - v)q_{n-2}(z) + b_{n-1}z + b_n, \qquad (2.18)$$

where

$$q_{n-2}(z) = b_0 z^{n-2} + b_1 z^{n-1} + \cdots + b_{n-3}z + b_{n-2},$$

and the $\{\, b_j \mid j = 0, 1, \ldots, n \,\}$ are real. Hence

$$p_n(x + iy) = (b_{n-1}x + b_n) + ib_{n-1}y,$$

and if $x + iy$ is a root of $p_n(z)$, we have

$$p_n(z) \equiv (z^2 - uz - v)q_{n-2}(z),$$

and $q_{n-2}(z)$ is the deflated polynomial.

Comparing coefficients in (2.18), we find

$$\begin{aligned} a_0 &= b_0, \\ a_1 &= b_1 - ub_0, \\ a_j &= b_j - ub_{j-1} - vb_{j-2}, & j = 2, 3 \ldots, n-1, \\ a_n &= b_n - vb_{n-2}, \end{aligned}$$

which suggests the Horner-like algorithm

$$\begin{aligned} b_0 &= a_0, \\ b_1 &= a_1 + ub_0, \\ b_j &= a_j + ub_{j-1} + vb_{j-2}, & j = 2, 3, \ldots, n-1, \\ b_n &= a_n + vb_{n-2}. \end{aligned}$$

Hence

$$p_n(x + iy) = R + iS = (xb_{n-1} + b_n) + iyb_{n-1},$$

yielding a fast algorithm for the evaluation of R and S in (2.17). For the derivatives we note that

$$p_n'(z) = (z^2 - uz - v)q_{n-2}'(z) + (2z - u)q_{n-2}(z) + b_{n-1},$$

giving

$$p'_n(x + iy) = (2x - u + 2iy)q_{n-2}(x + iy) + b_{n-1}$$
$$= 2iyq_{n-2}(x + iy) + b_{n-1}.$$

Let

$$q_{n-2}(z) = (z^2 - uz - v)t_{n-4}(z) + c_{n-3}z + c_{n-2},$$

where

$$t_{n-4}(z) = c_0 z^{n-4} + c_1 z^{n-3} + \cdots + c_{n-5}z + c_{n-4}.$$

Equating coefficients we obtain

$$
\begin{aligned}
c_0 &= b_0, \\
c_1 &= b_1 + uc_0, \\
c_j &= b_j + uc_{j-1} + vc_{j-2}, \qquad j = 2, 3, \ldots, n - 3, \\
c_{n-2} &= b_{n-2} + vc_{n-4},
\end{aligned}
$$

which gives

$$q_{n-2}(x + iy) = (c_{n-3}x + c_{n-2}) + iyc_{n-3}.$$

Hence

$$p'_n(x + iy) = (b_{n-1} - 2c_{n-3}y^2) + i2y(c_{n-3}x + c_{n-2}) = R_x + iS_x,$$

and since $p'_n(x + iy) = R_x + iS_x$ (see Copson [20], p. 40) this enables R_x and S_x to be evaluated efficiently.

EXERCISE 2.14   Write a program which implements the complex version of Horner's algorithm.

2.9   The NAG polynomial solvers

The C02 chapter contains two routines for computing the roots of polynomials. C02ADF attempts to find all the roots of a polynomial with complex coefficients, whilst C02AEF does the same for the real coefficient case. If only some of the real roots are required the use of a C05 routine (see Section 2.4) might be more appropriate.

Both C02ADF and C02AEF use the method of Grant and Hitchins [39] which is based on the Newton iteration described in Section 2.8. The solution of the system (2.13) is equivalent to the minimization of the quantity $\Phi = R^2(x, y) + S^2(x, y)$, and an appropriate search is incorporated into the iteration to generate an initial root estimate. Note that in the complex coefficient case the complex roots do not generally appear in conjugate pairs and, hence, must be factored out one at a time. An attempt is made in both routines to evaluate the roots to as high an accuracy as possible to avoid problems with deflation.

Since the use of the two routines is so similar, we confine our attention to the slightly simpler C02AEF routine which is probably of more practical use. A typical call is of the form

```
CALL CO2AEF(A,N,REZ,IMZ,TOL,IFAIL)
```

Note that the degree of the polynomial, n, is given by $N - 1$, N being used to dimension the arrays. The array A should contain the coefficients such that $A(1)$ is set to the coefficient of z^n, $A(2)$ to the coefficient of z^{n-1}, and so on. (CO2AEF was written to the Fortran 66 standard when arrays had to be indexed from 1.) The roots are returned in two arrays; REZ contains the real parts of the computed roots, and IMZ the imaginary parts. Do not forget to declare the type of IMZ explicitly, otherwise it will default to an INTEGER array! The jth root is thus given by $REZ(J) + i IMZ(J)$. The use of two arrays is necessary for portability because the Fortran standard does not define a double-precision complex data type (see Section 1.11). On exit N is used, in a rather unnatural way, to signify how many roots have been successfully computed. Thus if, on input, the value of N is N_1 and on output it is N_2, then the number of roots successfully computed is $N_1 - N_2 + 1$; i.e., $N_2 - 1$ roots have *not* been found. A value of 1 means that all roots have been determined. Only in exceptional circumstances, which we discuss below, will N be other than 1 on exit. Beware that on exit the polynomial coefficients will also have been destroyed. The parameter TOL is used to determine the accuracy to which roots are computed; the recommended value is that returned by XO2AJF. The use of values of TOL in excess of this will cause a loss of accuracy in the computed roots. If, on input, TOL is less than this value, it will reset to XO2AJF internally.

A failure with IFAIL = 1 means that one of the input parameters is incorrect. A value of IFAIL = 2 means that the routine has been unable to obtain all the roots of the polynomial. This is generally due to detection of a *saddle point*; a point where both the first and second derivatives of the, possibly deflated, polynomial are zero. In this situation the soft error option may be useful, since this error need not be catastrophic, and, provided additional information can be supplied, the computation may be restarted. On exit the value of N gives the number of coefficients in the deflated polynomial which caused the problem. To restart the calculations it is necessary just to provide a new starting value in $REZ(1) + i IMZ(1)$ and to recall CO2AEF. Note that the values of all other parameters, including N, should remain unchanged.

As an example, consider the polynomial equation $x^{15} + 1 = 0$. For this problem CO2AEF returns with IFAIL = 2 and $N = 16$; that is, the algorithm has run into difficulties with the original polynomial. Restarting the computation at the point $-1 + 0i$ (an obvious root) all roots are found without further difficulties.

Finally, we note that when computing the roots of a polynomial it is advisable to perturb the coefficients slightly, say by a few units of the order of accuracy to which they are known, and to resolve. This will help to detect whether or not the root positions are ill-conditioned with respect to the defining coefficients.

EXERCISE 2.15 How well does CO2AEF perform on polynomials of the form $(x+1)^r$ for $r = 2, 3, \ldots$? Explain.

EXERCISE 2.16 Use CO2AEF to compute the roots of the tenth-order polynomial

$$
\begin{aligned}
p_{10}(x) = \; & x^{10} - 55x^9 + 1320x^8 - 18150x^7 + 157773x^6 \\
& - 902055x^5 + 3416930x^4 - 8409500x^3 \\
& + 12753576x^2 - 10628640x + 3628800.
\end{aligned}
$$

Change the coefficient of x^2 to 12753756 and repeat. Explain.

2.10   Systems of nonlinear equations

We now consider the problem of finding an approximate solution to a system of m simultaneous, nonlinear equations in m unknowns of the form

$$
\begin{aligned}
f_1(x_1, x_2, \ldots, x_m) &= 0 \\
f_2(x_1, x_2, \ldots, x_m) &= 0 \\
&\;\;\vdots \\
f_m(x_1, x_2, \ldots, x_m) &= 0.
\end{aligned}
$$

This system may be written in vector form as

$$
\mathbf{f}(\mathbf{x}) = \mathbf{0}. \tag{2.19}
$$

It is assumed that the functions f_i are

- continuous and differentiable so that the Jacobian matrix (2.15) exists;

- independent (the Jacobian is singular if they are linearly dependent); if this is not the case (2.19) will have an infinite number of solutions and the routines to be discussed below will fail. It may be the case that more than one solution exists anyway. For example,

$$
\begin{aligned}
x_1(x_1 + x_2) + 2 &= 0 \\
x_2^2 - 9 &= 0
\end{aligned}
$$

has two solutions, $(-2, 3)^T$ and $(-1, 3)^T$. In the case of multiple solutions the NAG routines generally find that closest to the user provided starting value.

All the routines provided in the C05 chapter are based on the same hybrid algorithm, proposed by Powell [65], which is a combination of Newton's method, and the method of *steepest descent*. This algorithm has excellent theoretical convergence properties and, in practice, convergence is obtained even from starting values a long way from the solution.

We have already seen, Section 2.8, how Newton's method may be applied to a system of equations. At each step an update vector $\mathbf{u}_k$ is computed using

$$J_k \mathbf{u}_k = -\mathbf{f}(\mathbf{x}_k),$$

and the next iterate $\mathbf{x}_{k+1}$ is then formed as $\mathbf{x}_k + \mathbf{u}_k$. The steepest descent method recasts the solution of (2.19) as the problem of finding the minimum value of

$$F(\mathbf{x}) = \sum_{i=1}^{m} f_i^2(\mathbf{x}) = \mathbf{f}^T(\mathbf{x})\mathbf{f}(\mathbf{x}). \qquad (2.20)$$

Successive approximations

$$\mathbf{x}_{k+1} = \mathbf{x}_k - \lambda_k \mathbf{g}_k,$$

are formed where

$$\mathbf{g}_k = 2 \sum_{i=1}^{m} f_i(\mathbf{x}_k) \frac{\partial f_i(\mathbf{x}_k)}{\partial x_i}$$

is the *gradient* of $\mathbf{f}$, and λ_k is chosen so that $\mathbf{x}_{k+1}$ minimizes F in the direction $-\mathbf{g}_k$ from $\mathbf{x}_k$. However, in practice, since this procedure may involve a considerable amount of work, it is usually considered sufficient to locate a point $\mathbf{x}_{k+1}$ so that $\|F(\mathbf{x}_{k+1})\| < \|F(\mathbf{x}_k)\|$. The hybrid algorithm uses a linear combination of the two *search directions*, $\mathbf{u}_k$ and $\mathbf{g}_k$. Full details of the implementation and of the convergence proofs are in Powell [65].

The NAG library has four routines in the C05 chapter designed specifically for solving the system (2.19), and many more in the E04 chapter for solving the recast problem (2.20). The C05 routines are C05NBF, C05NCF, C05PBF and C05PCF. The main choice facing the user is whether or not to provide a function to compute, explicitly, the elements of the Jacobian for a given $\mathbf{x}_k$. If practicable such a routine should be supplied since accurate evaluation of J_k will increase the reliability of the method. If the user chooses not to supply a routine, the partial derivatives are approximated using the *finite difference approximations*

$$\frac{\partial f_i}{\partial x_j} = \frac{f_i(x_1, \ldots, x_j + h_j, \ldots, x_m) - f_i(x_1, \ldots, x_j, \ldots, x_m)}{h_j},$$

$$i, j = 1, 2, \ldots, m \qquad (2.21)$$

for some suitable step lengths h_j. The routines

- C05PBF/PCF — use a user supplied routine to compute the Jacobian,

- C05NBF/NCF — approximate the Jacobian using finite differences.

C05NBF and C05PBF are easy-to-use routines; the user provides just enough information to define the problem and the accuracy required of the solution. The other two routines provide an experienced user with the ability to 'tune' the algorithm. They allow a more efficient solution to some problems at the expense of a longer parameter list. For the majority of problems the routines C05NBF and C05PBF should prove adequate.

The difficulties of coding a function to compute the Jacobian matrix for a general nonlinear system should not be underestimated. A routine, C05ZAF, is available to assist in checking for possible coding errors in the user provided function. Consistency checks are performed by approximating the partial derivatives using a finite difference approximation of the form (2.21), and comparing these values with those obtained directly from the user's code. Programs do exist (for example, JAKEF in Toolpack – see Appendix C) to generate automatically a Fortran subprogram for computing the Jacobian from a Fortran 66 routine for evaluating the function.

By considering a simple example we illustrate the use of C05PBF and C05ZAF, some of the problems of obtaining a solution, and how to check the accuracy of the returned solution vector. Our example consists of the pair of equations

$$
\begin{aligned}
0.001x_1 - x_2 + 6 &= 0 \\
x_1 x_2 - 72000 &= 0.
\end{aligned}
\tag{2.22}
$$

This system has two solutions, at $(6000, 12)$ and $(-12000, -6)$. C05PBF requires a user supplied routine to compute both the function vector and the Jacobian matrix. This must be of the form

> SUBROUTINE FUN(N,X,FVEC,FJAC,LDFJAC,IFLAG)

where N denotes the number of equations in the system, and the vector X, of length N, contains the point at which information is required. The parameter IFLAG, set by the calling routine, indicates whether a function evaluation (IFLAG = 1) or a Jacobian evaluation (IFLAG = 2) is being requested. In the case of a function evaluation, the values should be returned in the array FVEC which must, therefore, be of length N. The result of a Jacobian evaluation should be returned in the two-dimensional array FJAC whose leading dimension in the calling routine is given by LDFJAC $\geq$ N. On exit the element FJAC(I, J) should contain $\partial f_i / \partial x_j$ for $i, j = 1, 2, \ldots, N$. Note that when computing function values in the array FVEC the array FJAC should not be changed and vice versa. Under no circumstances should the contents of the vector X be altered by FUN. The value of IFLAG may be set negative by the user in order to abort the computation; for example, if X contains illegal values. In such a case this value is returned via IFAIL. Code 2.10 shows the code defining the equations and Jacobian of example (2.22).

Before proceeding with the call to C05PBF we should use C05ZAF to check FUN for consistency. C05ZAF uses reverse communication and a call takes the form

> CALL C05ZAF(M,N,X,FVEC,FJAC,LDFJAC,XP,FVECP,MODE,ERR)

```
      SUBROUTINE FUN(N,X,FVEC,FJAC,LDFJAC,IFLAG)
      INTEGER IFLAG,LDFJAC,N
      DOUBLE PRECISION FJAC(LDFJAC,N),FVEC(N),X(N)
*
      IF (IFLAG.EQ.1) THEN
*
*.. FUNCTION EVALUATION
         FVEC(1) = -0.001D0*X(1)+X(2)-6.0D0
         FVEC(2) = X(1)*X(2)-72000.0D0
*
*.. JACOBIAN EVALUATION (IFLAG = 2)
      ELSE
         FJAC(1,1) = -0.001D0
         FJAC(1,2) = 1.0D0
         FJAC(2,1) = X(2)
         FJAC(2,2) = X(1)
      ENDIF
      END
```

Code 2.10   Example SUBROUTINE FUN for C05ZAF.

Although at first sight its use may appear a little complicated, the effort required is often amply repaid by the pinpointing of errors which would otherwise be difficult to locate. The following steps are required:

(1) Choose a 'suitable' evaluation point, X, at which the consistency check is to take place. Here, suitable means that the values obtained by evaluating the function at this point should not be significantly affected by rounding or cancellation errors. In general terms it is best to avoid small values for the components of X and, in particular, zero values should not be used.

(2) Call C05ZAF with MODE set to one. The routine will return, in the array XP a neighbouring point to X which is suitable for computing the finite difference approximations.

(3) Use the routine, FUN, being tested to compute

   • $f(X)$ in FVEC,

   • $f(XP)$ in FVECP,

   • $J(X)$ in FJAC.

(4) Call C05ZAF a second time with MODE set to two.

(5)   Check the values returned in the vector ERR. A value of 1.0 in ERR(I) denotes that the gradients associated with the ith component of f are correct, whilst a returned value of zero signifies that those gradients are incorrect. Values of ERR(I) between zero and unity are less decisive; a value in the range $[0.5, 1.0)$ is almost certainly correct, values less than 0.5 could be caused either by errors in the code, or by rounding or cancellation errors. In these circumstances, it is well worth rerunning the test routine with a different starting vector X.

We note that C05ZAF may be used to check user supplied routines used by the differential equation routines in the D02 chapter (see Chapter 6), and by the optimization routines in the E04 chapter. This latter use explains the need for two arguments, M and N, to define the number of function components and the number of variables respectively. For use with the C05 and D02 routines M will always be equal to N.

The sequence of steps is illustrated in Code 2.11; this program is quite general in that only the value in the parameter statement and the input data need to be changed for different user supplied functions.

```
      PROGRAM CHECK
      CHARACTER*4 STARS
      INTEGER I,IFLAG,IST,MODE,LENG,N,NIN,NOUT
      DOUBLE PRECISION ONE
      PARAMETER (N = 2,NIN = 5,NOUT = 6,ONE = 1.0D0)
      DOUBLE PRECISION ERR(N),FJAC(N,N),FVEC(N),
     +                 FVECP(N),X(N),XP(N)
      EXTERNAL C05ZAF,FUN
      INTRINSIC LEN,INT
*
*.. DRIVER FOR JACOBIAN CHECKER C05ZAF
      STARS = '****'
      LENG = LEN(STARS)
*
*.. USE FREE FORMAT READ FOR THE EVALUATION POINT
*.. STOP AT END OF DATA
10    READ(NIN,*,END = 30)(X(I),I = 1,N)
*
*.. CALL C05ZAF WITH MODE=1 TO OBTAIN A SUITABLE
*.. NEIGHBOUR OF THE INPUT POINT
      MODE = 1
      CALL C05ZAF(N,N,X,FVEC,FJAC,N,XP,FVECP,MODE,ERR)
*
*.. EVALUATE F(X), F(XP) AND J(X) USING THE TEST ROUTINE
*.. NOTE THE ARRAY FJAC IS A DUMMY AND SHOULD NOT BE
```

```
*.. ASSIGNED TO WHEN EVALUATING THE FUNCTION
      IFLAG = 1
      CALL FUN(N,X,FVEC,FJAC,N,IFLAG)
      CALL FUN(N,XP,FVECP,FJAC,N,IFLAG)
*
*.. SIMILARLY DON'T ASSIGN TO FVEC WHEN COMPUTING THE
*.. JACOBIAN MATRIX
      IFLAG = 2
      CALL FUN(N,X,FVEC,FJAC,N,IFLAG)
*
*.. CALL CO5ZAF FOR THE CONSISTENCY CHECK (MODE=2)
      MODE = 2
      CALL CO5ZAF(N,N,X,FVEC,FJAC,N,XP,FVECP,MODE,ERR)
*
*.. LOOK AT THE ERR VALUES - THE MORE STARS THE MORE LIKELY
*.. THE INCONSISTENCY IN THAT COMPONENT
      DO 20 I = 1,N
         IF(ERR(I).GT.ONE-ONE/LENG)THEN
            WRITE(NOUT,100)I,ERR(I)
         ELSE
            IST = INT(ERR(I)*LENG)+1
            WRITE(NOUT,100)I,ERR(I),STARS(IST:LENG)
         ENDIF
20    CONTINUE
100   FORMAT(1X,I4,F10.4,A)
      GOTO 10
30    CONTINUE
      END
```

Code 2.11   Example driver program for CO5ZAF.

Once the problem defining function has been checked we may proceed to use CO5PBF with some degree of confidence. A call to this routine takes the form

```
      CALL CO5PBF(FUN,N,X,FVEC,FJAC,LDFJAC,XTOL,
     +            WA,LWA,IFAIL)
```

where N denotes the number of equations in the system defined by the routine FUN whose construction has been described above. Remember that FUN must be declared in an EXTERNAL statement in the calling unit. On entry the vector X must contain an initial guess at the solution vector; on exit it will contain the final estimate of the solution. The user can control the accuracy to which the solution is determined by setting XTOL appropriately. The convergence criterion used by the algorithm is based

Starting values		Number of evaluations	
x_1	x_2	$f(x)$	$J(x)$
1	1	1575	21
4831	15	248	6
5800	13	6	1

Figure 2.11   Cost of solving (2.22) for various starting values.

purely on a relative error test. Normally a value of 10^{-k} will ensure that the components of X are determined correct to k significant digits; however, there is a danger that the smaller components of X will be subject to high relative errors although this is minimized by the very rapid convergence properties of the algorithm near a solution. Additional care should be taken to check the accuracy of solutions which lie close to the origin. The recommended value of XTOL is SQRT(X02AFF(DUMMY)), although this should obviously be increased if the problem definition contains inherent errors (for example, in experimentally determined parameters) which are larger than this value.

On successful exit the vector FVEC, of length at least N, contains the value of the function at the final point X (the *residuals*). Workspace arrays FJAC (dimensioned (LDFJAC, p), where LDFJAC is the leading dimension as defined in the calling routine, and both LDFJAC and p are at least N) and WA (of length at least LWA = N $\times$ (N + 13)/2) return information which may be used to analyse the quality of the final solution.

To illustrate the importance of good starting values, Figure 2.11 shows how the amount of computation, measured in the number of function and Jacobian evaluations, required to solve (2.22) decreases as the starting vector gets closer to the solution. In all cases the solution with positive components was returned correct to at least 9 significant digits.

The algorithms implemented by the C05 routines perform most efficiently when the equations are scaled to make all the components at the solution of similar magnitude. This is obviously impossible if the user has absolutely no idea where the solution is; however in most practical applications the solution components are known to at least an order of magnitude, and even scaling at this level can lead to a more efficient solution. To illustrate the effect of scaling, we re-solve problem (2.22), with $\hat{x}_1 = x_1/1000$ and $\hat{x}_2 = x_2/10$. This gives the system

$$\hat{x}_1 - 10\hat{x}_2 + 6 = 0$$
$$\hat{x}_1\hat{x}_2 - 7.2 = 0.$$

Starting at $(0,0)$, that is, with no significant digits in either component, C05PBF requires only 14 function and 2 Jacobian evaluations. As before we obtain at least 9 significant digits. For large systems the computation time is dominated by function and Jacobian evaluations; anything that can reduce the number of times the user supplied routine is called is therefore worthwhile.

CO5PBF may fail for a number of reasons. We have already seen that it is possible for the user to abort the calculation by returning a negative value of IFLAG from a call to FUN. This is either passed back to the unit calling CO5PBF via IFAIL, or it is printed in the NAG error message. Inconsistent input values are signalled by IFAIL = 1. A value of IFAIL = 2 is used as a safety valve to control the number of function evaluations required to obtain a solution; if the user is happy to carry on, the computation may be restarted from the current value of X. Note that in order to restart without rerunning the program, it is necessary to use the soft error option (see Section 1.9) and to test the returned value of IFAIL. IFAIL = 3 signifies that the accuracy requirements imposed by the user, via XTOL, are too strict; the results are those which would have been obtained if XTOL had been set to X02AJF(DUMMY), the machine precision.

Finally a returned value of IFAIL = 4 indicates that the algorithm has run into difficulties. There may be a number of reasons for this:

- the system does not possess a zero,

- the solution is very close to the origin,

- the algorithm is 'lost'.

Printing the returned values of X and FVEC will certainly give some indication of whether the solution lies close to zero. In such a case it may be possible to scale the problem. However, it is not obvious how to distinguish between the remaining conditions. Various possible approaches are

- use another starting value,

- monitor the X, and the associated FVEC, values that are passed to the function by inserting WRITE statements into FUN,

- attempt further analysis of the problem; in general this is likely to be very difficult for anything other than small problems.

Functions without a solution often show convergence to local or global minima, whilst lost iterations may produce X values which show no sign of convergence. In neither situation will the residuals get very small. Cases of non-convergence are rare, especially if the Jacobian is being evaluated by the user.

On successful exit we should check both the accuracy of the solution and the conditioning of the problem at the solution. An approximation to the error, $\mathbf{e}$, in the final solution may be obtained by solving the system of linear equations

$$J(\mathbf{x}_{sol})\mathbf{e} = -\mathbf{f}(\mathbf{x}_{sol}), \qquad (2.23)$$

where x_{sol} is the final solution vector.

From the conditioning point of view, we are interested in the effect of the uncertainty in the specification of the function components on the computed solution. We assume that each f_i is changed by an amount p_i (this could, for example, quantify the errors in the parameters defining the system), and look at the effect that this change has on the solution. The approximate change, c, in x_{sol} is given as the solution of

$$J(x_{sol})c = -p. \tag{2.24}$$

Both (2.23) and (2.24) could be solved directly, using $O\left(n^3\right)$ arithmetic operations, by the methods and routines to be given in Chapter 4. However, as a by-product of the solution of the nonlinear system, C05PBF returns the Jacobian matrix as the product of two matrices Q and R, where Q is orthogonal and R is upper triangular (see also Section 5.7). (It may be shown that any nonsingular matrix may be uniquely written in this form, see Stewart [78], p. 214.) We may thus write $Jx = b$ as $QRx = b$ and since $Q^TQ = I$, the unit matrix, we have $Rx = Q^Tb$. Since R is upper triangular both the formation of the right-hand side and the solution of the resulting triangular system of linear equations require $O\left(n^2\right)$ operations (see Section 4.3). This is an order of magnitude more efficient than using the general solvers. Code 2.12 illustrates a call to C05PBF and the additional calls necessary to obtain the error and conditioning information. This code may easily be adapted for use with C05NBF.

```
        PROGRAM CHECK
        INTEGER LWA,N,NOUT
        DOUBLE PRECISION HUNDRED,ONE,ZERO
        PARAMETER (NMAX = 2,LWA = (NMAX*(NMAX+13))/2,
     +            NOUT = 6,ZERO = 0.0D0,HUNDRED = 100.0D0,
     +            ONE = 1.0D0)
        DOUBLE PRECISION E(NMAX),EPS(NMAX),FJAC(NMAX,NMAX),
     +            FVEC(NMAX),WA(LWA),X(NMAX)
        DOUBLE PRECISION F06EJF,FUN,MCHEPS,X02AJF,XTOL
        INTEGER I,IFAIL,N
        INTRINSIC SQRT
        EXTERNAL C05PBF,F06EJF,F06PAF,F06PLF,FUN,X02AJF
*
*.. SET THE STARTING VALUES
        N = 2
        X(1) = 5800.0D0
        X(2) = 13.0D0
        MCHEPS = X02AJF(ZERO)
        XTOL = SQRT(MCHEPS)
```

```
            IFAIL = 0
            CALL CO5PBF(FUN,N,X,FVEC,FJAC,N,XTOL,WA,
     +                  LWA,IFAIL)
*
*.. ONLY GET HERE IF NO ERROR
*.. OUTPUT RESULTS
            WRITE(NOUT,100)F06EJF(N,FVEC,1)
            WRITE(NOUT,101)
            WRITE(NOUT,102)(X(I),I = 1,N)
100         FORMAT(1X,'FINAL RESIDUAL NORM = ',D12.4)
101         FORMAT(1X,'FINAL APPROXIMATE SOLUTION')
102         FORMAT(1X,3D16.4)
*
*.. CHECK ERROR IN THE RETURNED APPROXIMATION X BY SOLVING
*..        J(X) E = -F
*.. THE QR FACTORIZATION OF J(X) IS AVAILABLE SO IT'S FASTER
*.. TO SOLVE
*..        R E = -Q(TRANSPOSE)F
*.. WHERE R IS UPPER TRIANGULAR IN PACKED FORM IN WA(6N+1)
*.. TO WA(LWA) AND Q(TRANSPOSE)F IS IN WA(N+1) ... WA(2N).
*.. NOTE. THE FINAL VECTOR HAS THE WRONG SIGN
            DO 20 I = 1,N
              E(I) = -WA(N+I)
20          CONTINUE
*
*.. NOW SOLVE THE LINEAR SYSTEM
            CALL F06PLF('L','N','N',N,WA(6*N+1),E,1)
            WRITE(NOUT,103)F06EJF(N,E,1)
            WRITE(NOUT,104)
            WRITE(NOUT,102)(E(I),I = 1,N)
103         FORMAT(1X,'NORM OF ERROR VECTOR = ',D12.4)
104         FORMAT(1X,'INDIVIDUAL ELEMENTS OF ERROR VECTOR')
*
*.. COMPUTE THE SENSITIVITY OF THE APPROXIMATE SOLUTION TO
*.. CHANGES IN THE FUNCTION VECTOR AT THE FINAL SOLUTION.
*
*.. FOR THIS EXAMPLE WE SET THE ERROR VECTOR TO BE 100 TIMES
*.. THE MACHINE EPSILON TIMES THE FUNCTION VALUES AT THE
*.. FINAL SOLUTION.
*.. THIS SHOULD BE ADJUSTED TO REFLECT THE EFFECT OF ANY
*.. KNOWN ERRORS, FOR EXAMPLE IN THE DEFINING PARAMETERS.
            DO 30 I = 1,N
              E(I) = -HUNDRED*MCHEPS*FVEC(I)
              EPS(I) = ZERO
30          CONTINUE
```

```
*
*.. FORM EPS=Q(TRANSPOSE)*-E
      CALL F06PAF('T',N,N,-ONE,FJAC,NMAX,E,1,
     +              ZERO,EPS,1)
*
*.. SOLVE R E = EPS
      CALL F06PLF('L','N','N',N,WA(6*N+1),E,1)
*
*.. OUTPUT RESULTS
      WRITE(NOUT,105)F06EJF(N,E,1)
      WRITE(NOUT,106)
      WRITE(NOUT,102)(E(I),I = 1,N)
105   FORMAT(1X,'NORM OF SENSITIVITY VECTOR = ',D12.4)
106   FORMAT(1X,'INDIVIDUAL ELEMENTS OF SENSITIVITY VECTOR')
      END
```

Code 2.12   Code to check error and conditioning of solution vector
from C05PBF.

EXERCISE 2.17   We wish to solve a system of nonlinear equations of the form (2.19) where f_1 is dependent only on $\{x_1, x_2\}$, f_i on $\{x_{i-1}, x_i, x_{i+1}\}$ for $i = 2, 3, \ldots, n - 1$ and f_n on $\{x_{n-1}, x_n\}$. Such systems occur in practice in the numerical solution of nonlinear partial differential equations. Show that it is only necessary to perform four function evaluations to approximate the Jacobian of such a system using the finite difference formula (2.21). Extend this method to the computation of general banded and sparse Jacobians.

EXERCISE 2.18   Write a subroutine to implement Newton's method for solving a system of nonlinear equations assuming that a function will be provided for calculating the Jacobian. Compare the efficiency of your routine with C05PBF using the number of function and Jacobian evaluations as a measure. Study carefully the convergence properties of the two methods by using different starting vectors; pay special attention to initial vectors a long way from the solution.

2.11 Summary

The routines described in this chapter of the book are contained in the C02 (zeros of polynomials) and C05 (roots of one or more transcendental equation) chapters of the NAG library. Considering polynomial equations first, we observe that the C02 routines attempt to locate all roots. For the case of complex coefficients, use C02ADF; if the coefficients are all real, use C02AEF (Section 2.9). If only one root is required and its approximate position is known, it may be more sensible to use an appropriate C05 routine.

For locating the real roots of a single real-valued equation, the 'direct communication' routines C05ADF, C05AGF and C05AJF are available (Section 2.4). If a bracket containing a root is known use C05ADF; otherwise, if a good approximation to a root is known, use C05AJF. Finally, if no such information is available, use C05AGF. In addition to these three basic routines, the 'reverse communication' routines C05AVF, C05AXF and C05AZF are available (Section 2.4). These latter codes implement the same methods as the direct communication routines but permit more sophisticated use. In consequence they require greater care in the setting up of the call sequence. If a bracket containing a root is known, use C05AZF; otherwise, if a good approximation to a root is known use C05AXF. If no such information is available, use C05AVF (which attempts to determine a bracket for a root), followed by C05AZF.

If the user requires to locate a real root of a system of real-valued equations the routines C05NBF, C05NCF, C05PBF and C05PCF are available (Section 2.10). If the user is brave enough to determine and program the Jacobian matrix he can call C05PBF, or its more sophisticated form C05PCF. A prior call of C05ZAF (Section 2.10) can be used to check the user defined Jacobian. Less adventurous users will be content to call C05NBF, or its comprehensive version C05NCF, where the Jacobian is estimated internally.

Chapter 3
Quadrature

In this chapter we consider methods for estimating integrals defined over finite, semi-infinite and infinite ranges. We

- show how polynomial approximation can be used to give integral estimates;

- introduce the idea of orthogonal polynomials and indicate their importance to integral approximation;

- outline the adaptive approach which aims to return an approximation correct to some specified tolerance.

For a general integrand the NAG routines we describe are based on Gauss rules. We shall also consider a number of special routines for calculating certain commonly occurring integrals using Chebyshev approximations.

3.1 What is quadrature?

In this chapter we are concerned with ways of estimating the definite integral

$$I = \int_a^b f(x)\,dx, \qquad (3.1)$$

that is, the area bounded by the curve $y = f(x)$ and the x-axis. We distinguish three different ranges of integration

- *finite* : both a and b are finite,

- *semi-infinite* : one of $a = -\infty$ or $b = \infty$,

- *infinite* : $a = -\infty$ and $b = \infty$.

If the integrand f is known as a continuous function, it may be possible to evaluate the integral analytically using one of the standard techniques such as substitution or integration by parts. However when analytic integration is not possible (e.g., f may only be known at a discrete set of

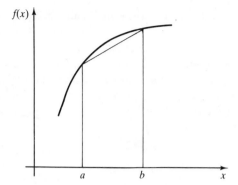

Figure 3.1   Basic trapezium rule.

points), or is difficult, a numerical approach to the problem may be necessary. Basically, we approximate f by, say, a polynomial and integrate analytically the approximating function instead. Our estimate of I will be of the form

$$R = \sum_{i=1}^{n} w_i f_i \qquad (3.2)$$

where $f_i = f(x_i)$ and the x_is are the *abscissae* or *integration points* with corresponding *weights* w_i. R is referred to as a *numerical integration* or *quadrature rule*.

A simple example of a quadrature rule is the well-known *trapezium rule*. If we know the values of $f(a)$ and $f(b)$ we can determine a straight line which passes through the coordinates $\{(a, f(a)), (b, f(b))\}$, and take as our integral approximation the area bounded by this straight line, the x-axis and the lines $x = a$ and $x = b$ (see Figure 3.1). Let $p_1(x)$ be the equation of our straight line; then

$$p_1(x) = \frac{x - b}{a - b} f(a) + \frac{x - a}{b - a} f(b).$$

This may easily be verified by noting that $p_1(a) = f(a)$ and $p_1(b) = f(b)$. The function $p_1(x)$ is an *interpolating polynomial*, i.e., it is a polynomial which passes exactly through a set of coordinates (cf. linear interpolation, Section 2.3).

Formally we should now integrate $p_1(x)$ over $[a, b]$ to derive the trapezium rule. However since the area of a trapezium is given by the product of the average length of the two parallel sides and the distance between them, we may immediately write

$$R = \frac{b - a}{2} (f(a) + f(b)).$$

Hence, in the notation of (3.2), we have $w_1 = w_2 = \frac{1}{2}(b - a)$ and $x_1 = a$, $x_2 = b$.

We observe that for the trapezium rule the quadrature weights are easily computable functions of a and b, the limits of integration. This will not necessarily be the case for other rules which we shall consider if a direct approach to the estimation of (3.1) is taken. To overcome this problem we choose to define a rule in terms of some *standard* interval and transform our integral before attempting to approximate it. For example, suppose that rule (3.2) is valid for the interval $[-1, 1]$, then, using the transformation

$$z = \frac{2x - a - b}{b - a}, \tag{3.3}$$

(3.1) becomes

$$I = \int_{-1}^{1} f\left(\frac{(b-a)z + a + b}{2}\right) \frac{b-a}{2}\, dz,$$

so that the transformed rule is

$$R = \sum_{i=1}^{n} \frac{b-a}{2} w_i f\left(\frac{(b-a)z_i + a + b}{2}\right).$$

We see that the weights are simple multiples of the standard weights, and the integration points are easily obtained from the original z_is. This transformation is only valid provided both a and b are finite.

For semi-infinite integrals we note that a simple transformation can be used to map the given range onto $[0, \infty)$. A further transformation is then required to convert the integral to one over a finite range (see Exercise 3.1). In Section 3.10 we shall look at special rules for the standard semi-infinite range $[0, \infty)$, and also for $(-\infty, \infty)$. In these, and other, cases we shall consider *weighted integrals* of the form

$$I = \int_{a}^{b} g(x) f(x)\, dx.$$

Our quadrature rule will again take the form (3.2), that is, a linear combination of f values, where f now forms only part of the integrand; the remaining part, g, being absorbed into the weights w_i.

Weighted integration is often used in the estimation of integrals for which the integrand is unbounded at one or more points. For example, consider

$$I = \int_{-1}^{1} \frac{1}{\sqrt{1 - x^2}} f(x)\, dx.$$

Clearly $1/\sqrt{1 - x^2}$ is singular at $x = \pm 1$, i.e., the denominator is zero at these points. If a quadrature rule is used which includes either of these values as integration points computational difficulties will result. Note that this has no significance whatsoever with respect to the existence, or otherwise, of the integral. One solution is to use a rule which only requires evaluation of the non-singular part, f, of the integrand; the singular part is integrated out analytically.

In considering numerical integration algorithms we distinguish between the situations where

- values of the function f are only available at a fixed set of points. The integration points are thus fixed and we seek weights which make the rule as accurate as possible in some sense;

- f may be evaluated at any chosen point within the range of integration. In this case we have further degrees of freedom since we may choose the abscissa points to make the rule 'optimal'.

In either case we will want to be able to say something about the error, E, in the quadrature rule which we formally define as $E = I - R$. In general it will not be possible to determine E exactly, although we may be able to produce either an estimate of, or an upper bound on, its value. An alternative way of measuring the accuracy of a quadrature rule is to determine its degree of precision. We define a rule to have *degree of precision* n if it integrates any polynomial of degree n exactly but $E \neq 0$ for a polynomial of degree $n + 1$. If f is polynomial-like we can hope that a rule with a higher degree of precision will provide a more acceptable estimate of the integral. Thus, in subsequent discussions, we aim to determine the weights and, if possible, the points so as to optimize the accuracy of the rule in this sense. Note that if a rule has degree of precision at least zero (which we will take to be a basic requirement), the sum of the weights is equal to $b - a$.

EXERCISE 3.1   Given a rule of the form (3.2) which is valid for the interval $[-1, 1]$, show that, by using the transformation $z = 1 - 2/(x + 1 - a)$,

$$I = \int_a^\infty f(x)\,dx$$

becomes

$$I = \int_{-1}^1 f\left(\frac{1+z}{1-z} + a\right) \frac{2}{(1-z)^2}\,dz$$

which may be approximated by

$$R = \sum_{i=1}^n \frac{2w_i}{(1-z_i)^2} f\left(\frac{1+z_i}{1-z_i} + a\right).$$

What problems may occur if the integration points include the end-points of the interval?

EXERCISE 3.2   Show that

$$\int_a^\infty \phi(x)\,dx = \int_b^c \phi\{f(t)\} f'(t)\,dt,$$

where $f(b) = a$ and $f(c) = \infty$. Hence, using $f(t) = t/(1-t)$ rewrite

$$\int_0^\infty \frac{1}{(1+x)^m}\,dx$$

for some c and b as an integral over a finite range.

3.2   Integration of interpolating polynomials

Let $\{\,x_i \mid i = 1, 2, \ldots, n\,\}$ be a set of points at which function values $\{\,f_i \mid i = 1, 2, \ldots, n\,\}$ are available. Then if $f(x)$ is the underlying function (i.e., the function which, when evaluated at $x = x_i$, gives the value f_i) we may write

$$\int_{x_1}^{x_n} f(x)\,dx = \int_{x_1}^{x_2} f(x)\,dx + \int_{x_2}^{x_3} f(x)\,dx + \cdots + \int_{x_{n-1}}^{x_n} f(x)\,dx. \quad (3.4)$$

We can then use the trapezium rule to approximate the integral over the subinterval $[\,x_i, x_{i+1}\,]$ for $i = 1, 2, \ldots, n-1$ to yield the *composite* trapezium rule estimate

$$R = \frac{x_2 - x_1}{2}\{f_1 + f_2\} + \frac{x_3 - x_2}{2}\{f_2 + f_3\} + \cdots$$
$$+ \frac{x_n - x_{n-1}}{2}\{f_{n-1} + f_n\}. \quad (3.5)$$

It can be shown (see Exercise 3.4 and Johnson and Riess [46], p. 302) that the error, $E^{(i)}$, associated with using the trapezium rule to approximate the integral over $[\,x_i, x_{i+1}\,]$ is

$$E^{(i)} = -\tfrac{1}{12}\left(x_{i+1} - x_i\right)^3 f''(\xi_i), \quad (3.6)$$

where $\xi_i \in (x_i, x_{i+1})$. The total error in the quadrature rule (3.5) is simply the sum of terms of the form (3.6), i.e.,

$$E = I - R = -\sum_{i=1}^{n-1} \tfrac{1}{12}\left(x_{i+1} - x_i\right)^3 f''(\xi_i). \quad (3.7)$$

Now, if f is a constant or straight line function, $f''(x) = 0$ for all x and hence the rule is exact. Since the second derivative of higher degree polynomials is not identically zero we conclude that the rule has degree of precision one. Result (3.7) may be useful in other ways; if a bound on the magnitude of f'' can be determined it may be possible to bound $|E|$.

It is possible to increase the degree of precision of the integration rule by improving the local approximation to the underlying function. One way is to construct a kth degree polynomial which passes through $k + 1$ distinct points. Thus, for example, we could choose to approximate the integrand over successive groups of three points by interpolating quadratic polynomials. If the points are equally spaced then on integrating these local approximations we obtain *Simpson's rule* which has degree of precision three (see Exercise 3.6).

Generating rules with higher degree of precision is straightforward; for each subinterval $[x_i, x_{i+k}]$ we introduce the polynomial $p_{k,i}(x)$ which interpolates at the points $\{(x_j, f_j) \mid j = i, i+1, \ldots, i+k\}$. By substitution, we may verify that $p_{k,i}(x)$ takes the form

$$p_{k,i}(x) = \sum_{j=i}^{i+k} l_j(x) f_j, \tag{3.8}$$

where

$$l_j(x) = \frac{(x - x_i) \cdots (x - x_{j-1})(x - x_{j+1}) \cdots (x - x_{i+k})}{(x_j - x_i) \cdots (x_j - x_{j-1})(x_j - x_{j+1}) \cdots (x_j - x_{i+k})}. \tag{3.9}$$

Equation (3.8) is known as the *Lagrange* form of the interpolating polynomial. For any $x \in [x_i, x_{i+k}]$, the error in the approximation is given by

$$q_i(x) \equiv f(x) - p_{k,i}(x) = (x - x_i)(x - x_{i+1}) \cdots$$
$$(x - x_{i+k}) f^{(k+1)}(\eta_i)/(k+1)! \tag{3.10}$$

where $\eta_i \in (x_i, x_{i+k})$. Integration of (3.9) yields the weights for the new rule which, for the general case, has degree of precision k, and provided $n - 1$ is divisible by k we can approximate (3.1) using a composite version of this rule. Integration of (3.10) yields an expression for the error.

In the case of equally spaced points, when the rules are known as *Newton–Cotes formulae*, we get a bonus; the degree of precision is k if k is odd, but $k+1$ if k is even. The use of Newton–Cotes rules with high degree of precision requires some caution as the quadrature weights become large and of mixed sign leading to the possibility of loss of significance when the summation (3.2) is performed. There is also nothing to recommend the use of a sequence of Newton–Cotes rules of increasing order to obtain high accuracy approximations (see the example in Davis and Rabinowitz [23], pp. 80–81). In Section 3.3 we shall see that if the integration points have special values, then the degree of precision of the rule may be increased to $2k + 1$ and that for such rules all weights are guaranteed to be positive.

EXERCISE 3.3

(a) Show that for equally spaced abscissa points, i.e., $x_{i+1} - x_i = h$, for all i, the composite trapezium rule, (3.5), becomes

$$R = h \sum_{i=1}^{n}{}'' f_i,$$

where the double prime on the summation sign indicates that the first and last terms are to be halved.

(b)   Confirm that the method has degree of precision one by using this formula to compute

$$\int_0^1 x^k \, dx,$$

for $k = 0, 1, 2$.

(c)   Write a subroutine TRAP which successively halves the value of h until two approximations agree to within a user supplied relative tolerance. Make sure that no function evaluations are computed twice at the same point.

EXERCISE 3.4   Define $F(t) = \int_{x_i}^t f(x) \, dx$. Using the Taylor series expansion (1.9) show that, for $x_i \leq x \leq x_{i+1}$,

$$F(x_{i+1}) = F(x_i) + hF'(x_i) + \tfrac{1}{2}h^2 F''(x_i) + \cdots,$$

and

$$f(x) = f_i + (x - x_i)f_i' + \tfrac{1}{2}(x - x_i)^2 f_i'' + \cdots,$$

where $h = x_{i+1} - x_i$. Hence show that the leading error term in the trapezium rule approximation is of the form $-h^3 f_i''/12$.

EXERCISE 3.5   Compute approximations to

$$\int_0^2 x^3 \, dx$$

using the trapezium rule for successively smaller values of h. Plot the absolute errors in the approximation against h. Explain what happens when h gets very small.

EXERCISE 3.6   Let $p_{2,i-1}(x)$ be the polynomial interpolating the data $\{(x_j, f_j) \mid j = i-1, i, i+1\}$, with the x_js equally spaced, distance h apart. Show that

$$\int_{x_{i-1}}^{x_{i+1}} p_{2,i-1}(x) \, dx = \tfrac{1}{3}h(f_{i-1} + 4f_i + f_{i+1}).$$

This is known as Simpson's rule. Using it to integrate the functions $f(x) \equiv x^k$ for $k = 0, 1, 2, 3, 4$ deduce that this method has degree of precision three. For the composite Simpson's rule for equally spaced abscissa points, write a subroutine SIMP to the specification of TRAP in Exercise 3.3. Note that at each bisection you should only evaluate the function at the newly introduced points.

3.3   Rules with high degree of precision

Next we consider rules of the form (3.2) where both the weights, w_i, and the abscissa points, x_i, are free parameters, and we attempt to choose values for them to define integration rules with the highest possible degree of precision. For an n-point rule we have $2n$ parameters (n weights and n points) available and we might hope to select values for these to produce a rule of degree of precision $2n - 1$. There are certain restrictions on our choice of values

n	x_i	w_i	Degree of precision
1	0	2	1
2	$\pm\,0.577350$	1	3
3	0 $\pm\,0.774597$	8/9 5/9	5
4	$\pm\,0.339981$ $\pm\,0.861136$	0.652145 0.347855	7

Figure 3.2   Gauss–Legendre quadrature rules.

- all integration points must lie in the range of integration,
- we would prefer all the weights to be of the same sign.

Without loss of generality we restrict the discussion to the interval $[-1, 1]$ (remembering that the transformation (3.3) can be used if necessary). We consider the derivation of a two-point rule of the form

$$\int_{-1}^{1} f(x)\,dx \approx w_1 f_1 + w_2 f_2.$$

Such a rule will have degree of precision three if we impose the four conditions

$$\begin{aligned}
\int_{-1}^{1} 1\,dx &\equiv 2 = w_1 + w_2, \\
\int_{-1}^{1} x\,dx &\equiv 0 = w_1 x_1 + w_2 x_2, \\
\int_{-1}^{1} x^2\,dx &\equiv \tfrac{2}{3} = w_1 x_1^2 + w_2 x_2^2, \\
\int_{-1}^{1} x^3\,dx &\equiv 0 = w_1 x_1^3 + w_2 x_2^3,
\end{aligned}$$

that is, if we ensure that the rule integrates 1, x, x^2 and x^3 exactly. We need to solve a nonlinear system of four equations in the four unknowns, x_1, x_2, w_1 and w_2. By substitution it can be seen that the required solution is $w_1 = w_2 = 1$, $x_1 = -x_2 = -1/\sqrt{3}$. We are fortunate in that both weights are positive and both integration points lie in $[-1, 1]$. To confirm that the rule has degree of precision exactly three we observe that

$$\int_{-1}^{1} x^4\,dx \equiv \frac{2}{10} \neq w_1 x_1^4 + w_2 x_2^4 \equiv \frac{2}{9}.$$

The rule we have derived is referred to as the two-point *Gauss–Legendre rule* and it is possible to obtain rules with higher degree of precision using this approach; Figure 3.2 lists the points and weights we would derive for the first few cases. Note the symmetry. Unfortunately, the resulting systems of nonlinear equations become more difficult to solve as we increase n, and hence we consider an alternative approach which, although not as straightforward as that outlined above, does allow us

- to show that all weights are positive, and all points lie in $(-1,1)$,
- to obtain a bound on the error in quadrature.

The starting point is the *Hermite interpolating polynomial*

$$p_{2n-1}(x) = \sum_{i=1}^{n} h_i(x)f_i + \sum_{i=1}^{n} \bar{h}_i(x)f_i', \qquad (3.11)$$

where $f_i' \equiv f'(x_i)$,

$$\begin{aligned} h_i(x) &= \big(1 - 2(x - x_i)l_i'(x_i)\big)l_i^2(x), \\ \bar{h}_i(x) &= (x - x_i)l_i^2(x), \end{aligned}$$

and $l_i(x)$ is defined by (3.9). By direct substitution it can be verified that $p_{2n-1}(x)$ interpolates f and its first derivative on the point set $\{\, x_i \mid i = 1, 2, \ldots, n \,\}$, i.e.,

$$\left. \begin{aligned} p_{2n-1}(x_i) &= f_i \\ p_{2n-1}'(x_i) &= f_i' \end{aligned} \right\} \quad i = 1, 2, \ldots, n.$$

Now, the error in interpolation is defined to be $q_{2n-1}(x) = f(x) - p_{2n-1}(x)$, and it can be shown that

$$q_{2n-1}(x) = \frac{1}{(2n)!} \prod_{i=1}^{n} (x - x_i)^2 f^{(2n)}(\xi), \qquad (3.12)$$

where ξ is some point in the open interval spanned by x and x_1, x_2, ..., x_n (see Ralston and Rabinowitz [67], p. 72). Observe that (3.12) is valid at each x_i, since at these points we know that $q_{2n-1}(x) = 0$. Furthermore, the $2n^{th}$ derivative of a polynomial of degree $2n - 1$ is zero and hence the interpolation process is exact for such a function.

To proceed we now substitute (3.11) for the integrand in (3.1) to obtain the quadrature rule

$$R = \sum_{i=1}^{n} \left(\int_{-1}^{1} h_i(x)\,dx \right) f_i + \sum_{i=1}^{n} \left(\int_{-1}^{1} \bar{h}_i(x)\,dx \right) f_i'.$$

By integrating (3.12) we may also show that the quadrature error is given by

$$E = I - R = \frac{1}{(2n)!} f^{(2n)}(\eta) \int_{-1}^{1} \prod_{i=1}^{n} (x - x_i)^2 \, dx, \qquad \eta \in (x_1, x_n),$$

(Ralston and Rabinowitz [67], p. 100). At first sight this does not appear to be very promising. In order to use the rule we apparently require both function and derivative values at the interpolation points; we have doubled the degree of precision at twice the cost. However a judicious choice for the interpolation points ensures that the integrals

$$\int_{-1}^{1} \bar{h}_i(x)\, dx \tag{3.13}$$

are all zero, avoiding the need for derivative values. Our next task is to investigate just what this choice of points should be.

EXERCISE 3.7   Find the weights and points so that the quadrature rule

$$\sum_{i=1}^{2} w_i f(x_i) \approx \int_{-\pi/2}^{\pi/2} \sin(x) f(x)\, dx$$

is exact for $f(x) = 1,\, x,\, x^2,\, x^3$.

3.4   Orthogonal polynomials and Gauss rules

Before proceeding we need to discuss briefly some properties of *orthogonal polynomials* and explain how these may be used in the derivation of high-order quadrature rules.

Consider the sequence of polynomials $\{Q_i(x) \mid i = 0, 1, \ldots, n\}$, where Q_i is of degree i. These polynomials are termed *orthogonal* on the interval $[a, b]$ with respect to a positive weight function $g(x)$ if

$$\int_{a}^{b} Q_i(x) Q_j(x) g(x)\, dx = 0, \qquad i \neq j.$$

If, in addition,

$$\int_{a}^{b} Q_i^2(x) g(x)\, dx = 1$$

the polynomials are termed *orthonormal*. For example, if $g(x) = 1$ and $[a, b] = [-1, 1]$ it may be verified that the *Legendre polynomials*

$$1,\ x,\ x^2 - \frac{1}{3},\ x^3 - \frac{3x}{5}, \ldots$$

are orthogonal.

Orthogonal polynomials possess the following three properties which are of relevance to our discussion of Gaussian quadrature:

(1) $Q_i(x)$ has i simple roots in the interval (a, b), so that if the interval lies on the real axis all roots are real.

(2) They satisfy a three-term recurrence relation; that is, there exist co-
     efficients α_i, β_i and γ_i such that

$$Q_{i+1}(x) = (\alpha_i x - \beta_i)Q_i(x) - \gamma_i Q_{i-1}(x).$$

This provides an efficient means of generating sequences of orthogonal
polynomials.

(3) They form a basis for all polynomials of degree at most n. That is,
     for any polynomial $p_n(x)$ of degree at most n, there exist coefficients
     $\{\,\omega_i \mid i = 0, 1, \ldots, n\,\}$ such that

$$p_n(x) = \sum_{i=0}^{n} \omega_i Q_i(x).$$

It follows that if we extend the sequence (using, say, the three term
recurrence relation) then

$$\int_a^b p_n(x)Q_m(x)\,dx = 0, \qquad m > n.$$

The connection between the derivation of high-order quadrature formulae and orthogonal polynomials can now be established. We choose the
points of the Hermite interpolating polynomial (3.11) to be the roots of the
n^{th} degree Legendre polynomial, which must all lie in $(-1, 1)$. Apart from a
multiplicative factor, $(x - x_i)l_i(x)$ is equal to the nth Legendre polynomial,
whence

$$\int_{-1}^{1} \bar{h}_i(x)\,dx = \int_{-1}^{1} (x - x_i)l_i(x)l_i(x)\,dx = 0$$

by using the orthogonality condition and noting that $l_i(x)$ is of degree $n-1$.
The weights in this rule are defined by

$$\begin{aligned} w_i &= \int_{-1}^{1} h_i(x)\,dx \\ &= \int_{-1}^{1} l_i^2(x)\,dx - 2l_i'(x_i)\int_{-1}^{1} (x - x_i)l_i(x)l_i(x)\,dx. \end{aligned}$$

and we deduce that

$$w_i = \int_{-1}^{1} l_i^2(x)\,dx > 0.$$

Quadrature rules whose integration points are the zeros of the Legendre polynomials are referred to as *Gauss–Legendre* rules. The points and
weights corresponding to $n = 1$, 2, 3 and 4 are given in Figure 3.2, that is
the methods outlined here and in Section 3.3 generate the same rules.

A straightforward implementation of Gauss–Legendre quadrature is given by the subroutine GAUSS in Code 3.1. It returns in RES an approximation to a user defined integral over the finite range [A,B], the integrand being specified via the FUNCTION parameter FUN. Note that this must appear in an EXTERNAL statement in any program unit which calls GAUSS. N defines the order of the Gauss–Legendre rule to be used; values of N equal to 1, 2, 3, 4 and 5 only are permitted, although this restriction may be lifted in an obvious manner. (Gauss–Legendre rules have been extensively tabulated; see, for example, Stroud and Secrest [80].) Observe that we only need to store half the points and weights of the rule. This follows from the fact that, before transformation, they are symmetrically distributed about the origin; that is, if $\bar{x}$ is a Gauss point with associated weight $\bar{w}$ then $-\bar{x}$ is also a Gauss point and has the same weight as $\bar{x}$.

```
      SUBROUTINE GAUSS(FUN,A,B,N,RES,IFAIL)
*
*.. ROUTINE GAUSS RETURNS IN RES AN APPROXIMATION TO
*.. THE INTEGRAL FROM A TO B (WITH B .GT. A) OF FUN USING
*.. A GAUSS RULE OF ORDER N WHERE N MAY BE 1,2,3,4 OR 5.
*..
*.. THE USER SUPPLIED FUNCTION FUN IS OF THE FORM
*..        DOUBLE PRECISION FUNCTION FUN(X)
*..        DOUBLE PRECISION X
*..        .........
*..
*.. ON RETURN
*..        IFAIL = 0  ON A SUCCESSFUL EXIT
*..        IFAIL = 1  IF N IS NOT ONE OF 1,2,3,4,5
*..        IFAIL = 2  IF A.GT.B
      DOUBLE PRECISION HALF,ZERO
      INTEGER MAXELT,MAXRUL
      PARAMETER (MAXELT = 9, MAXRUL = 5)
      PARAMETER (ZERO = 0.0D0,HALF = 0.5D0)
      DOUBLE PRECISION A,ABSC(MAXELT),B,FUN,RES,
     +               TRANS1,TRANS2,W(MAXELT),
     +               X,YPOS,YNEG
      INTEGER END,I,IFAIL,N,PTR(MAXRUL),START
      EXTERNAL FUN
*
*.. STORE THE WEIGHTS AND POSITIVE ABSCISSAE IN
*.. DESCENDING ORDER
      DATA W /
*
*.. 1-POINT RULE
```

```
      +             2.0D0,
*
*.. 2-POINT RULE
      +             1.0D0,
*
*.. 3-POINT RULE
      +             0.55555 55555 55556D0, 0.88888 88888 88889D0,
*
*.. 4-POINT RULE
      +             0.34785 48451 37454D0, 0.65214 51548 62546D0,
*
*.. 5-POINT RULE
      +             0.23692 68850 56189D0, 0.47862 86704 99366D0,
      +             0.56888 88888 88889D0/
      DATA ABSC /
*
*.. 1-POINT RULE
      +             0.0D0,
*
*.. 2-POINT RULE
      +             0.57735 02691 89626D0,
*
*.. 3-POINT RULE
      +             0.77459 66692 41483D0, 0.0D0,
*
*.. 4-POINT RULE
      +             0.86113 63115 94053D0, 0.33998 10435 84856D0,
*
*.. 5-POINT RULE
      +             0.90617 98459 38664D0, 0.53846 93101 05683D0,
      +             0.0D0                              /
*
*.. PTR IS A POINTER ARRAY GIVING THE START POSITIONS OF
*.. THE ABSCISSAE AND WEIGHTS IN THE ARRAYS
*.. ABSC AND W FOR EACH RULE
      DATA PTR /1,2,3,5,7/
*
*.. CHECK INPUT DATA
      RES = ZERO
      IF(A.GT.B)THEN
        IFAIL = 2
        RETURN
*
      ELSE IF(N.GT.MAXRUL .OR. N.LT.1)THEN
        IFAIL = 1
```

```
         RETURN
      ENDIF
      IFAIL = 0
*
*.. FORM THE APPROXIMATION
*..
*.. GENERATE THE LINEAR TRANSFORMATION FROM -1,1 TO A,B
      TRANS1 = (A+B)*HALF
      TRANS2 = (B-A)*HALF
      START = PTR(N)
      IF(N.EQ.1)THEN
         END = 0
      ELSE
         END = START+(N-2)/2
         DO 10 I = START,END
            X = ABSC(I)
            YPOS = TRANS1+TRANS2*X
            YNEG = TRANS1-TRANS2*X
            RES = RES+W(I)*(FUN(YPOS)+FUN(YNEG))
10       CONTINUE
      ENDIF
*
*.. IF N IS ODD THEN THE LAST POINT IS ZERO
      IF(MOD(N,2).EQ.1)THEN
         RES = RES+W(END+1)*FUN(TRANS1)
      ENDIF
      END
```

Code 3.1   Subroutine GAUSS.

Gauss–Legendre rules are members of a wide family of quadrature rules known simply as Gauss rules. Other rules in the family are generated by changing the range of integration and/or the weight function and we give further examples of Gauss rules in Section 3.10. However, unless specifically designated otherwise, we use the term Gauss rule to mean a Gauss–Legendre rule.

EXERCISE 3.8   The polynomials

$$1, 2x, 4x^2 - 2, 8x^3 - 12x, \ldots,$$

are known as the *Hermite* polynomials. Verify that these may be generated from

$$H_n(x) = (-1)^n e^{x^2} \frac{d^n}{dx^n} e^{-x^2}$$

(the *Rodriguez formula*) and determine the next member of the sequence. Let $g(x) = e^{-x^2}$ and deduce that $g^{(n)}(x) = (-1)^n e^{-x^2} H_n(x)$. Hence show that the Hermite polynomials satisfy $H_{n+1}(x) = 2x H_n(x) - H_n'(x)$ and that the leading coefficient of H_n is 2^n. Prove that the Hermite polynomials are orthogonal on the interval $(-\infty, \infty)$ with respect to $g(x) = e^{-x^2}$. By what factor should the Hermite polynomials be multiplied to generate an orthonormal set? (Hint: $\int_{-\infty}^{\infty} g(x) = \sqrt{\pi}$).

3.5   The Kronrod and Patterson schemes

When using a quadrature rule of the form (3.2) we need to select a value for n and this must be done with some care. Too small a value may result in an inaccurate integral approximation being computed, whilst too large a value may produce a very accurate estimate but at considerable cost. In either case we have no way of determining the accuracy of the value obtained by a single call.

Conceptually, the simplest method of obtaining an integral approximation to a specified accuracy is to apply rules of increasing order to the whole interval until two successive values agree to within the requested tolerance, that is, $|R_m - R_n| < \epsilon$ where $m > n$ and R_i is the result obtained by applying an i-point quadrature rule. Such an algorithm is termed a *global automatic quadrature method*. In the interests of efficiency it is important that, where possible, we reuse available information when computing the next approximation in the sequence (see Exercises 3.3 and 3.6). It would seem natural to generate quadrature estimates from Gauss rules of increasing order, but, unfortunately, these rules will have no abscissae in common (with the exception of the mid-point of the range). This means that when moving from a lower-order rule to a higher-order rule all function values used to compute the lower-order approximation are discarded.

Kronrod [52] suggests a scheme based on Gauss–Legendre quadrature which reuses function values at the cost of sacrificing maximal degree of precision. The basic idea is to start with an n-point Gauss rule G_n, say, and then add a further $n+1$ points to produce a $(2n+1)$-point rule, K_{2n+1}, which has degree of precision $3n + 1$ if n is even and $3n + 2$ if n is odd (instead of the optimal degree of precision of $4n + 1$). Note that although the rules G_n and K_{2n+1} share n points in common, the corresponding weights will be different. To proceed to yet higher degrees of precision we could add a further $(2n + 1) + 1 = 2n + 2$ points, and so on. Points and weights corresponding to the choice $n = 7$, 10, 15, 20, 25 and 30 are given by Piessens et al. [63].

Patterson [61] advocates the sequence of Gauss type rules which uses 1, 3, 7, 15, 31, 63 and 127 points, each containing all points of its predecessor. An n-point rule in the sequence has degree of precision $(3n+1)/2$ for $n > 1$. This forms the basis of D01ARF for which a call takes the form

```
      CALL D01ARF(A,B,FUN,RELACC,ABSACC,MAXRUL,IPARM,
   +              ACC,ANS,N,ALPHA,IFAIL)
```

A and B define the range of integration (which is transformed to $[-1, 1]$ before application of the Patterson rules) and FUN defines the integrand. The role of ALPHA is discussed in Section 3.7; here we simply ensure that it is an array of at least 390 elements, but the choice IPARM = 0 means that they are not used. If P_n is the integral estimate produced by the nth rule in the Patterson sequence, the routine terminates successfully when $|P_{n+1} - P_n| < \max(\text{ABSACC}, \text{RELACC} \times |P_{n+1}|)$. The value of P_{n+1} is returned via ANS and $|P_{n+1} - P_n|$ via ACC. A purely absolute (relative) error criterion can be enforced by setting RELACC (ABSACC) to zero. MAXRUL is used to restrict the number of rules taken in the sequence should convergence not be achieved. If its value is not in the range 1–9 a value of 9 will be assumed. This means that the routine will extend the Patterson sequence by up to two terms where necessary, and that the maximum number of function calls made is 511. On exit, N gives the number of function calls actually made. An exit from D01ARF with IFAIL = 1 means that it has not been possible to satisfy the user's accuracy criteria. Subdividing the range of integration in an appropriate manner and then using D01ARF to integrate over each subinterval in turn may improve manners. Ideally we would like to have a routine which automatically did the subdivision for us, and our attention is next turned to how this might be achieved.

EXERCISE 3.9   Show that the first k Patterson approximations and the first k Gauss rules in the sequence $\{G_{2^i} \mid i = 0, 1, 2, \ldots\}$ may be evaluated using the same number of function evaluations. Compare the degree of precision of the first 6 approximations in each sequence.

3.6   Adaptive quadrature

The problem with the algorithms described in the previous section is their global approach. If an integrand is badly behaved in one small region then the order of the approximation is increased over the whole interval. For such integrands it is likely that the global approach will use many more function evaluations than are actually necessary to attain the required accuracy. It would appear more efficient to increase the order of the approximation only in those regions of the range of integration where the function is badly behaved, and to use fewer points in regions where the function is less troublesome. This is the philosophy behind *adaptive quadrature* algorithms which attempt to sample the function depending on the underlying behaviour of each particular integrand. Figure 3.3 illustrates how an adaptive routine samples the integrand; each mark on the x-axis represents a function evaluation at that point. Notice how the points

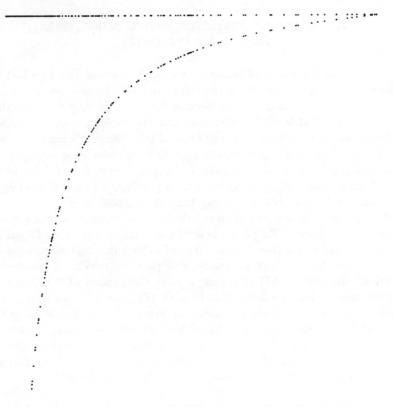

Figure 3.3   Sampling of function values by an adaptive routine.

are packed together in the region where the function is changing rapidly, and are relatively sparse in regions of little action.

The logic of adaptive quadrature is basically very simple. Below we present a skeletal pseudo-code which attempts to highlight the key features

while (termination criterion not satisfied) do

(1)   choose the next interval I to operate on,

(2)   subdivide I into I_1 and I_2 and for each half compute a local quadrature estimate and a local error estimate,

(3)   update the data structure used to store subinterval information, replacing information about I with the newly calculated information for I_1 and I_2,

endwhile.

We see that as the algorithm proceeds the range of integration is divided into a number of subranges, and for each subrange both an integral approximation and an error estimate are computed. Only those subintervals for which the error estimate is considered to be too large are selected in stage (1).

Rice [69] postulates that there are several million adaptive quadrature algorithms which fit into this framework. This paper, and Rice [68], contain discussions of several different strategies for choosing the next subinterval to divide, and the data structures necessary to store the information about each subdivision. We concentrate on the methods and data structures used by the NAG adaptive routines, many of which are based on QUADPACK codes; see Piessens et al. [63] for complete details.

We first look at the data structure used to keep track of the subinterval information, and comment that this needs to allow both for the rapid selection of the next subinterval to be divided and for easy updating. Suppose that m subdivisions have already taken place, so that $[a, b]$ has been partitioned such that $a = x_1 < x_2 < \cdots < x_m < x_{m+1} = b$.

For each subinterval we need to store

- the left and right end-points a_i and b_i,

- the local quadrature estimate q_i,

- the local error estimate e_i.

This information is stored in four arrays (see Figure 3.4) in descending order of e_i. When we update these arrays it is more efficient to keep the list *ranked* by using an integer array of pointers rather than by sorting blocks of real numbers each time we change the list (q.v. Exercise 4.8). This requires an additional integer workspace array. Figure 3.4 indicates that the interval $[a_3, b_3]$ has the largest local error estimate associated with it, $[a_6, b_6]$ the next largest, and so on. It is the interval pointed to by p_1 which is selected as the next candidate for subdivision. The a_i, b_i, q_i and e_i are computed for each half of the chosen interval; one set, corresponding to the larger error estimate, overwrites the values pointed to by p_1 and the other set is added to the bottom of the list. The pointer array is then updated to preserve the ordering with respect to the size of the error estimates. Other strategies are possible and Rice ([68] and [69]) gives an extensive discussion of possible alternatives.

To complete the definition of our adaptive algorithm we examine how the values of the quadrature approximation, q_i, and its associated error estimate, e_i, are obtained for each subinterval. The majority of the adaptive NAG routines obtain a first estimate of the integral over $[a_i, b_i]$ using an n-point Gauss rule G_n. The Kronrod extension, K_{2n+1}, is then used to provide what is hoped to be a better approximation, and the difference is taken as an estimate of the local error in G_n, i.e., $e_i = |K_{2n+1} - G_n|$. Note that since G_n and K_{2n+1} share n points only $2n + 1$ function evaluations are made. Since K_{2n+1} is expected to be more accurate than G_n we take $q_i = K_{2n+1}$ and, hence, e_i is usually an overestimate of the error in K_{2n+1}.

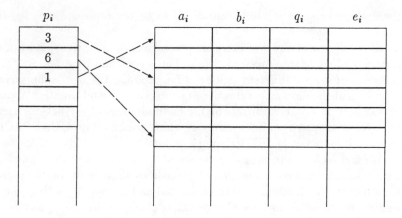

Figure 3.4   Adaptive quadrature data structure.

To compensate for this a number of heuristics are introduced into the algorithm. Details of the actual error estimate used may be found in Piessens et al. [63].

The strategy for the successful termination of our algorithm is that the process halts when the sum of the magnitudes of the local errors is less than some user specified tolerance. A failure termination will occur if

- a particular subinterval becomes too small (that is, successive bisections have not yielded a suitably small error estimate),

- the number of subdivisions is too large.

These conditions may occur if the integrand is badly behaved and/or the user tolerance is too small. The termination strategy used in the NAG routines also detects whether rounding errors are preventing termination. This may occur if the tolerance is set too small, or because of the existence of large rounding errors in the computation of the integrand, or errors in any empirical data used to define the integrand. A complete discussion may be found in Malcolm and Simpson [56].

If the smallest subinterval has the largest associated error estimate then prior to bisection an attempt may be made to reduce the error in the integral by applying an extrapolation process known as the ϵ-algorithm. (We observe that the condition just described may be symptomatic of a singularity in the integrand.) Very simply, we have a sequence of values $\{\, q_k \mid k = 1, 2, \ldots, n \,\}$ which are known to converge to a value q. It is assumed that the error in each q_k is of the form

$$e_k = q_k - q = c_1 \lambda_1^k + c_2 \lambda_2^k + \cdots + c_j \lambda_j^k,$$

where c_1, c_2, ..., c_j are constants and $1 < \lambda_1 < \lambda_2 < \cdots < \lambda_j$. The ϵ-algorithm takes combinations of the q_k so as to eliminate leading terms in the error expansion, thus yielding an approximation to q which is more accurate than any of the q_k. For more details see Wynn [84].

The adaptive algorithm just described is implemented in D01AJF which uses the pair of rules G_{10}, K_{21}. This is a general purpose integrator and may be used effectively even when the integrand possesses end-point singularities. The call sequence is

```
      CALL D01AJF(FUN,A,B,ABSACC,RELACC,ANS,ACC,W,
    +              LW,IW,LIW,IFAIL)
```

The purpose of FUN, A, B, ABSACC and RELACC are as for D01ARF (Section 3.5). The final approximation is returned via ANS and on successful exit ACC contains an estimate of the absolute error in ANS. In general ACC is an upper bound for

$$\left| \int_A^B f(x)\, dx - \text{ANS} \right|.$$

The array W, of length LW, is used to store the a_i, b_i, q_i and e_i shown in Figure 3.4. The maximum number of subintervals allowed is therefore $LW/4 = maxsubs$. For most problems a value for LW in the range $800 - 2000$ (allowing $200 - 500$ subdivisions) is likely to be adequate. The pointers are stored in the array IW, of length LIW where $\text{LIW} \geq maxsubs/2 + 2$. On exit IW(1) contains the amount of workspace actually used in the approximation of the integral. This may be used to measure the efficiency of the routine for a particular problem, by calculating the number of subintervals required, IW(1)/4. The number of integrand evaluations performed could also be measured by inserting a counter in a common block within the user supplied integrand function and the calling unit.

There are a number of error indicators that can be returned by D01AJF. Their general vagueness reflects the major problem with quadrature, namely that by sampling the integrand at a finite number of points it is usually not possible either to guarantee an approximation correct to some specified tolerance or to say whether the value of the integral is determinable or not. Apart from the case IFAIL = 6 (invalid input parameters) some idea of the state of play when an error is flagged may be obtained by selecting the soft error option and inspecting the values of ANS and ACC. The value IFAIL = 2 or 4 indicates that it was not possible to achieve the requested accuracy. Values of 1 or 3 usually indicate abnormal integrand behaviour, assuming that a sufficient amount of workspace has been made available. There are two ways of proceeding should either error occur. We could simply increase the amount of workspace and/or reduce the accuracy condition. However we should really attempt to analyse the form of the integrand in order to discover precisely what is the difficulty. (Ideally we should attempt such an analysis even before attempting to use a quadrature routine.) If the difficulty is local (for example, an internal singularity) then splitting the range at that point and calling D01AJF twice, once for each subrange, may solve the problem.

The NAG library also contains a number of adaptive routines for one-dimensional quadrature which take into account specific features of the integrand. It is to be expected that when used in an appropriate context they are likely to be more reliable, and more efficient, than the general purpose integrators discussed earlier. However, generally the counter argument also holds, and hence before attempting to use one of these routines it is advisable to confirm its suitability by analysing the form of the integrand carefully.

We have already observed that D01AJF is capable of dealing with integrands possessing end-point singularities. For integrands with a number of singularities, discontinuities, etc., within the range of integration the use of D01ALF is recommended. This does little more than subdivide the range using points of difficulty indicated by the user, and then applies D01AJF to each subinterval in turn.

For highly oscillatory integrands there are two routines available. If the integrand may be factored as $g(x)f(x)$ where $g(x) = \cos(\omega x)$ or $\sin(\omega x)$ then D01ANF may be used. This permits end-point singularities in f. If the integrand is not of this specific form and contains no singularities then D01AKF is the recommended choice. However if the integrand does contain end-point singularities, use of the general purpose integrator D01AJF is suggested. Two other routines which deal with integrands which may be factored as $g(x)f(x)$ are

- D01AQF, for *Hilbert transforms* with

$$g(x) = (x - c)^{-1}$$

where c is a real constant, and

- D01APF, with

$$g(x) = (b - x)^c (x - a)^d \ln(b - x) \ln(x - a)$$

where $c, d > -1$.

For infinite or semi-infinite intervals, D01AMF is available. The range of integration is first transformed to $[0, 1]$ and an approach similar to that of D01AJF is then used.

Before leaving adaptive quadrature we make brief mention of D01AHF. For each sub-interval this routine computes the Patterson sequence which forms the basis of D01ARF until convergence is achieved. If the sequence does not converge then the sub-interval is further subdivided. For most applications where an adaptive routine is required D01AJF is likely to prove satisfactory. However, if the integrand contains internal singularities then the use of D01AHF may be more appropriate.

3.7 Series approximation and quadrature

Occasionally we are interested in computing a number of integrals of the form

$$\int_c^d f(x)\,dx,$$

for several subranges, $[c,d]$, of $[a,b]$. Rather than regard each problem separately we look at the possibility of replacing the integrand using a series approximation and integrating this analytically.

We begin by transforming $[a,b]$ onto $[-1,1]$ and then look for a polynomial approximation to f of the form

$$f(x) \approx \sum_{i=0}^m \alpha_i P_i(x), \qquad (3.14)$$

where $P_i(x)$ is the Legendre polynomial of degree i. Now

$$\int_{-1}^1 f(x)P_j(x)\,dx = \sum_{i=0}^n \alpha_i \int_{-1}^1 P_i(x)P_j(x)\,dx$$

and using the orthogonality property of the Legendre polynomials the coefficients in our series approximation are immediately given by

$$\alpha_i = \frac{\int_{-1}^1 P_i(x)f(x)\,dx}{\int_{-1}^1 P_i^2(x)\,dx}. \qquad (3.15)$$

If the P_is are generated using the three-term recurrence relation

$$P_i(x) = \frac{2i-1}{i} x P_{i-1} - \frac{i-1}{i} P_{i-2}(x),$$

then the result

$$\int_{-1}^1 P_i^2(x)\,dx = \frac{2}{2i-1}$$

holds. The numerator in (3.15) may be evaluated numerically. In particular we may use the Patterson sequence, in which case it can be shown that m in (3.14) should be at most $(3n-1)/4$ if an n-point rule is to be used. Note that the same evaluations of the integrand f are used in the calculation of each α_i.

Now,

$$\int_c^d f(x)\,dx \approx \sum_{i=0}^m \alpha_i \int_c^d P_i(x)\,dx$$

and the Legendre polynomials may be integrated analytically, to yield a method which has degree of precision m. Both the evaluation of the polynomial coefficients, and the evaluation of the integral over $[c,d]$ of the polynomial approximation, may be achieved using appropriate calls to D01ARF.

The first is as described in Section 3.5, only now we set `IPARM` $= 1$. The coefficients $\{ \alpha_i \mid i = 0, 1, \ldots, m \}$ are returned in the first $m + 1$ positions of `ALPHA`. It is imperative that the contents of this array should remain unaltered between a call to `D01ARF` with `IPARM` $= 1$ and a subsequent call with `IPARM` $= 2$. Note that with `IPARM` $= 1$, `D01ARF` also computes the definite integral of `FUN` over $[A, B]$, and returns this value in `ANS` in the normal way.

The second call, with `IPARM` $= 2$, takes the form

```
    CALL D01ARF(C,D,FUN,RELACC,ABSACC,MAXRUL,IPARM,
  +              ACC,ANS,N,ALPHA,IFAIL)
```

where $[C, D]$ is a subrange of $[A, B]$. (`IFAIL` is set to 4 if this is not the case.) On successful termination `ANS` returns the required integral approximation. Note that `FUN`, `RELACC`, `ABSACC`, `MAXRUL`, `ACC` and `N` have no significance for a call with `IPARM` $= 2$. Nevertheless, it is essential that compatible parameters are used, and, in particular, that `FUN` is declared in an `EXTERNAL` statement within the calling unit.

3.8 Special integrals

A number of commonly occurring mathematical functions may be defined as integrals, where the function parameters are in either the range of integration or in the definition of the integrand. For example, the error function *erf* is defined by

$$erf(x) = \frac{2}{\sqrt{\pi}} \int_0^x e^{-t^2} \, dt \qquad (3.16)$$

and the sine integral by

$$Si(x) = \int_0^x \frac{\sin(u)}{u} \, du.$$

Although the general purpose routines from the D01 chapter could be used to compute these integrals, it is possible to produce more efficient methods of approximation for specific integrands. A number of these specialized routines are available in the S chapter of the NAG library. The majority use a *Chebyshev series* (see Section 5.3) to approximate the integrals, although a number of different series expansions are required to cover the complete range of possible parameters for any particular integral. Chebyshev approximations take the form

$$f(x) = g(x) \sum_{r=0}^{n}{}' a_r T_r(x) \qquad (3.17)$$

where n, the order of the approximation, varies depending on

- different functions,

- different ranges of the arguments of a particular function,

- different implementations of the library due to differences in precision and the floating-point arithmetic.

The approximated function is often treated as a product of two functions where the auxiliary function, g, is used to extract 'difficulties' from the underlying function (for example, zeros, singularities or asymptotes) in a way similar to that used for generating special Gaussian quadrature rules (see Section 3.10). The remaining function should then be well-behaved and can be approximated by a rapidly convergent Chebyshev expansion. A function $t(x)$ is used to map the general region of interest $[a, b]$ into the range $[-1, 1]$ required by the Chebyshev approximation.

There are a number of reasons for choosing the Chebyshev series approach for a multi-machine implementation:

- the coefficients, a_r, in (3.17) are relatively easy to generate;

- once generated very accurately, the coefficients may be truncated to suit different machine precisions;

- the magnitude of the first neglected term in the Chebyshev expansion gives a good indication of the maximum truncation error;

- the evaluation of the series (3.17) can be performed in an efficient and numerically stable way, for details see Section 5.3.

For many of the special functions, Section 10 of the routine document contains an analysis of the error in the computed approximation. This information is frequently presented in graphical form. We define E and Δ as the absolute errors in the function value and argument, and ϵ and δ as the corresponding relative errors. Using the Taylor series expansion (1.9), we then have

$$f(x + \Delta) = f(x) + \Delta f'(x) + \text{higher order terms},$$

whence

$$E = |f(x + \Delta) - f(x)| \approx \Delta |f'(x)|$$

or

$$\frac{E}{\Delta} \approx |f'(x)|.$$

Similarly we may obtain

$$\frac{\epsilon}{\delta} \approx \left| \frac{x f'(x)}{f(x)} \right| \qquad (3.18)$$

and

$$\frac{E}{\delta} \approx |x f'(x)|.$$

If we assume the only error in the argument is caused by its representation as a floating-point number then δ is bounded and independent of x. For example, using a 55-bit mantissa we have $\delta \approx 2^{-55}$. We then know that $\Delta = x\delta$ is also bounded but is dependent on x. E/δ, E/Δ and ϵ/δ are termed *amplification factors* and indicate how an error in the argument affects the accuracy of the computed approximation. Generally the behaviour of (3.18) is the most useful since it gives an estimate of the number of incorrect digits in the returned result. For example, $\lceil \log_2(\epsilon/\delta) \rceil$ gives the number of bits lost, where $\lceil x \rceil$ is the smallest integer greater than x. There are cases (for example, near the zero of a function) where considering absolute errors is more appropriate. The predicted error amplifications are usually a good measure of the actual error when the amplification factors are large, say around 10. For factors around unity the actual errors are usually dominated by rounding errors in the computer arithmetic and the predicted errors tend to be underestimates.

As an example we look at the complementary error function

$$erfc(x) = \frac{2}{\sqrt{\pi}} \int_x^\infty e^{-u^2} \, du$$

which is implemented as routine S15ADF. Figure 3.5 shows a plot of the function which approaches 2 asymptotically as $x \to -\infty$ and to zero as $x \to \infty$. The implementation of S15ADF contains two machine dependent constants, x_{low} and x_{hi}, chosen so that, to within rounding error, $erfc(x) = 2$ if $x < x_{low}$ and $erfc(x) = 0$ if $x > x_{hi}$. The values of x_{low} and x_{hi} are given in the implementation dependent documentation (see Section 1.8); on the VAX 11/780 the values are $x_{low} = -4.5$ and $x_{hi} = 9.5$. For values of x in the range (x_{low}, x_{hi}) the amplification factor (3.18) is given by

$$\frac{\epsilon}{\delta} = \left| \frac{2xe^{-x^2}}{\sqrt{\pi} \, erfc(x)} \right|$$

and its behaviour is illustrated in Figure 3.6. For large negative values of x the factor is $\sim xe^{-x^2}/\sqrt{\pi}$ and accuracy is limited principally by machine precision. Around zero the factor behaves as $2x/\sqrt{\pi}$ and again, since x is small, the effects of rounding errors are more important than approximation errors. For positive x the amplification factor increases steadily, and for large positive values it acts like $2x^2$. However since

$$\frac{E}{\delta} \approx \left| \frac{2xe^{-x^2}}{\sqrt{\pi}} \right|$$

absolute accuracy is guaranteed for all values of x. For x positive and increasing the true function tends to zero and, although the number of correct bits in the mantissa of the returned result decreases, the magnitude is always accurately determined.

The following integrals are available as special function routines:

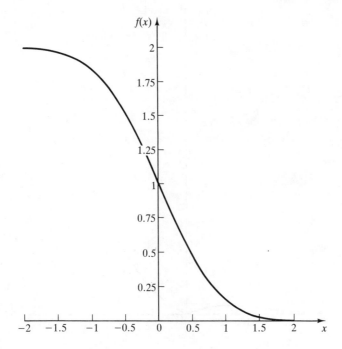

Figure 3.5   Plot of $f(x) = erfc(x)$.

- exponential integral (S13AAF)

$$E_1(x) = \int_x^\infty \frac{e^{-u}}{u}\, du, \quad x > 0,$$

- cosine integral (S13ACF)

$$Ci(x) = \gamma + \ln(x) + \int_0^x \frac{\cos(u) - 1}{u}\, du, \quad x > 0,$$

- sine integral (S13ADF)

$$Si(x) = \int_0^x \frac{\sin(u)}{u}\, du,$$

- cumulative normal distribution function (S15ABF)

$$P(x) = \frac{1}{\sqrt{2\pi}} \int_{-\infty}^x e^{-\frac{1}{2}u^2}\, du,$$

- complement of cumulative normal distribution function (S15ACF)

$$Q(x) = \frac{1}{\sqrt{2\pi}} \int_x^\infty e^{-\frac{1}{2}u^2}\, du,$$

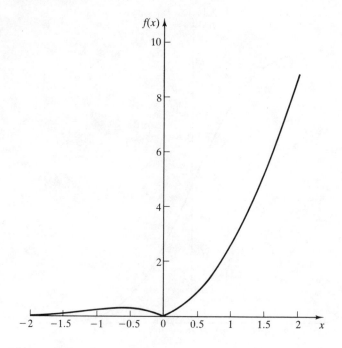

Figure 3.6   Amplification factor $f(x) = \epsilon/\delta$ for S15ADF.

- complementary error function (S15ADF)

$$erfc(x) = \frac{2}{\sqrt{\pi}} \int_x^\infty e^{-u^2} \, du,$$

- error function (S15AEF)

$$erf(x) = \frac{2}{\sqrt{\pi}} \int_0^x e^{-u^2} \, du,$$

- Dawson's integral (S15AFF)

$$F(x) = e^{-x^2} \int_0^x e^{t^2} \, dt,$$

- Fresnel integral (S20ACF)

$$S(x) = \int_0^x \sin\left(\frac{\pi}{2} t^2\right) \, dt,$$

- Fresnel integral (S20ADF)

$$C(x) = \int_0^x \cos\left(\frac{\pi}{2} t^2\right) \, dt,$$

- degenerate symmetrized elliptic integral of the first kind (S21BAF)

$$R_C(x,y) = \frac{1}{2} \int_0^\infty \frac{dt}{\sqrt{(t+x)(t+y)}}, \quad x \geq 0, \, y > 0,$$

- symmetrized elliptic integral of the first kind (S21BBF)

$$R_F(x,y,z) = \frac{1}{2} \int_0^\infty \frac{dt}{\sqrt{(t+x)(t+y)(t+z)}},$$

where $x, y, z \geq 0$ and at most one is zero,

- symmetrized elliptic integral of the second kind (S21BCF)

$$R_D(x,y,z) = \frac{3}{2} \int_0^\infty \frac{dt}{\sqrt{(t+x)(t+y)(t+z)^3}},$$

where $x, y \geq 0$ with at most one zero, and $z > 0$,

- symmetrized elliptic integral of the third kind (S21BDF)

$$R_J(x,y,z,\rho) = \frac{3}{2} \int_0^\infty \frac{dt}{(t+\rho)\sqrt{(t+x)(t+y)(t+z)}},$$

where $x, y, z \geq 0$ with at most one zero, and $\rho > 0$.

The definitions of the elliptic integrals (S21) are given in their symmetrized forms for which more stable numerical algorithms exist than for the traditional canonical Legendre forms. These canonical forms may easily be computed in terms of the symmetrized forms:

- elliptic integral of the first kind

$$
\begin{aligned}
F(\phi \mid m) &= \int_0^\phi \frac{d\theta}{\sqrt{1 - m\sin^2\theta}} \\
&= \int_0^{\sin\phi} \frac{dt}{\sqrt{(1-mt^2)(1-t^2)}} \\
&= \sin(\phi) R_F(\cos^2\phi, 1 - m\sin^2\phi, 1),
\end{aligned}
$$

- elliptic integral of the second kind

$$
\begin{aligned}
E(\phi \mid m) &= \int_0^\phi \sqrt{1 - m\sin^2\theta} \, d\theta \\
&= \int_0^{\sin\phi} \sqrt{\frac{1-mt^2}{1-t^2}} \, dt \\
&= \sin(\phi) R_F(\cos^2\phi, 1 - m\sin^2\phi, 1) \\
&\quad - \tfrac{m}{3}\sin^3\phi \, R_D(\cos^2\phi, 1 - m\sin^2\phi, 1),
\end{aligned}
$$

- elliptic integral of the third kind

$$\begin{aligned}
\Pi(n; \phi \mid m) &= \int_0^\phi \frac{d\theta}{(1 - n\sin^2\theta)\sqrt{1 - m\sin^2\theta}} \\
&= \int_0^{\sin\phi} \frac{d\theta}{(1 - nt^2)\sqrt{(1 - mt^2)(1 - t^2)}} \\
&= \sin(\phi)R_F(\cos^2\phi, 1 - m\sin^2\phi, 1) \\
&\quad - \tfrac{n}{3}\sin^3\phi R_J(\cos^2\phi, 1 - m\sin^2\phi, 1, 1 + n\sin^2\phi),
\end{aligned}$$

- complete elliptic integral of the first kind

$$K(m) = F\left(\tfrac{\pi}{2} \mid m\right) = \int_0^{\pi/2} \frac{d\theta}{\sqrt{1 - m\sin^2\theta}} = R_F(0, 1 - m, 1),$$

- complete elliptic integral of the second kind

$$\begin{aligned}
E(m) &= E\left(\tfrac{\pi}{2} \mid m\right) = \int_0^{\pi/2} \sqrt{1 - m\sin^2\theta}\, d\theta \\
&= R_F(0, 1 - m, 1) - \tfrac{m}{3}R_D(0, 1 - m, 1).
\end{aligned}$$

More details of the symmetrized forms may be found in Carlson [14] which, along with Abramowitz and Stegun [1], is a goldmine of information on the properties and forms of many of the special functions which arise in applied mathematics.

EXERCISE 3.10   Tabulate the error function (3.16) for $x = 0.0$ to 2.0 in steps of 0.1 using (i) S15AEF; (ii) D01AJF with a relative tolerance of 10^{-4}. Compare the accuracy of the results obtained from D01AJF with those of S15AEF; explain. Compare the execution times of the two routines.

EXERCISE 3.11   Write routines to compute the Legendre forms of the elliptic integrals. Check the results obtained against published tables (e.g., Abramowitz and Stegun [1]).

EXERCISE 3.12   Show that the cumulative normal distribution function and its complement can be evaluated using appropriate calls to S15ADF.

EXERCISE 3.13   We wish to evaluate $f(x) = \ln x/(x - 1)$. What computational problems may arise if f is computed naively for values of x close to one? Compare the accuracy and efficiency of the following methods for computing f for such values of x

- the intrinsic log function,
- the approximation, suitably truncated,

$$\ln(x) = 2\left[\frac{x - 1}{x + 1} + \frac{1}{3}\left(\frac{x - 1}{x + 1}\right)^3 + \frac{1}{5}\left(\frac{x - 1}{x + 1}\right)^5 + \cdots\right],$$

$$f_i$$
$$\qquad f[\,x_i x_{i+1}\,]$$
$$f_{i+1} \qquad\qquad\qquad f[\,x_i x_{i+1} x_{i+2}\,]$$
$$\qquad f[\,x_{i+1} x_{i+2}\,] \qquad\qquad\qquad\qquad f[\,x_i x_{i+1} x_{i+2} x_{i+3}\,]$$
$$f_{i+2} \qquad\qquad\qquad f[\,x_{i+1} x_{i+2} x_{i+3}\,]$$
$$\qquad f[\,x_{i+2} x_{i+3}\,]$$
$$f_{i+3}$$

Figure 3.7   Divided difference table.

- the identity (see Carlson [14] for details)

$$\frac{\ln x - \ln y}{x - y} = R_C\left(\left(\frac{x+y}{2}\right)^2, xy \right)$$

and then using S21BAF to compute f directly.

3.9   Cubic interpolation

We now investigate the problem of approximating an integral when values of the integrand are available only at a predetermined, fixed set of points.

We begin by considering the cubic polynomial, $p_{3,i}(x)$, which interpolates at the distinct coordinates $\{(x_j, f_j) \mid j = i,\, i+1,\, i+2,\, i+3\}$. There are a number of ways of determining the form of this unique polynomial; from a theoretical standpoint the Lagrange form (3.8) is often convenient, however, for ease of computation we choose to work with the *divided difference* formulation.

We define $f[\,x_j x_{j+1}\,]$, the *first divided difference* of f with respect to x_j and x_{j+1}, as

$$f[\,x_j x_{j+1}\,] = \frac{f_{j+1} - f_j}{x_{j+1} - x_j}.$$

From this we define, recursively, the kth divided difference of f with respect to $x_j, x_{j+1}, \ldots, x_{j+k}$ as

$$f[\,x_j x_{j+1} \cdots x_{j+k}\,] = \frac{f[\,x_{j+1} x_{j+2} \cdots x_{j+k}\,] - f[\,x_j x_{j+1} \cdots x_{j+k-1}\,]}{x_{j+k} - x_j}.$$

Figure 3.7 shows how differences based on the points x_i, x_{i+1}, x_{i+2} and x_{i+3} may be computed. Starting with the function values we compute the first divided differences to form the second column — these values are used to compute the second differences (column 3) and finally the third difference is determined. We leave to an exercise (Exercise 3.14) a simple proof that $p_{3,i}(x)$ may be written as

$$
\begin{aligned}
p_{3,i}(x) = & \tfrac{1}{2}\{(f_{i+1} + f_{i+2}) + (2x - x_{i+1} - x_{i+2})f[x_{i+1}x_{i+2}] \quad\quad (3.19)\\
& + (x - x_{i+1})(x - x_{i+2})(f[x_ix_{i+1}x_{i+2}] + f[x_{i+1}x_{i+2}x_{i+3}])\\
& + (x - x_{i+1})(x - x_{i+2})(2x - x_i - x_{i+3})f[x_ix_{i+1}x_{i+2}x_{i+3}]\}.
\end{aligned}
$$

Now, provided $n - 1$ is divisible by 3, we may write

$$
\int_a^b f(x)\,dx \approx \int_{x_1}^{x_4} p_{3,1}(x)\,dx + \int_{x_4}^{x_7} p_{3,4}(x)\,dx + \cdots + \int_{x_{n-3}}^{x_n} p_{3,n-3}(x)\,dx,
$$

where each subinterval approximation may be integrated exactly. However the restriction on n is not in line with our desire for general purpose algorithms and an alternative approach, proposed by Gill and Miller [36] is to express the integral, I, in the form (3.4), and then approximate the integral over each subinterval, $[x_i, x_{i+1}]$, by

$$
R^{(i)} = \int_{x_i}^{x_{i+1}} p_{3,i-1}(x)\,dx,
$$

that is, integrate over $[x_i, x_{i+1}]$ the cubic polynomial which interpolates f at $\{x_j \mid j = i-1, i, i+1, i+2\}$. Assuming that values f_0 and f_{n+1} are not available we note that special consideration must be given to integration over $[x_1, x_2]$ and $[x_{n-1}, x_n]$; we return to this later.

We introduce the notation

$$
\begin{aligned}
h_j &= x_{j+1} - x_j,\\
s &= (x - x_i)/h_i,\\
e_j &= h_j/h_i - 1.
\end{aligned}
$$

Then

$$
\begin{aligned}
p_{3,i-1}(x) &\equiv p_{3,i-1}(x_i + sh_i)\\
&= \tfrac{1}{2}(f_i + f_{i+1}) + (s - \tfrac{1}{2})h_i f[x_ix_{i+1}]\\
&\quad + \tfrac{1}{2}h_i^2 s(s-1)(f[x_{i-1}x_ix_{i+1}] + f[x_ix_{i+1}x_{i+2}])\\
&\quad + h_i^2 s(s-1)(sh_i + \tfrac{1}{2}(h_{i-1} - h_i - h_{i+1}))f[x_{i-1}x_ix_{i+1}x_{i+2}]\\
&= (1-s)f_i + sf_{i+1}\\
&\quad + \tfrac{1}{2}h_i^2 s(s-1)(f[x_{i-1}x_ix_{i+1}] + f[x_ix_{i+1}x_{i+2}])\\
&\quad + h_i^3 s(s-1)(s - \tfrac{1}{2} + \tfrac{1}{2}(e_{i-1} - e_{i+1}))f[x_{i-1}x_ix_{i+1}x_{i+2}].
\end{aligned}
$$

Now

$$
\int_{x_i}^{x_{i+1}} p_{3,i-1}(x)\,dx = h_i \int_0^1 p_{3,i-1}(x_i + sh_i)\,ds,
$$

and we also have

$$\int_0^1 ds = 1; \qquad \int_0^1 s\,ds = \tfrac{1}{2}; \qquad \int_0^1 s(s-1)\,ds = -\tfrac{1}{6};$$

$$\int_0^1 s(s-1)(s-\tfrac{1}{2})\,ds = 0.$$

Thus

$$\int_{x_i}^{x_{i+1}} p_{3,i-1}(x)\,dx \qquad\qquad\qquad\qquad (3.20)$$

$$= \tfrac{1}{2}h_i(f_i + f_{i+1}) - \tfrac{1}{12}h_i^2\big(f[\,x_{i-1}x_ix_{i+1}] + f[\,x_ix_{i+1}x_{i+2}]$$
$$+ h_i(e_{i-1} - e_{i+1})f[\,x_{i-1}x_ix_{i+1}x_{i+2}]\big).$$

Having derived the quadrature formula (3.20), we should now attempt to quantify the error to be incurred. From (3.10) we have

$$f(x) - p_{3,i-1}(x) = (x - x_{i-1})(x - x_i)(x - x_{i+1})(x - x_{i+2})f^{(4)}(\eta_i)/24$$

where $\eta_i \in (x_{i-1}, x_{i+2})$. Since $(x - x_{i-1})(x - x_i)(x - x_{i+1})(x - x_{i+2})$ does not change sign in $[\,x_i, x_{i+1}]$ it follows, from the intermediate value theorem (Conte and de Boor [19], p. 25), that the error in (3.20) may be written

$$E^{(i)} = \tfrac{1}{24}f^{(4)}(\xi_i)\int_{x_i}^{x_{i+1}}(x - x_{i-1})(x - x_i)(x - x_{i+1})(x - x_{i+2})\,dx$$

$$= \tfrac{1}{24}h_i f^{(4)}(\xi_i)\int_0^1 (sh_i + h_{i-1})sh_i(s-1)h_i\big((s-1)h_i - h_{i+1}\big)\,ds$$

$$= \tfrac{1}{24}h_i^5 f^{(4)}(\xi_i)\int_0^1 s(s-1)(s+1+e_{i-1})(s - e_{i+1} - 2)\,ds$$

where $\xi_i \in (x_{i-1}, x_{i+2})$ and we see that the rule has degree of precision three. Now

$$s(s-1)(s+1+e_{i-1})(s - e_{i+1} - 2)$$
$$= s(s-1)(s^2 - se_{i+1} - s - e_{i+1} - 2 + se_{i-1} - e_{i+1}e_{i-1} - 2e_{i-1})$$
$$= s(s-1)\big((s-2)(s+1) + (s - \tfrac{1}{2})(e_{i-1} - e_{i+1})$$
$$\qquad - \tfrac{3}{2}(e_{i-1} + e_{i+1}) + e_{i-1}e_{i+1}\big),$$

and since

$$\int_0^1 s(s-1)(s-2)(s+1)\,ds = \tfrac{11}{30},$$

we have

$$E^{(i)} = \tfrac{1}{24}\Big(\tfrac{11}{30} - \tfrac{1}{6}\big(e_{i-1}e_{i+1} - \tfrac{3}{2}(e_{i-1} + e_{i+1})\big)\Big)h_i^5 f^{(4)}(\xi_i). \qquad (3.21)$$

If an upper limit on the magnitude of $f^{(4)}$ is available, (3.21) can be used to determine a bound on the error in the rule. In the absence of such a bound we can estimate the error in the following way. Inspection of the definition of $f[x_i x_{i+1}]$ reveals that this quantity can be used as a local approximation to f' in the vicinity of the range of integration (cf. the relationship between the secant and Newton iterations in Section 2.3). It follows that

$$f[x_{i-1} x_i x_{i+1}] = \frac{f[x_i x_{i+1}] - f[x_{i-1} x_i]}{x_{i+1} - x_{i-1}},$$

or the average of this and $f[x_i x_{i+1} x_{i+2}]$ may be used to estimate f'' in (x_{i-2}, x_{i+2}). Continuing in this manner, we have that an approximation to $f^{(4)}(\xi_i)$ is given by either $f[x_{i-2} x_{i-1} x_i x_{i+1} x_{i+2}]$, $f[x_{i-1} x_i x_{i+1} x_{i+2} x_{i+3}]$, or an average of the two. Having computed such an estimate we may add the value given by (3.21) to the rule approximation to yield, we hope, an improved integral approximation.

We now return to the problem of integration over the first and last subintervals. Integrating over $[x_1, x_2]$ the cubic interpolating polynomial based on the point set $\{x_1, x_2, x_3, x_4\}$ gives

$$\int_{x_1}^{x_2} f(x)\, dx =$$
$$h_1\Big\{f_1 + \tfrac{1}{2} h_1 f[x_1 x_2] - \tfrac{1}{6} h_1^2 f[x_1 x_2 x_3] + h_1^3 \big(\tfrac{1}{4} + \tfrac{1}{6} e_2\big) f[x_1 x_2 x_3 x_4]\Big\}$$
$$- \tfrac{1}{24} h_1^5 f^{(4)}(\xi_1)\big(\tfrac{19}{30} + \tfrac{2}{3} e_2 + \tfrac{1}{4} e_3 + \tfrac{1}{6} e_2(e_2 + e_3)\big).$$

If $\{x_{n-3}, x_{n-2}, x_{n-1}, x_n\}$ is chosen as the set of interpolation points the result

$$\int_{x_{n-1}}^{x_n} f(x)\, dx =$$
$$h_{n-1}\Big\{f_n - \tfrac{1}{2} h_{n-1} f[x_{n-1} x_n] - \tfrac{1}{6} h_{n-1}^2 f[x_{n-2} x_{n-1} x_n]$$
$$- h_{n-1}^3 \big(\tfrac{1}{4} + \tfrac{1}{6} e_{n-2}\big) f[x_{n-3} x_{n-2} x_{n-1} x_n]\Big\}$$
$$- \tfrac{1}{24} h_{n-1}^5 f^{(4)}(\xi_n)\big(\tfrac{19}{30} + \tfrac{2}{3} e_{n-2} + \tfrac{1}{4} e_{n-3}$$
$$+ \tfrac{1}{6} e_{n-2}(e_{n-2} + e_{n-3})\big)$$

may be derived.

The method described above is implemented in routine D01GAF of the NAG library. A typical call takes the form

```
CALL D01GAF(X,Y,N,ANS,ER,IFAIL)
```

Here the N coordinates are supplied via the arrays X and Y. On exit ANS contains the computed estimate of the integral. Note that the error estimate, ER, is automatically added in.

Suppose we use D01GAF to estimate

$$I_j = \int_0^1 x^j\, dx, \qquad j = 0, 1, 2, \ldots$$

using function values at the points 0.0, 0.0625, 0.2225, 0.525, 0.9995 and 1.0. Then, as expected, DO1GAF integrates I_0, I_1, I_2 and I_3 exactly, allowing for rounding errors. It also integrates I_4 exactly because adding in the error estimate, ER, increases the degree of precision to four. I_5 is not integrated exactly; DO1GAF returns the value 0.16541 and the difference between this and the exact value, $1/6$, is 0.126×10^{-2}. We observe that in this case the magnitude of ER, 0.788×10^{-2}, is a bound for the error. This, however, is not always the case. If we now use function values at the points 0.0, 0.2, 0.4, 0.6, 0.8, 1.0, the degree of precision of the rule is increased to 5 due to the equal spacing of the points. The computed estimate of $I_6 = \frac{1}{7}$ is 0.14307 and this is in error by 0.210×10^{-3}. The returned value for ER is -0.219×10^{-3} illustrating that this is an estimate and not a bound on the error in integration.

EXERCISE 3.14 By substituting $x = x_i$, x_{i+1}, x_{i+2} and x_{i+3}, verify that $p_{3,i}(x)$ defined by (3.19) interpolates f on the points x_i, x_{i+1}, x_{i+2} and x_{i+3}.

3.10 Generalized Gaussian quadrature

Before leaving quadrature mention should be made of a number of non-adaptive routines which require the user to have some detailed knowledge of the form of the integrand. They are based on generalizations of and extensions to the Gauss rules introduced in Section 3.4. However, the difficulty of

- recognizing the correct routine to use,

- selecting a suitable value for n, the number of points,

should not be underestimated. Further the user cannot specify an error tolerance nor is any error estimate returned and hence it can be difficult to derive any real confidence in the results produced. At the very least it would be necessary to make two calls using different values of n. In most instances if accuracy and reliability are of prime importance an adaptive routine should be used. Nevertheless the correct use of one of these generalized routines can lead to a significant reduction in the number of function calls made and the effort needed to classify the problem appropriately can be worthwhile. In particular their use is clearly suitable for problems in which the rule is exact for the given integrand.

We consider the weighted integral

$$I = \int_a^b g(x)f(x)\,dx,$$

where g describes the predominant shape of the integrand and $g(x) \geq 0$ on the interval of integration. We approximate the component f only using a Hermite interpolating polynomial. Proceeding as before, we choose as our integration points the roots of the nth orthogonal polynomial on the range of integration with respect to the weight g. The requirement of derivative values is thus once again avoided and we obtain the rule

$$R = \sum_{i=1}^{n} w_i f_i,$$

where

$$w_i = \int_a^b g(x) l_i^2(x)\, dx$$

and the $\{l_i(x) \mid i = 1, 2, \ldots, n\}$ are defined by (3.9). Such a rule has degree of precision $2n - 1$.

The most commonly used Gauss rules (after Gauss–Legendre) are

(1) Gauss–Jacobi

   weight: $g(x) = (b - x)^c (x - a)^d$
   range: any finite range $[a, b]$
   restrictions: $c, d > -1, b > a$
   applications: integrands with end-point singularities
   precision: exact for $f(x) = x^i$, $i = 0, 1, \ldots, 2n - 1$;

(2) Gauss–Exponential

   weight: $g(x) = |x - (a + b)/2|^c$
   range: any finite range $[a, b]$
   restrictions: $c > -1, b > a$
   applications: integrands with mid-point singularities
   precision: exact for $f(x) = x^i$, $i = 0, 1, \ldots, 2n - 1$;

(3) Gauss–Laguerre

   weight: $g(x) = (x - a)^c e^{-bx}$
   range: the semi-infinite range $[a, \infty)$
   restrictions: $c > -1, b > 0$
   applications: integrands with possible left-hand end-point singularities and asymptotic behaviour like e^{-bx}
   precision: exact for $f(x) = x^i$, $i = 0, 1, \ldots, 2n - 1$;

(4) Gauss–Hermite

   weight: $g(x) = (x - a)^c e^{-b(x-a)^2}$
   range: the entire real axis, $(-\infty, \infty)$
   restrictions: $c > -1, b > 0$
   applications: integrands with possible left-hand end-point singularities and asymptotic behaviour like e^{-bx^2}
   precision: exact for $f(x) = x^i$, $i = 0, 1, \ldots, 2n - 1$;

(5)   Gauss–Rational

   weight: $g(x) = |x - a|^c / |x + b|^d$

   range: the semi-infinite range $[a, \infty)$

   restrictions: $c > -1$, $d > c + 1$, $a + b > 0$

   applications: integrands with inverse power rates of decay

   precision: exact for $f(x) = (x + b)^{-i}$, $i = 0, 1, \ldots, 2n - 1$.

The points and weights for these rules have been extensively tabulated and may be found in, for example, Stroud and Secrest [80].

3.11   The D01 direct routines

The NAG library contains a number of routines which calculate an integral estimate using the direct application of a Gauss rule. D01BAF approximates integrals of the form

$$\int_a^b f(x)\, dx.$$

The first argument of D01BAF is a subroutine name and indicates the type of Gauss rule to be used. The choices available are

Gauss–Hermite: D01BAW

Gauss–Laguerre: D01BAX

Gauss–Rational: D01BAY

Gauss–Legendre: D01BAZ

It should be noted that the implemented rules are not quite as general as those defined in the previous section, for example, the Gauss–Hermite, Gauss–Laguerre and Gauss–Rational rules only permit $c = 0$.

   A call to D01BAF takes the form

```
RES = D01BAF(D01BAZ,A,B,N,FUN,IFAIL)
```

if Gauss–Legendre quadrature is required. Note that since D01BAZ is itself a SUBROUTINE, it must appear, along with FUN, a user supplied function defining the whole integrand, in an EXTERNAL statement in the program unit which calls D01BAF. Only rules for which N = 1–6, 8, 10, 12, 14, 16, 20, 24, 48 and 64 are available. If N is not one of these then IFAIL is set to 1. If this happens when the soft error option has been selected (see Section 1.9) then the value returned by D01BAF is derived from the $\bar{N}$-point rule, where $\bar{N}$ is the largest value for which $\bar{N} \leq$ N, and such a rule is available. The values of A and B are used for two purposes:

Rule	Integration range	Weight function
Gauss–Legendre	$[A, B]$	1
Gauss–Laguerre	$[A, \infty)$ if $B > 0$ $(-\infty, A]$ if $B < 0$	e^{-Bx}
Gauss–Hermite	$(-\infty, \infty)$	$e^{-B(x-A)^2}$
Gauss–Rational	$[A, \infty)$ if $B > 0$ $(-\infty, A]$ if $B < 0$	–

Figure 3.8   Use of A and B parameters to DO1BAF.

(1) they determine the range of integration;

(2) they are used to adjust the free parameters in the underlying weight functions so as to obtain as close a match to the integrand as possible.

Using DO1BAF the user provided function FUN needs to specify the complete integrand. The integration is performed exactly, except for the effects of rounding errors, for functions of the form

DO1BAW: $f(x) = e^{-B(x-A)^2} p_{2n-1}(x),$

DO1BAX: $f(x) = e^{-Bx} p_{2n-1}(x),$

DO1BAY:

$$f(x) = \sum_{i=2}^{2n+1} \frac{c_i}{(x + B)^i}$$
$$= \frac{1}{(x + B)^{\hat{N}}} \sum_{i=0}^{2n-1} \alpha_i(x + B)^i$$

where $n \leq \hat{N}$,

DO1BAZ: $f(x) = p_{2n-1}(x),$

where $p_{2n-1}(x)$ is a polynomial of degree at most $2n - 1$.

Where appropriate, the parameters should be chosen to fit one of these forms as closely as possible. A good choice of B is often difficult to obtain and, unfortunately, an unsatisfactory choice of this value may lead to a poor approximation.

If the user is not satisfied with the restrictions imposed by DO1BAF (the restricted choice for n, the value of c for Gauss–Hermite or Gauss–Rational), the points and weights for the general forms of the five Gauss rules discussed in Section 3.10, plus Gauss–Legendre, may be generated using a call to DO1BCF; DO1FBF can then be used to determine the corresponding integral approximation. The values computed by DO1BCF are

Rule	Range	Weight parameters	Restrictions
Legendre	$[A, B]$	None	$B > A$
Jacobi	$[A, B]$	$(B - x)^C (x - A)^D$	$C + D + 2 \leq GMAX$ $C, D > -1;\ B > A$
Exponential	$[A, B]$	$\left\| x - \frac{A+B}{2} \right\|^C$	$C > -1;\ B > A$
Laguerre	$[A, \infty);\ B > 0$	$\|x - A\|^C e^{-Bx}$	$C > -1;\ B \neq 0$ $C + 1 \leq GMAX$
Hermite	$(-\infty, \infty)$	$\|x - A\|^C e^{-B(x-A)^2}$	$C > -1;\ B > 0$ $C + 1 \leq 2\,GMAX$
Rational	$[A, \infty);\ A > -B$ $(-\infty, A];\ A < -B$	$\|x - A\|^C / \|x + B\|^D$	$C > -1$ $D > C + 1$ $A + B \neq 0$

Figure 3.9   Parameter choices for D01BCF.

returned in one of two forms, normal or adjusted. For *normal weights* the integrand is treated as being separable and the rule defined by

$$R = \sum_{i=1}^{n} w_i f(x_i)$$

is used to approximate

$$\int_a^b g(x) f(x)\, dx.$$

For *adjusted weights* the integrand is treated as a whole and the situation is equivalent to that in D01BAF. In essence then, for adjusted weights the rule is taken as a linear combination of evaluations of the entire integrand, including the weight function g. For normal weights g has been absorbed in the w_is and the rule involves evaluation of the f part of the integrand only.

Because of the multiplicity of rules available and the diversity of parameters and ranges, the documentation for D01BCF can be difficult to come to terms with. To aid understanding of the way the routine works we list in Figure 3.9 some details of the purpose of the arguments A, B, C and D. GMAX is a machine dependent constant whose value is the largest integer such that $\Gamma(GMAX) = (GMAX - 1)!$ can be computed without overflow; this may be found in the implementation dependent documentation (see Section 1.8). On the VAX 11/780 the value of GMAX is 34.

The call sequence for D01BCF takes the form

```
CALL D01BCF(ITYPE,A,B,C,D,N,WEIGHT,ABSCIS,IFAIL)
```

n	CPU time (seconds)	$\left\lvert 2 - \sum_{i=1}^{n} w_i \right\rvert$
50	2.5	0.78×10^{-14}
100	9.5	0.18×10^{-13}
150	21.0	0.27×10^{-13}
200	37.3	0.34×10^{-13}
250	55.0	0.45×10^{-13}

Figure 3.10   Points and weights computed by D01BCF.

where the magnitude of ITYPE specifies the required rule. ITYPE $= 0$ gives Gauss–Legendre, ITYPE $= 1$ gives Gauss–Jacobi, and so on. (See Figure 3.9 for the order of the remaining rules.) The sign of ITYPE dictates whether normal (positive) or adjusted (negative) weights are to be computed (there is no such distinction for Gauss–Legendre), and N specifies the number of weights and points to be returned. Note that all the weights and integration points are returned in the arrays WEIGHT and ABSCIS respectively, regardless of any symmetry that may exist. Thus for $[A,B] = [-1,1]$ the Gauss–Legendre rules will include both the positive and negative abscissa points. The algorithm used to determine the points is an $O\left(N^3\right)$ process (see Golub and Welsch [37]). Large values of N are thus likely to require a significant amount of computation. Furthermore, there is an increasing loss of accuracy as $N \to \infty$. Figure 3.10 shows the relative performance of D01BCF when used to compute points and normal weights for the Gauss–Legendre rules with $[A,B] = [-1,1]$. Since the rules integrate $f(x) = 1$ exactly, we have $\int_{-1}^{1} dx = 2 = \sum_{i=1}^{n} w_i$ and the third column indicates how well this condition is satisfied. Clearly if the required weight functions and number of points are catered for by D01BAF, it is more efficient to use this routine since the points and weights are stored, and therefore not computed at each call. If an M-point Gauss–Legendre approximation is required and M is not in the available set it is generally more efficient to use D01BAF with a value of $N > M$ which is in the set, rather than use D01BCF to compute the weights and points for the M-point rule. Use of D01BCF in these circumstances would be considered only if the integrand were very expensive to evaluate.

For Gauss–Exponential and Gauss–Hermite rules with adjusted weights the value of N passed to D01BCF should be chosen to be even. If N is odd and C $\neq 0$ then either the exact integrand value or the exact weight for the middle point is infinite. In both cases the contribution to the approximation is indeterminate. Such a choice of N and C results in IFAIL being given the value 6. The routine sets IFAIL equal to 5 if one or more of the computed weights has underflowed. If the soft fail option is chosen these weights are returned as zero and they may produce a usable integral approximation. The failure may be prevented either by reducing the value of

N or, if appropriate, changing from normal to adjusted weights. If the computation of a weight would result in a value larger than $rmax$, the largest representable floating-point number (whose value may be determined from X02ACF), then IFAIL is set to 4 and the weight set equal to $rmax$ if the soft fail option is chosen. The generated rule is not suitable for estimating the integral and a further call will be necessary, either with a different value of N, or by changing from normal to adjusted weights. Error flag values of 2 or 3 indicate that inappropriate values of N, ITYPE, A, B, C or D have been chosen. Finally, a halt with IFAIL = 1 occurs if the algorithm used to compute the points, which is iterative in nature, fails to converge. This rarely happens and, when it does, is usually circumvented by adjusting N.

Once a set of points and weights has been successfully computed the function D01FBF may be used to calculate an integral approximation. The routine is designed to cope with multi-dimensional integration (up to 20 dimensions) and in the function call

$$RES = D01FBF(NDIM,NPTVEC,N,WEIGHT,ABSCIS,FUN,IFAIL)$$

we set the number of dimensions NDIM to one and the number of integration points NPTVEC(1) to N. As usual, the user supplied function FUN is used to define the integrand and its form depends on whether normal or adjusted weights are being used; hence great care must be taken if a decision is made to switch between the two types. Note that since multiple integration is allowed by D01FBF, FUN now has two arguments, the first being NDIM, the number of dimensions, and the second being a vector of size NDIM indicating the points of evaluation. For example, consider the approximation of

$$\int_0^\infty x^2 e^{-x}\, dx$$

using a Gauss–Laguerre rule with normal weights. In the call to D01BCF we choose values for ITYPE, A, B and C of 3, 0, 1 and 0 respectively and, in the call to D01FBF, FUN would be of the form

```
DOUBLE PRECISION FUNCTION FUN (NDIM,X)
INTEGER NDIM
DOUBLE PRECISION X(NDIM)
FUN = X(1)**2
END
```

For adjusted weights, the assignment to FUN would take the form

```
FUN = X(1)**2*EXP(-X(1))
```

and B would now be assigned a value of 0 and/or ITYPE set to -3 prior to the call of D01BCF. In both cases the value given to D is immaterial.

D01FBF can also be used in conjunction with D01BBF which stores internally points and adjusted weights for Gauss rules of order 1, 2, 3, 4, 5, 6, 8, 10, 12, 14, 16, 20, 24, 32, 48, and 64. The fact that D01BBF does not recompute the points and weights means that, say, it is usually more efficient to use a 48-point rule supplied by D01BBF than a corresponding 46-point rule for which the points and weights are freshly generated by D01BCF.

3.12 Summary

We know that we cannot guarantee an accurate approximation to the integral of an arbitrary given function by sampling it at a finite number of points. It is always possible to construct an integrand which will cause a sampling method to generate results in error by an arbitrary quantity without returning an error indicator. An error flag generally means that the routine considers the problem to be intractable using the standard strategy. Recent testing (Berntsen [9]) has shown that the NAG routines D01AJF and D01AKF are both very reliable and efficient on a wide range of problems. However, different classes of problems require different methods of approximation, and hence different routines, if we are to keep the success rate as high as possible. Almost invariably, failure to classify a problem correctly will lead to a less than optimal method being used. This may lead to incorrect or, at best, inaccurate results being produced.

Having correctly classified his problem the user may still make life difficult for both himself and the chosen routine by a poor choice of the error tolerance. In general users tend to overestimate the number of significant digits they require. Rarely is it necessary to generate results to machine precision and the routines in the NAG library do have safety devices in the code to prevent a runaway number of function evaluations in search of an impossible accuracy goal. On the other hand it is advisable not to set the accuracy sights too low since this may cause the integrator to stop too soon and deliver a grossly inaccurate result.

For one-off integrations where the result is required to a prescribed accuracy an adaptive routine is the best approach. The price paid for having the routine discover the number of function evaluations actually required for the desired accuracy is generally around three times the number we would have used had we known what order rule to use in the first place. For most problems this is a modest overhead.

Routines for numerical integration are contained in the D01 chapter of the NAG library. However, it should be noted that special routines exist for a number of commonly occurring integrals (e.g., $erf(x)$, Fresnel integrals, etc.) and the S chapter should be investigated first to determine whether the user's problem is, or can easily be converted to, one of the forms covered there (Section 3.8).

There are a large number of D01 routines available, many of which perform a similar task. We have concentrated here on the routines specifically designed to deal with one-dimensional integration, plus multi-dimensional integration routines which may be used when multi means 1. We note that the question of routine selection is largely governed by the form of the integrand. An inappropriate routine may return a value which is inaccurate, or even bad, without any real indication that this is the case. Hence some preliminary investigation into the properties of the integrand is essential if any sensible use of these routines is to be made.

If the integrand is not known as a continuous function, but only at a discrete set of points, the routine to use is D01GAF (Section 3.9). (Alternatively, a routine from the E02 chapter could be used, and the result integrated. Of particular relevance in this respect are the least squares routines E02BDF and E02AJF which integrate cubic spline and polynomial approximations to data (Section 5.6 and Section 5.3).)

For integrands which can be evaluated for any value of the independent variable and if the range of integration is finite D01AJF adapts to the form of the integrand and its use is recommended as a general purpose integrator since it can cope with algebraic and logarithmic end-point singularities, and the user may specify an acceptable accuracy requirement (Section 3.6). If the function is known to be difficult at a finite number of points within the interval, the adaptive process may be assisted by using D01ALF and providing the routine with these break points (Section 3.6). For integrands which are oscillatory but have no singularities D01AKF is likely to prove more efficient. If the integrand can be factored in a particular manner, the appropriate use of D01APF, D01AQF or D01ANF is likely to give superior results (Section 3.6). These routines are specifically designed to handle known singularities of a particular form in the integrand which would cause considerable difficulties to a general integrator. If the integrand is known to possess singularities other than those types covered by D01APF and D01AQF, then either D01AJF or D01AHF should be used (Section 3.6). For infinite integrals, D01AMF is available.

An alternative to D01AJF is the non-adaptive routine D01AHF whose use may be appropriate when the integrand contains internal singularities (Section 3.7). A further, non-adaptive, routine, D01ARF, also attempts to satisfy a user specified accuracy condition by considering a sequence of rules of increasing order. It has the advantage that it may additionally be used to compute indefinite integrals (Section 3.5 and Section 3.7).

In exceptional circumstances the routine D01BAF may be used to compute an approximation using an n-point Gauss rule, where n is user defined (Section 3.11). The range of integration may be finite, semi-infinite or infinite. The integral estimate returned is, for sufficiently large n, likely to be acceptable, provided that the integrand is reasonably smooth. This estimate will have been determined as a linear combination of the evaluations of the integrand, and the weights and evaluation points used can be obtained by a call of D01BBF. A subsequent call of the multi-dimensional integrator D01FBF can then be used to form the sum (Section 3.11). The range of integrands which can be dealt with adequately can be extended considerably if D01BBF is replaced by D01BCF, but the use of this routine imposes a considerable onus on the user to choose parameters appropriately in defining the form of his integrand (Section 3.11).

Chapter 4
Linear Equations

In this chapter we consider the solution of a linear system of equations. We are primarily concerned with the real coefficient case, but we shall have something to say about the extension of the methods to complex problems. We

- describe the basic Gaussian elimination process for computing a solution;

- consider a simple variation based on pivoting which is designed to reduce error build-up;

- develop the method as a decomposition of the coefficient matrix;

- show how an improvement in accuracy in the solution may be obtained using iterative refinement;

- apply the method to certain special types of coefficient matrix.

4.1 Introduction

Systems of linear equations occur frequently in numerical problem solving, both directly (for example, analysing circuits by applying Kirchoff's Laws), and indirectly (for example, as the result of using finite difference techniques to obtain an approximate solution to a partial differential equation). We restrict our discussion to the case when the number of equations and unknowns is the same, and express the system as

$$A\mathbf{x} = \mathbf{b}, \qquad (4.1)$$

where A (the *coefficient matrix*) is a given $n \times n$ matrix, with $(i, j)^{th}$ element a_{ij}, and $\mathbf{b} = (b_1, b_2, \ldots, b_n)^T$ is a *right-hand side vector*. The vector $\mathbf{x} = (x_1, x_2, \ldots, x_n)^T$ is to be determined and is thus referred to as the *solution vector*. The entries of A and $\mathbf{b}$ may be complex, in which case $\mathbf{x}$ may be complex. However, if all entries of A and $\mathbf{b}$ are real then the elements of the solution vector will also be real. Throughout the following we shall assume that all systems are real unless otherwise stated. However, many of

149

the techniques we outline for the solution of real problems can be extended with little, or no, modification to the complex case.

Systems of the form (4.1) may differ in size from a handful of equations to large systems containing thousands, or even tens of thousands, of equations. There is also a broad spectrum to the structure of A, from full complex matrices, with no symmetry at one extreme, through symmetric and banded matrices to triangular (or even diagonal) matrices at the other. In general the very large systems which are solvable by current direct techniques tend to be either very sparse (most of the elements are zero), or highly structured (often the non-zero elements are concentrated in bands), or both.

We may show that if the inverse, A^{-1}, of the coefficient matrix exists then a solution exists and is unique. However (4.1) should never be solved by forming A^{-1} explicitly, and then $A^{-1}\mathbf{b}$, because of the computational cost and the problems of rounding errors.

The condition that A is *invertible* or *nonsingular* (i.e., A^{-1} can be calculated) is equivalent to the following statements

- no one equation in the system can be expressed as a linear combination of the others,

- the coefficient matrix has full *rank*, i.e., rank n,

- the columns (rows) of the coefficient matrix are *linearly independent*,

- the *determinant* of A is non-zero,

- A has no zero eigenvalue, i.e. $\lambda = 0$ is not a solution of $A\mathbf{x} = \lambda\mathbf{x}$.

We would expect that robust numerical software should detect and report a singular coefficient matrix. The major problem is to determine what effect rounding errors have on the computation and this requires a detailed analysis of the underlying algorithms. Further the problem as posed may be ill-conditioned (q.v. Section 4.7); that is, small changes in either the elements of the coefficient matrix, or in the elements of the right-hand side, may cause relatively large changes in the elements of the computed solution. These small changes may occur either as the result of experimental error, or through approximating exact values by floating-point numbers.

All the methods we study in this chapter can be classed as *direct*, at least in their basic form. This means that the solution is calculated in a predeterminable number of arithmetic operations. This quantity is dependent on the structure of the coefficient matrix and for special types of coefficient matrix it is possible to obtain methods two orders of magnitude faster than for the general case. Iterative methods are also available; these are similar in algorithmic procedure to the root-finding techniques discussed in Chapter 2 and the interested reader is referred to Ralston and Rabinowitz [67], p. 440, or Hageman and Young [40].

4.2 Matrix definitions

As we have already remarked, the efficiency of many matrix algorithms may be improved, often dramatically, if A exhibits some special structure. Before using numerical routines implementing these methods it is therefore important to attempt to classify the problem under consideration according to the properties of the coefficient matrix. Hence some basic definitions are necessary before we can proceed.

A *diagonal* matrix is one possessing non-zero entries on the diagonal only. Formally, we may write

$$a_{ij} = \begin{cases} d_i, & \text{if } i = j; \\ 0, & \text{otherwise;} \end{cases} \qquad i, j = 1, 2, \ldots, n. \qquad (4.2)$$

An important member of this class is the *unit* matrix, I_n, which is defined by (4.2) with $d_i = 1$ for all i. For any $n \times n$ matrix A, $AI_n = I_n A = A$.

Two types of matrix which we find very useful in the solution of linear equations are *upper* and *lower triangular* matrices. These have non-zero elements, on and below the diagonal (lower triangular), and on and above the diagonal (upper triangular) only. If, in addition, the diagonal elements are all one, a triangular matrix is further classified as unit lower or unit upper triangular.

Another structure which occurs very frequently in practice is a *banded* structure. Here all the non-zero elements are confined to a band around the leading diagonal. We define the *band width* to be k if $a_{ij} = 0$ for $|i - j| > k$. For example, if $k = 1$ we obtain the tridiagonal matrix

$$A = \begin{pmatrix} a_{11} & a_{12} & & & 0 \\ a_{21} & a_{22} & a_{23} & & \\ & \ddots & \ddots & \ddots & \\ & & a_{n-1,n-2} & a_{n-1,n-1} & a_{n-1,n} \\ 0 & & & a_{n,n-1} & a_{nn} \end{pmatrix}.$$

For the above types of matrix it is clearly wasteful of computer space for the zero entries to be stored explicitly. An alternative is to store the non-zero elements in *compact form*. For example,

- a diagonal matrix may be stored as an n-vector viz. $(d_1, d_2, \ldots, d_n)$,

- an upper (lower) triangular matrix has at most $(n^2 + n)/2$ non-zero elements, and may be stored in a vector by columns (rows). For example, a lower triangular matrix may be represented by the vector $\mathbf{c}$ with components $c_k = a_{ij}$ for $k = i(i - 1)/2 + j$, that is, $\mathbf{c}^T = (a_{11}, a_{21}, a_{22}, a_{31}, \ldots, a_{nn})$,

- a banded matrix may be stored in a two-dimensional array of dimensions $(n, 2k + 1)$, e.g., for a tridiagonal matrix

$$
T = \begin{pmatrix}
* & a_{11} & a_{12} \\
a_{21} & a_{22} & a_{23} \\
\vdots & \vdots & \vdots \\
a_{n-1,n-2} & a_{n-1,n-1} & a_{n-1,n} \\
a_{n,n-1} & a_{nn} & *
\end{pmatrix},
$$

where $*$ indicates that the array element is not part of the original tridiagonal array.

It is not only patterns of zeros which allow the efficiency of a general systems solver to be improved, and some further classification is in order. A real matrix is said to be *symmetric* if $a_{ij} = a_{ji}$, for all i and j. Again, storage may be conserved if the upper, or lower, triangle of A is represented by a vector of length $(n^2 + n)/2$. A tridiagonal symmetric matrix may be represented by an $n \times 2$ matrix, or by two vectors, one of length n for the diagonal, and the other of length $n - 1$ for the off-diagonals.

We term a real symmetric matrix, A, *positive definite* if $\mathbf{x}^T A \mathbf{x} > 0$ for all non-zero vectors, $\mathbf{x}$. This is not a very helpful definition since it provides little indication of how the positive definiteness, or otherwise, may be determined. Equivalent definitions which improve the situation slightly are

- all eigenvalues of A are real and positive,

- the leading submatrices, A_j of A, of orders $j = 1, 2, \ldots, n$ all have positive determinants. A_j consists of the first j rows and columns of A,

- the matrix A may be written as $W^T W$ where W is a non-singular matrix.

Unfortunately, even these definitions involve considerable computational effort if they are to be used to determine whether or not a given matrix is positive definite. However, the following is a sufficient (but not necessary) condition for a matrix to be positive definite, and is easily checked

- a matrix is positive definite if it is symmetric, has positive diagonal entries, and is *diagonally dominant*, that is,

$$
|a_{ii}| \geq \sum_{\substack{j=1 \\ j \neq i}}^{n} |a_{ij}|, \qquad \forall i,
$$

where the inequality must be strict for at least one value of i.

Finally we look briefly at *sparse* matrices. The term sparse is reserved for matrices of very large order (usually several hundreds upwards) which have a high percentage of zero elements. The exact percentage of zeros required for a sparse classification is vague and often depends on

(1) whether or not there may be any other structure, e.g., the matrix is tridiagonal;

(2) what the matrix is to be used for;

(3) how the non-zero elements are represented; this may also depend on (2).

A symmetric tridiagonal matrix whose diagonal elements are all equal and whose off-diagonal elements are all equal may be represented by just two real numbers, independently of the order of the matrix. For performing a matrix/vector multiply the order of the problem solvable would be constrained only by the storage available for the vectors. On the other hand $3n$ non-zero elements randomly positioned in a large matrix would require somewhat more information to be stored in order to define the matrix. For example, the numerical value and two integers giving the row and column indices define the crudest data structure we could use. For more details of possible storage strategies see Exercise 4.4 and George and Liu [35]. A matrix which is not classed as sparse will be regarded as being *full*.

The above definitions of matrix types have all been applied to real matrices. Although diagonal, triangular, banded and sparse matrices may be defined in exactly the same way for the complex case, we need to change the definitions of symmetry and positive definiteness to cope with complex elements.

We define a complex matrix, A, to be *Hermitian* if $a_{ij} = \bar{a}_{ji}$ for all i and j, where $\bar{a}_{ji}$ denotes the complex conjugate of a_{ji}. We may then define the Hermitian form, $\mathbf{x}^H A \mathbf{x}$, where $\mathbf{x}^H = (\bar{x}_1, \bar{x}_2, \ldots, \bar{x}_n)$, which is real if A is Hermitian. The concepts of positive definiteness carry over to Hermitian forms.

EXERCISE 4.1   Show that the product of two lower (upper) triangular matrices is also lower (upper) triangular.

Write two subroutines

(a) which takes two lower triangular matrices stored in two-dimensional arrays and computes their product, taking into account the triangular form to minimize the number of arithmetic operations;

(b) which uses the compact form $\mathbf{c}^T = (a_{11}, a_{21}, a_{22}, a_{31}, \ldots, a_{nn})$ and produces the product in a compact form.

Compare the execution time of the two methods for increasing values of n. What is the minimum amount of additional workspace required if the product is to overwrite one of the input matrices?

EXERCISE 4.2   A sparse vector may be stored as a pair of vectors, one real vector, $\mathbf{v}$, which stores the non-zero elements and the other, an integer array, $\mathbf{ind}$, which stores the corresponding index. For example, to store $\mathbf{x}^T = (x_1, 0, \ldots, 0, x_{47}, 0, \ldots, 0, x_{93}, 0, \ldots, 0)$ we would use $\mathbf{v}^T = (x_1, x_{47}, x_{93})$, and $\mathbf{ind}^T = (1, 47, 93)$.

Write a routine which, given two vectors $\mathbf{v}$ and $\mathbf{w}$ representing n-vectors $\mathbf{x}$ and $\mathbf{y}$ with respective index vectors $\mathbf{indv}$ and $\mathbf{indw}$, forms the inner product $\mathbf{x}^T\mathbf{y} = \sum_{i=1}^{n} x_i y_i$ using the minimum number of multiplications. What problems would there be implementing a routine to form the sum of two such vectors?

EXERCISE 4.3   There are a number of routines in the F01 chapter of the library which perform basic matrix manipulations. In particular F01CSF forms the matrix–vector product Ax where A is a symmetric matrix whose lower triangle is stored by rows in a one-dimensional array. Compare the execution time required to form the matrix product AB where A is symmetric and B is a general $n \times n$ matrix for various values of n by using on each column of B

- F01CSF,

- F01CKF, which requires A to be stored in full.

EXERCISE 4.4   Consider the following strategy for storing a sparse $n \times n$ matrix:

- store all the non-zero elements in a vector $\mathbf{v}$;

- store the respective column subscripts in a vector $\mathbf{c}$;

- store a vector of row pointers, $\mathbf{r}$, such that the non-zero values in row i are the elements v_{r_i+1} to $v_{r_{i+1}}$. Note that if $r_i + 1 > r_{i+1}$ then row i contains no non-zero elements.

For example,

$$\begin{aligned}
\mathbf{v}^T &= (a_{15}, a_{23}, a_{42}, a_{73}, a_{92}, a_{93}, a_{95}), \\
\mathbf{c}^T &= (5, 3, 2, 3, 2, 3, 5), \\
\mathbf{r}^T &= (0, 1, 2, 2, 3, 3, 3, 4, 4, 7).
\end{aligned}$$

Note that in general $\mathbf{r}$ is of length $n + 1$, and, unlike the given example, will usually be much shorter than either $\mathbf{v}$ or $\mathbf{c}$.

Write a package of routines using this structure to provide useful matrix and vector operations. For example, you should provide routines for

- a matrix–matrix add and multiply,

- a matrix–vector multiply,

- multiplying all elements of a matrix by a constant.

Pay careful attention to the form of the user interface. Some routines to help the user to set up the internal structures may also be useful.

4.3   The solution of triangular systems

We start by looking at one of the simplest structures that the coefficient matrix may take as far as a non-trivial computational algorithm is concerned. Recall that a triangular matrix is one in which all the elements either below or above the principal diagonal are zero. We recall that a matrix is lower (upper) triangular if $a_{ij} = 0$, for all $j > i$ (for all $i > j$). We consider the upper triangular system

$$
\begin{pmatrix}
u_{11} & u_{12} & \cdots & u_{1n} \\
 & u_{22} & \cdots & u_{2n} \\
 & & \ddots & \vdots \\
 & u_{n-1,n-1} & u_{n-1,n} \\
0 & & & u_{nn}
\end{pmatrix}
\begin{pmatrix}
x_1 \\ x_2 \\ \vdots \\ x_{n-1} \\ x_n
\end{pmatrix}
=
\begin{pmatrix}
c_1 \\ c_2 \\ \vdots \\ c_{n-1} \\ c_n
\end{pmatrix} . \quad (4.3)
$$

We see immediately that the solution to the nth equation is $x_n = c_n/u_{nn}$. The $(n-1)^{st}$ equation now has only one unknown, x_{n-1}, which is calculated by $x_{n-1} = (c_{n-1} - u_{n-1,n} x_n)/u_{n-1,n-1}$, and so on. This leads to the general form for the ith equation

$$
x_i = \frac{1}{u_{ii}} \left(c_i - \sum_{j=i+1}^{n} u_{ij} x_j \right) . \quad (4.4)
$$

Assuming that the coefficient matrix is stored in a two-dimensional array, U, and the right-hand side vector and the solution vector are one-dimensional arrays, B and X, respectively, then the code fragment Code 4.1 solves the upper triangular system (4.3).

```
*.. VISIT EACH ROW (EQUATION)
      DO 20 I = N,1,-1
*
*.. COLLECT THE SUM
        SUM = 0.0D0
        DO 10 J = I+1,N
          SUM = SUM+U(I,J)*X(J)
10      CONTINUE
*
*.. SOLVE THE ITH EQUATION
        X(I) = (B(I)-SUM)/U(I,I)
20    CONTINUE
```

Code 4.1   Backward substitution.

The scheme outlined above is referred to as *backward substitution*. For systems in which the coefficient matrix is lower triangular, the process of *forward substitution* may be defined in an analogous manner. Now the unknowns are determined in the order x_1, x_2, ..., x_n.

It can be readily seen that the determinant of a lower (or upper) triangular matrix is equal to the product of the entries on the leading diagonal. It is very easy, therefore, to detect a mathematical singularity; if any diagonal element is identically zero then the matrix is singular. The matrix could be computationally singular if a diagonal element is so small that it underflows when stored as a floating-point number or if, to the known accuracy of the coefficients, a diagonal element is not distinguishable from zero.

We next consider the computational cost of backward substitution. We assume that each basic *floating-point operation (flop)* is of equivalent cost. From Code 4.1 we see that the computation of the x_is requires approximately $2n$ flops, whilst the calculation of the sums requires $0+2+4+\cdots+2(n-1)$ flops, giving a total cost of $n(n-1)+2n = n(n+1)$ flops. The process, therefore, has an $O\left(n^2\right)$ operation count.

EXERCISE 4.5   Chapter F06 of the NAG library contains a number of routines for solving triangular systems of equations. For example, `F06PJF` can be used to solve systems of the form $A\mathbf{x} = \mathbf{b}$ or $A^T\mathbf{x} = \mathbf{b}$ where A is either upper or lower triangular. In some applications it is more efficient to store the matrix in a vector $\mathbf{u}$ of length $n \times (n-1)/2$ where, for an upper triangular matrix

$$\mathbf{u} = \left(u_{11}, u_{12}, \ldots, u_{1n}, u_{22}, u_{23}, \ldots, u_{nn}\right)^T.$$

`F06PLF` does an equivalent job to `F06PJF` except that it operates on the packed form of the array.

Compare the execution times of the two routines as the order of the system is increased.

4.4   Elementary row operations and elimination

For a general linear system of equations we observe that the solution is unaffected by the following *elementary operations*.

- the multiplication of an equation by a constant,

- a change in the order of the equations,

- the addition of a multiple of one equation to another.

We now look at how we might devise an algorithm for computing the solution of (4.1) which makes use of these operations. The basic idea is the same as that used for solving small systems by hand.

For example, consider the system

$$\begin{aligned}
x_1 &- x_2 &+ x_3 &= 4 \\
x_1 &- 2x_2 &- 4x_3 &= -5 \\
x_1 &+ 2x_2 &- x_3 &= -3
\end{aligned}.$$

Subtracting the first equation from the second and third gives

$$\begin{aligned}
x_1 &- x_2 &+ x_3 &= 4 \\
&- x_2 &- 5x_3 &= -9 \\
&3x_2 &- 2x_3 &= -7
\end{aligned}.$$

Adding three times the new second equation to the new third equation gives

$$\begin{aligned}
x_1 &- x_2 &+ x_3 &= 4 \\
&- x_2 &- 5x_3 &= -9 \\
&&- 17x_3 &= -34
\end{aligned}.$$

We have now reduced our original matrix to upper triangular form and using backward substitution we obtain $x_3 = 2$, $x_2 = -1$ and $x_1 = 1$.

We now formalize the method for a general system of order n. Each row is used in turn to eliminate the coefficients directly beneath its diagonal element. We look at the state of the system as the kth equation is about to be used in this process

$$A^{(k)} = \begin{pmatrix}
a_{11}^{(1)} & a_{12}^{(1)} & \cdots & \cdots & \cdots & a_{1n}^{(1)} \\
& a_{22}^{(2)} & \cdots & \cdots & \cdots & a_{2n}^{(2)} \\
& & \ddots & & & \vdots \\
& & & a_{kk}^{(k)} & \cdots & a_{kn}^{(k)} \\
& & & \vdots & & \vdots \\
0 & & & a_{nk}^{(k)} & \cdots & a_{nn}^{(k)}
\end{pmatrix}, \quad
\mathbf{b}^{(k)} = \begin{pmatrix}
b_1^{(1)} \\
b_2^{(2)} \\
\vdots \\
b_k^{(k)} \\
\vdots \\
b_n^{(k)}
\end{pmatrix}.$$

The original coefficient matrix is denoted by $A^{(1)}$ and, if all goes well, $A^{(n)}$ will be a triangular matrix. The superscript on the individual elements denotes the last stage at which those elements took part in the elimination process. The kth row is known as the *pivotal row* and the first non-zero element in this row is termed the *pivotal* element or *pivot*. Similarly the kth column is referred to as the *pivotal column*. The reason for these names will become apparent soon. The block of elements superscripted k is called the *active block*.

The objective of the kth step is to use multiples of the pivotal row to eliminate the elements $a_{k+1,k}^{(k)}$, $a_{k+2,k}^{(k)}$, ..., $a_{nk}^{(k)}$. The code fragment given in Code 4.2 achieves this, with RMULT being the appropriate multiplier at each stage. Note that it is not necessary formally to zero the elements below the diagonal; zero values will be assumed.

```
*..VISIT EACH REMAINING ROW
      DO 20 I = K+1,N
*
*..CALCULATE THE REQUIRED ROW MULTIPLIERS
      RMULT = A(I,K)/A(K,K)
*
*..CALCULATE THE NEW ELEMENTS IN THE JTH ROW
      DO 10 J = K+1,N
         A(I,J) = A(I,J)-RMULT*A(K,J)
10       CONTINUE
*
*..DO THE SAME TO THE RIGHT-HAND SIDE
      B(I) = B(I)-RMULT*B(K)
20    CONTINUE
```

Code 4.2   Row elimination.

The application of this for $k = 1, 2, \ldots, n-1$ results in the upper triangular system

$$
\begin{pmatrix}
a_{11}^{(1)} & a_{12}^{(1)} & \cdots & \cdots & a_{1n}^{(1)} \\
& a_{22}^{(2)} & \cdots & \cdots & a_{2n}^{(2)} \\
& & \ddots & & \vdots \\
& & & & a_{nn}^{(n)}
\end{pmatrix}
\begin{pmatrix}
x_1 \\ x_2 \\ \vdots \\ x_n
\end{pmatrix}
=
\begin{pmatrix}
b_1^{(1)} \\ b_2^{(2)} \\ \vdots \\ b_n^{(n)}
\end{pmatrix}
\tag{4.5}
$$

being formed, whose solution (which can be determined using backward substitution, Section 4.3) is the same as that of the original system, since only elementary row operations have been used. Note that the superscripts in the final row are set to n for notational consistency; there are only $n-1$ elimination steps.

We now consider the computational cost of the above process, referred to as *Gaussian elimination*. The calculation of the row multipliers plus the modification of the right-hand side requires $3\sum_{k=1}^{n-1} k$ flops, but the majority of the arithmetic operations are performed in the innermost loop which requires $2\sum_{k=1}^{n-1} k^2$ flops. The total cost is thus $n(n-1)(2n-1)/3 + 3n(n-1)/2$ flops, i.e., the elimination process has an $O\left(n^3\right)$ operation count. Hence the reduction of the system to upper triangular form costs an order of magnitude more than the backward substitution process used to solve it.

Often we are interested in solving not just a single system, but many systems in which the coefficient matrix is the same, but the right-hand sides are different. We express this as $AX = B$ where X and B are $n \times m$

with columns $\mathbf{x}_i$ and $\mathbf{b}_i$, with $\mathbf{x}_i$ being the solution of the system $A\mathbf{x}_i = \mathbf{b}_i$. Applying Gaussian elimination to this *multiple right-hand side* problem is a straightforward extension. As A is converted to upper triangular form, the elimination process is applied to each right-hand side. When this is complete, the components of the columns of X are computed in parallel.

4.5 Pivoting

Gaussian elimination is computationally feasible unless at any stage we find a zero pivotal value. For example, the process applied to

$$\begin{pmatrix} 0 & 1 \\ 1 & 1 \end{pmatrix} \begin{pmatrix} x_1 \\ x_2 \end{pmatrix} = \begin{pmatrix} 1 \\ 2 \end{pmatrix}$$

breaks down. The system has solution $\mathbf{x}^T = (1, 1)$, and it is clear that a zero pivot does not necessarily mean that a solution to the system does not exist. We thus need to modify our algorithm to take account of this situation. To carry on the elimination process, we may exchange the offending pivotal row with any row within the active block which does not have a zero value in the pivotal column. If all elements in the pivotal column of the active block are zero, then the matrix is singular and we are left with no option but to abort the process and report the failure to the user.

The interchanging of rows is known as *partial pivoting*, and its use is not restricted to the situation outlined above. It turns out that from a numerical stability point of view it is undesirable to compute with relatively small pivots. This is because small pivots will lead to large multipliers which, in turn, tend to magnify both errors in the original data and accumulated rounding errors. We thus arrange that all multipliers which take part in the elimination procedure are less than one in magnitude by *always* choosing the pivotal row to be that row, within the active block, which has the pivot of largest magnitude.

Subroutine GSSELM, Code 4.3, solves a system of N equations of the form (4.1) using Gaussian elimination with partial pivoting. This implementation is designed for maximum efficiency on paged and pipelined machines. We saw in Section 1.6 that accessing Fortran two-dimensional arrays by rows can lead to thrashing and thereby increase the execution overheads of a program. Thus, although the Gaussian elimination algorithm is often described mathematically as proceeding by rows, it is actually implemented so that the elements within the active block are accessed by columns.

On entry the array A, dimensioned ($\text{IA} \geq \text{N}, \text{N}$), contains the coefficient matrix, and the N-vector B the right-hand side. During the transformation into upper triangular form rows are physically interchanged by the pivoting process. On successful exit ($\text{IFAIL} = 0$) the solution is in the vector B and A has been overwritten. $\text{IFAIL} = 1$ on exit signifies an illegal input value for N whilst a return value of 2 means that the input coefficient matrix, possibly affected by rounding errors, has been found to be singular.

```
      SUBROUTINE GSSELM(A,IA,N,B,IFAIL)
*
*.. ROUTINE TO IMPLEMENT GAUSSIAN ELIMINATION WITH
*.. PARTIAL PIVOTING FOR THE SYSTEM AX=B.
*.. THE SOLUTION VECTOR OVERWRITES THE RIGHT-HAND
*.. VECTOR.
*.. THIS CODE IS BASED ON THE LINPACK ROUTINES
*.. SGEFA AND SGESL AND OPERATES ON COLUMNS RATHER
*.. THAN ROWS WHERE POSSIBLE TO INCREASE EFFICIENCY
*.. ON PAGED MACHINES.
      INTEGER IA,IFAIL,N
      DOUBLE PRECISION A(IA,N), B(N)
      DOUBLE PRECISION EPS,ONE
      PARAMETER (EPS = 1.0D-8,ONE = 1.0D0)
      INTEGER I,J,K,L
      DOUBLE PRECISION T,TEMP
*
      DO 60 K = 1,N-1
*
*.. FIND PIVOT INDEX I.E. ELEMENT IN COL K
*.. OF LARGEST MAGNITUDE
      L = K
      TEMP = ABS(A(K,K))
      DO 10 I = K+1,N
        IF(ABS(A(I,K)).GT.TEMP)THEN
        L = I
        ENDIF
10      CONTINUE
*
*.. TEST FOR SINGULARITY
      IF(ABS(A(L,K)).LE.EPS)THEN
        IFAIL = 1
        RETURN
      ENDIF
*
*.. INTERCHANGE IF NECESSARY
      IF(L.NE.K)THEN
        T = A(L,K)
        A(L,K) = A(K,K)
        A(K,K) = T
```

```
            T = B(L)
            B(L) = B(K)
            B(K) = T
          ENDIF
*
*.. COMPUTE MULTIPLIERS
          T = -ONE/A(K,K)
          DO 20 I = K+1,N
            A(I,K) = T*A(I,K)
20        CONTINUE
*
*.. ROW ELIMINATION PERFORMED WITH COLUMN INDEXING
*.. FOR EFFICIENCY ON PAGED MACHINES
          DO 40 J = K+1,N
            T = A(L,J)
            IF(L.NE.K)THEN
              A(L,J) = A(K,J)
              A(K,J) = T
            ENDIF
            DO 30 I = K+1,N
              A(I,J) = A(I,J)+T*A(I,K)
30          CONTINUE
40        CONTINUE
          DO 50 I = K+1,N
            B(I) = B(I)+A(I,K)*B(K)
50        CONTINUE
60      CONTINUE
*
*.. BACKWARD SUBSTITUTION
        DO 80 K = N,1,-1
          B(K) = B(K)/A(K,K)
          T = -B(K)
          DO 70 I = 1,K-1
            B(I) = B(I)+T*A(I,K)
70        CONTINUE
80      CONTINUE
        IFAIL = 0
        END
```

Code 4.3 Gaussian elimination.

A theoretical improvement on partial pivoting by which the entire active block is searched for the largest pivot is known as *full pivoting*. The interchanges required by this scheme mean that the order of the x_is may need to be altered and this requires additional bookkeeping. In practice partial pivoting is usually considered to be adequate.

EXERCISE 4.6 The *coefficient of growth* in Gaussian elimination is defined to be $\max_{i,j} a_{ij}^{(i)} / \max_{i,j} a_{ij}$. Show that with row interchanges this is less than 2^{n-1}.

EXERCISE 4.7 Change the Gaussian elimination routine (Code 4.3) so that it operates on a set of M right-hand sides and produces a corresponding set of M solution vectors. What advantages are there in implementing the routine in this way?

EXERCISE 4.8 It is possible to implement Gaussian elimination with partial pivoting without physically interchanging the rows by using an integer array INDEX for indirect addressing. INDEX is initialized such that INDEX(I) = I and, when rows I and J are to be interchanged, the contents of INDEX(I) and INDEX(J) are exchanged. Using this scheme involves accessing all elements of the coefficient matrix and right-hand side via INDEX. Thus, when accessing the array element a_{ij} we must now refer to the Jth entry in the INDEX(I)th row.

Rewrite Code 4.3 to implement this strategy. Compare the execution times of the two versions for increasing values of N.

4.6   LU factorization and Crout's method

In this section we show how Gaussian elimination with partial pivoting is equivalent to the factorization of the original coefficient matrix, possibly with its rows permuted, into the product of a lower and a unit upper triangular matrix. This leads to *Crout's method* and an algorithm for solving linear systems which allows for better control of rounding errors.

The majority of the results from elementary linear algebra which are required in this section are given in Exercise 4.9 below.

First we need to look briefly at *permutation matrices*. A permutation matrix, P, of order n is an $n \times n$ matrix whose rows are a permutation of the rows of the identity matrix, I_n. PA is then a matrix formed by permuting the rows of A in the same way as the rows of I_n were permuted to form P.

For example, if

$$P = \begin{pmatrix} 1 & 0 & 0 \\ 0 & 0 & 1 \\ 0 & 1 & 0 \end{pmatrix} \text{ and } A = \begin{pmatrix} 1 & 2 & 3 \\ 4 & 5 & 6 \\ 7 & 8 & 9 \end{pmatrix} \text{ then } PA = \begin{pmatrix} 1 & 2 & 3 \\ 7 & 8 & 9 \\ 4 & 5 & 6 \end{pmatrix}.$$

When applying partial pivoting we solve the linear system $PA\mathbf{x} = P\mathbf{b}$. During the kth stage of the elimination process, we first select the pivotal row by interchanging the current row k with the row containing the element of maximum magnitude in column k. That is, we form

$$\tilde{A}^{(k)} = P_k A^{(k)},$$

where P_k is the requisite permutation matrix. We then eliminate the elements $\tilde{a}_{k+1,k}$, $\tilde{a}_{k+2,k}$, ..., $\tilde{a}_{nk}$ by adding multiples of the kth row of $\tilde{A}^{(k)}$ to rows $k+1$, $k+2$, ..., n. This is equivalent to the premultiplication of $\tilde{A}^{(k)}$ by the lower triangular matrix

$$M_k = \begin{pmatrix} 1 & & & & & & 0 \\ & \ddots & & & & & \\ & & 1 & & & & \\ & & & 1 & & & \\ & & & -m_{k+1,k} & 1 & & \\ & & & \vdots & & \ddots & \\ 0 & & & -m_{nk} & & & 1 \end{pmatrix}, \tag{4.6}$$

where $m_{jk} = \tilde{a}_{jk}^{(k)}/\tilde{a}_{kk}^{(k)}$ for $j = k+1$, $k+2$, ..., n.

Thus, provided the coefficient matrix is non-singular, we have, for a single step of the elimination process $A^{(k+1)} = M_k P_k A^{(k)}$, and the complete elimination process, with partial pivoting, may be written

$$A^{(n)} = M_{n-1} P_{n-1} M_{n-2} P_{n-2} \cdots M_1 P_1 A. \tag{4.7}$$

If we had known the required permutations in advance, we could have applied these to A and then used Gaussian elimination without interchanges. This implies that

$$\begin{aligned} A^{(n)} &= M_{n-1} M_{n-2} \cdots M_1 P_{n-1} P_{n-2} \cdots P_1 A \\ &= M_{n-1} M_{n-2} \cdots M_1 P A, \end{aligned}$$

where P is the required permutation matrix. Since the products of lower triangular matrices are also lower triangular and $A^{(n)}$ is upper triangular we may write

$$PA = LU,$$

where

$$L = \begin{pmatrix} 1 & & & 0 \\ l_{21} & 1 & & \\ \vdots & & \ddots & \\ l_{n1} & l_{n2} & \cdots & 1 \end{pmatrix} = \begin{pmatrix} 1 & & & 0 \\ m_{21} & 1 & & \\ \vdots & & \ddots & \\ m_{n1} & m_{n2} & \cdots & 1 \end{pmatrix},$$

$$U = \begin{pmatrix} u_{11} & u_{12} & \cdots & u_{1n} \\ & u_{22} & \cdots & u_{2n} \\ & & \ddots & \vdots \\ 0 & & & u_{nn} \end{pmatrix} = \begin{pmatrix} a_{11}^{(1)} & a_{12}^{(1)} & \cdots & a_{1n}^{(1)} \\ & a_{22}^{(2)} & \cdots & a_{2n}^{(2)} \\ & & \ddots & \vdots \\ 0 & & & a_{nn}^{(n)} \end{pmatrix},$$

and the $a_{ij}^{(k)}$ are as defined by (4.5), except that the elimination has been applied to PA rather than directly to A. The diagonal elements of U are thus the pivots of PA, and the off-diagonal elements of L are the row multipliers used during the elimination.

Assuming, for the moment, that no row exchanges are necessary we may write $A = LU$, and by comparing elements we have

$$a_{ij} = \sum_{r=1}^{\min(i,j)} l_{ir} u_{rj}, \qquad i,j = 1,2,\dots,n.$$

We now show how we may overwrite the original coefficient matrix with its LU factors. The first row of U is found simply since

$$u_{1j} = a_{1j}, \qquad j = 1,2,\dots,n,$$

and the first column of L is obtained from

$$a_{i1} = l_{i1} u_{11}, \qquad i = 2,3,\dots,n.$$

If we suppose that the first $k-1$ columns of L and the first $k-1$ rows of U are known then, noting that L has a unit diagonal, we obtain

$$u_{kj} = a_{kj} - \sum_{r=1}^{k-1} l_{kr} u_{rj}, \qquad j = k, k+1, \dots, n,$$

and similarly

$$l_{ik} = \frac{1}{u_{kk}} \left(a_{ik} - \sum_{r=1}^{k-1} l_{ir} u_{rk} \right), \qquad i = k+1, k+2, \dots, n,$$

This particular factorization procedure is known as *Doolittle's method*. An alternative approach is to generate $A = LU$ where U is a unit upper triangular matrix. Proceeding as above we may first compute column k of L using

$$l_{ik} = a_{ik} - \sum_{r=1}^{k-1} l_{ir} u_{rk}, \qquad i = k, k+1, \dots, n, \tag{4.8}$$

and similarly

$$u_{kj} = \frac{1}{l_{kk}} \left(a_{kj} - \sum_{r=1}^{k-1} l_{kr} u_{rj} \right), \qquad j = k+1, k+2, \dots, n. \tag{4.9}$$

This algorithm is called *Crout's method*. Once a_{ij} has been used in the computation of either l_{ij} or u_{ij} it is not required further in the calculation of the LU factors. We conclude that the elements of these factors may overwrite the coefficient matrix A as they are generated, with the diagonal of the upper triangular matrix being implicitly defined. We introduce pivoting into Crout's method by computing the complete k^{th} column of L and choosing the element of largest magnitude in that column as the divisor in (4.9).

The number of arithmetic operations required is still $\sim \frac{2}{3}n^3$ flops. However, the use of additional-precision arithmetic to accumulate the summations (*inner products*) in (4.8) and (4.9) can dramatically reduce the effects of rounding error. Crout's method is usually preferred because it is slightly more computationally convenient than Doolittle's method (see Exercise 4.10).

Having computed the LU factors by Crout's method it is relatively simple to solve the original linear system $Ax = b$, which has been transformed to $PAx = LUx = Pb$. We solve $Ly = Pb$ by forward substitution and thence obtain x by solving $Ux = y$ by backward substitution.

There is a further important advantage to Crout's method over direct Gaussian elimination. In the case of multiple right-hand sides, the LU factorization is performed once, and the backward and forward substitution process then applied to each right-hand side in turn. But note that it is no longer necessary to know the form of the right-hand sides before the elimination process begins. See Section 4.9 and Section 4.13 for some examples of the significance of this.

EXERCISE 4.9   Defining M_k by (4.6),

(a)   show that M_k may be written as $M_k = I_n - \mathbf{m}_k \mathbf{e}_k^T$ where

$$\mathbf{m}_k^T = (0, 0, \ldots, 0, m_{k+1,k}, m_{k+2,k}, \ldots, m_{nk}),$$

and $\mathbf{e}_k$ is the kth column of I_n;

(b)   show that $M_k^{-1} = I_n + \mathbf{m}_k \mathbf{e}_k^T$. Write out M_k^{-1} in full;

(c)   verify that $\hat{M} = M_1^{-1} M_2^{-1} \cdots M_{n-1}^{-1}$ is lower triangular, with elements

$$\hat{m}_{ii} = 1,$$
$$\hat{m}_{ij} = m_{ij}, \qquad i > j;$$

(d)   show that if $P_1, P_2, \ldots, P_{n-1}$ are permutation matrices, then their product is also a permutation matrix.

EXERCISE 4.10   Write a subroutine to implement the Crout factorization method with partial pivoting. Isolate the inner product routine as a separate module.

Compare the accuracy of the final solutions obtained using Crout's method with basic and additional precision accumulation of inner products.

Write a second routine to implement the Doolittle procedure with partial pivoting. Why is Crout's method to be preferred when the inner products are accumulated in additional precision?

4.7 Condition numbers

We now investigate the effect that small changes (perturbations) in the elements of the coefficient matrix and the right-hand side can have on the solution vector. We consider first a small change to $\mathbf{b}$ with A unchanged. Assume that replacing $\mathbf{b}$ by $\mathbf{b} + \delta\mathbf{b}$, where $\delta\mathbf{b}$ is small compared to $\mathbf{b}$, results in a change in the solution from $\mathbf{x}$ to $\mathbf{x} + \delta\mathbf{x}$. We may write

$$A\mathbf{x} = \mathbf{b}, \tag{4.10}$$

$$A(\mathbf{x} + \delta\mathbf{x}) = \mathbf{b} + \delta\mathbf{b}, \tag{4.11}$$

and we would like to know how large $\delta\mathbf{x}$ may be. Subtracting (4.10) from (4.11) we obtain $\delta\mathbf{x} = A^{-1}\delta\mathbf{b}$, whence, for a subordinate matrix norm, $\|\delta\mathbf{x}\| \leq \|A^{-1}\|\,\|\delta\mathbf{b}\|$. Equation (4.10) gives us $\|\mathbf{b}\| \leq \|A\|\,\|\mathbf{x}\|$, hence $\|\delta\mathbf{x}\|\,\|\mathbf{b}\| \leq \|A\|\,\|A^{-1}\|\,\|\delta\mathbf{b}\|\,\|\mathbf{x}\|$, and assuming $\mathbf{x}$ and $\mathbf{b}$ are not zero vectors we obtain

$$\frac{\|\delta\mathbf{x}\|}{\|\mathbf{x}\|} \leq \|A\|\,\|A^{-1}\|\,\frac{\|\delta\mathbf{b}\|}{\|\mathbf{b}\|} = \text{cond}\,(A)\,\frac{\|\delta\mathbf{b}\|}{\|\mathbf{b}\|}, \tag{4.12}$$

where $\text{cond}\,(A) = \|A\|\,\|A^{-1}\|$ is the *condition number* of A (q.v. Exercise 1.41).

We may interpret $\|\delta\mathbf{b}\|/\|\mathbf{b}\|$ as the relative uncertainty in the vector $\mathbf{b}$. For example, if the right-hand side vector were determined, experimentally, correct to 4 significant digits we could give $\|\delta\mathbf{b}\|/\|\mathbf{b}\|$ an upper bound of $\frac{1}{2}10^{-5}$ for a suitable choice of norm.

The inequality (4.12) bounds the relative uncertainty of the solution vector due to the uncertainty in the right-hand side. The closer the matrix A is to being singular the larger the value of $\text{cond}\,(A)$, and the larger the amplification of any errors in the right-hand side will be.

An analysis of the effects of a perturbation, δA, in the coefficient matrix on the solution vector gives

$$\frac{\|\delta\mathbf{x}\|}{\|\mathbf{x} + \delta\mathbf{x}\|} \leq \text{cond}\,(A)\,\frac{\|\delta A\|}{\|A\|},$$

which, provided $\|\delta A\|/\|A\|$ is small enough, gives

$$\frac{\|\delta\mathbf{x}\|}{\|\mathbf{x}\|} \leq \text{cond}\,(A)\,\frac{\|\delta A\|}{\|A\|}.$$

For further details see Strang [79] or Forsythe and Moler [29]. Once again we note that the condition number is the amplifying factor.

We say that a problem is ill-conditioned if $\text{cond}\,(A)$ is large, and well-conditioned if it is small. It is important to realize that, even for simple problems, it is possible for the value of $\text{cond}\,(A)$ to be enormous.

As a rule of thumb (Dongarra et al. [26]) $\log_{10}\big(\text{cond}\,(A)\big)$ is approximately the number of decimal digits lost in the computed solution vector.

Thus if $\text{cond}(A)$ is $O(10^{12})$ and the working precision of the computer is 14 decimal digits then only 2 significant digits in the computed solution vector should be taken as correct. Obviously this rule only applies provided a stable method of solution has been used.

4.8   Use of the NAG routine F04ARF

There are a large number of linear equation solvers available to users of the NAG library and decision trees are provided in the F04 chapter introduction to assist in the choice of routine. If you are not sure about the answer to any of the questions posed in the flow diagrams it is safer to take the *no* route. Although F04 largely consists of black-box routines, it is possible to obtain finer control over the solution of linear systems by performing the factorization and backward substitution phases of the solution separately. This mode of computation is discussed further in Section 4.13. We first concentrate on black-box routines to solve a general real system of the form (4.1).

Following the decision tree '*3.2(i) – Black-box routines for the unique solution of* $A\mathbf{x} = \mathbf{b}$' in the introduction to F04 two further decisions have to be made. First '*Are storage and time more important than accuracy?*' and secondly, whatever the answer to this is, '*Is B a single right-hand side?*'.

The NAG documentation uses the adjectives *accurate* and *approximate* to distinguish between solutions obtained with or without the use of iterative refinement (see Section 4.9). For most practical purposes the accuracy attained using routines returning *approximate* solutions will be satisfactory. Only if full machine precision is required, or if the system is suspected of being ill-conditioned (see Section 4.7), is the calculation of an *accurate* solution likely to be necessary.

Assume that we require a quick approximation for a single right-hand side (we will discuss the other options in Section 4.9 and Section 4.10). This leads to the routine F04ARF, which has the call sequence

```
CALL F04ARF(A,IA,B,N,X,WKSPCE,IFAIL)
```

The array A, dimensioned $(\text{IA}(\geq \text{N}), p(\geq \text{N}))$, is used to store the real coefficient matrix and the right-hand side is passed in the vector B of length at least N. The workspace array WKSPCE has a length of at least N. On successful exit, the vector X (declared length $\geq$ N) contains the approximate solution.

There are a number of further points worth mentioning. First, the original input coefficient matrix A is destroyed by this routine; on successful exit, A will contain the Crout LU decomposition of the original matrix. If the routine fails then the contents of A are, generally, meaningless. Secondly, strictly against the rules of standard Fortran 77, it is possible to

Order n	$\log_{10}\left(\mathrm{cond}\left(H_n^{-1}\right)\right)$	Number of significant decimal digits lost
2	1.4	0
3	2.9	0
4	4.5	0
5	6.0	1
6	7.5	1
7	9.0	5
8	10.5	6
9	12.0	9
10	13.5	8
11	15.1	12
12	16.6	13

Figure 4.1   Results obtained using FO4ARF to solve (4.13).

call FO4ARF with the parameters B and X the same. In this case the right-hand side vector is overwritten by the solution vector. Note that although declared to be real/double-precision, the vector WKSPCE is used to store integer pivotal information, a practice not to be copied! The only error detected by FO4ARF occurs when a small pivot is encountered. The routine assumes that the coefficient matrix is singular and halts with IFAIL $= 1$.

To study how accurate our final solutions are, we again turn to the Hilbert matrices (q.v. Section 1.12). Recall that the $(i, j)^{th}$ element of H_n, the $n \times n$ Hilbert matrix, is defined by $h_{ij} = (i + j - 1)^{-1}$. Hilbert matrices have a number of interesting properties:

- the elements of H_n^{-1}, $\hat{h}_{ij}$, are given by

$$\hat{h}_{ij} = \frac{(-1)^{i+j}(i + n - 1)!(j + n - 1)!}{(i + j - 1)(i - 1)!^2(j - 1)!^2(n - i)!(n - j)!},$$

which are integral for all values of i and j. See Knuth [48] for details of the derivation;

- the condition number of H_n increases rapidly with order (see Figure 4.1).

The exact solution of the system

$$H_n^{-1}\mathbf{x} = \mathbf{b}, \tag{4.13}$$

where $\mathbf{b}^T = (1, 0, \ldots, 0)$ is then given by $x_i = 1/i$, i.e., the first column of H_n.

For small values of n both the elements of H_n^{-1} and the right-hand side $\mathbf{b}$ may be calculated and stored exactly as floating-point numbers. Therefore any errors we see in the solution $\mathbf{x}$ are due solely to the amplification of rounding errors committed during the elimination and the forward and backward substitution processes.

For values of n from 2 to 12 we compute a solution using F04ARF and calculate the maximum number of decimal digits in error. These results, given at the end of Figure 4.1, may then be compared with the rule of thumb given in Section 4.7. From this table we observe that for well-conditioned to moderately ill-conditioned problems ($n \leq 7$) F04ARF is reasonably accurate and our rule of thumb tends to be somewhat pessimistic. For very ill-conditioned problems the routine produces progressively worse solution vectors *without giving any indication of the ill-conditioning*. For this type of problem the rule of thumb gives a very good indication of the observed numerical inaccuracies.

Unfortunately, at present, the NAG linear equation solvers do not return an estimate of the condition number of the coefficient matrix, although algorithms do exist which compute such estimates. Implementations of these are available in the LINPACK library of linear algebra software (Dongarra et al. [26]). The cost of the estimates is $O\left(n^2\right)$ flops and, in practice, these algorithms have been found to produce excellent order of magnitude estimates.

EXERCISE 4.11 Use the Crout factorization program of Exercise 4.10 to solve the system (4.13) for several values of n. Compare the accuracy of the final solution when basic and additional precision accumulation are used for the calculation of inner products.

4.9 Accurate solutions using F04ATF

It is sometimes necessary, and desirable, to be able to compute a very accurate solution to a system of linear equations and to be able to guarantee its accuracy. Iterative improvement (sometimes called *iterative refinement*) is a method whereby the accuracy of a solution obtained by a direct method is improved. Using this technique it is often possible to obtain a solution which is correct to the full precision of the floating-point arithmetic.

Suppose $\hat{\mathbf{x}}$ is an approximate solution to (4.1) obtained, for example, by Gaussian elimination. Defining the residual vector, $\mathbf{r}$, by $\mathbf{r} = \mathbf{b} - A\hat{\mathbf{x}}$, we note that if $\mathbf{z}$ is the *exact* solution of $A\mathbf{z} = \mathbf{r}$, we may form $\mathbf{x} = \hat{\mathbf{x}} + \mathbf{z}$, which is then the exact solution to the original system since $A\mathbf{x} = A\hat{\mathbf{x}} + A\mathbf{z} = A\hat{\mathbf{x}} + \mathbf{r} = \mathbf{b}$.

Skeel [76] derives error bounds associated with each cycle of iterative refinement for the cases where single or double precision are used in forming the residual. He shows that a single iteration using basic precision t, say, throughout is of benefit from a stability point of view. However, to obtain improved accuracy in the final solution it is essential to compute the residual in substantially higher precision, and it is usual to use about $2t$ significant digits.

We now define a computational algorithm:

(1) Compute **x** using a direct method, keeping the LU factors and the permutation vector.

(2) **Repeat**

(a) compute $\mathbf{r} = \mathbf{b} - A\mathbf{x}$ using additional-precision arithmetic,

(b) solve $A\mathbf{z} = \mathbf{r}$,

(c) set $\mathbf{x} = \mathbf{x} + \mathbf{z}$

Until ($\|\mathbf{z}\|$ is small enough *or* non-convergence suggests that A is too ill-conditioned for any further improvement to be achieved).

This may be regarded as a multiple right-hand side problem, although there is only a single right-hand side at each iteration. Direct Gaussian elimination would require decomposition of the coefficient matrix at each iteration, but if Crout's method is used the LU factorization is performed once only.

The termination criterion is complicated by the fact that it is possible for the coefficient matrix to be ill-conditioned enough to prevent convergence of the iterative refinement, but not ill-conditioned enough to be deemed computationally singular during the LU factorization. A good discussion of the implementation details of this algorithm can be found in Wilkinson and Reinsch [83].

The computational cost of the procedure is twofold. First there are the additional linear systems to solve in step (2b) of the algorithm. But since we already have the LU factorization of A available this is relatively cheap, $O\left(n^2\right)$, compared with the original $O\left(n^3\right)$ invested in the factorization. Secondly there is an additional storage overhead since we now need to store both the original matrix A, to compute the residuals, and the LU decomposition. Note that the LU factors cannot be used to compute the residuals because of rounding errors in their computation. We also need space to store the residual vector.

The additional storage requirements are reflected in the argument list of F04ATF, the routine we arrive at in the decision tree if we go for accuracy rather than time and storage. The call sequence is

```
CALL F04ATF(A,IA,B,N,X,AA,IAA,WKS1,WKS2,IFAIL)
```

The parameters A, IA, B, N and X are the same on input as for F04ARF (see Section 4.8). Note that, unlike F04ARF, this routine must not be called with the same name for parameters B and X, since both the original right-hand side and the computed solution need to be present to compute the residuals. The additional array AA, dimensioned $(\text{IAA}(\geq \text{N}), p(\geq \text{N}))$, is used to store the Crout LU factorization and A is now unchanged on exit.

The two workspace vectors, WKS1 and WKS2, both of length N, are for the permutation information and storage of the residuals.

The two non-zero IFAIL values distinguish between matrices which are found to be singular (possibly due to rounding error) during the factorization stage (IFAIL = 1), and those which are so ill-conditioned that the iterative refinement procedure will not converge (IFAIL = 2).

Using F04ATF to solve the system (4.13) for $n = 2, 3, \ldots, 12$ produces solution vectors which are correct to full machine precision. This demonstrates that the use of iterative refinement can be of benefit even for very ill-conditioned problems.

4.10 Multiple right-hand side routines

The NAG F04 chapter provides routines for computing both an approximate solution (F04AAF), and an accurate solution (F04AEF) for the case of multiple right-hand sides. In both cases the difference in the parameter list from the corresponding single right-hand side routine is the same. The call sequence for the F04ARF counterpart F04AAF is

CALL F04AAF(A,IA,B,IB,N,M,X,IX,WKSPCE,IFAIL)

The number of right-hand sides is given by M, and both the right-hand side parameter and the solution parameter are now arrays B and X, dimensioned (IB, p) and (IX, p) respectively, with $IB \geq N$, $IX \geq N$, and $p \geq M$. Each column of B will correspond to a different right-hand side vector and it follows that the solution vectors are stored as columns of X. All other parameters remain as described for F04ARF. As with F04ARF, A is destroyed during execution, and it is possible for the arrays B and X to be the same, in which case the right-hand side array is overwritten with the solution.

4.11 Cholesky decomposition

We now consider the case in which the coefficient matrix, A, is symmetric and positive definite (see Section 4.2). For such matrices a decomposition of the form $A = LL^T$ exists and it may be shown (Wilkinson [82]) that the factorization is numerically stable without pivoting. The entries in the lower triangular matrix L are given, for $i = 1, 2, \ldots, n$, by

$$l_{ij} = \left(a_{ij} - \sum_{k=1}^{j-1} l_{ik}l_{jk} \right) \Big/ l_{jj}, \qquad j = 1, 2, \ldots, i-1,$$

$$l_{ii} = \sqrt{a_{ii} - \sum_{k=1}^{i-1} l_{ik}^2}.$$

$$(4.14)$$

The positive definite condition ensures that the diagonal elements are the square roots of positive numbers. Since only one triangular matrix needs to be formed (L^T is implicitly available once L is formed) the method, known as *Cholesky factorization*, is inherently faster than the other schemes considered so far (and in particular Crout's method) although the operation count is still $O\left(n^3\right)$. For this reason the NAG library contains two routines, F04ABF and F04ASF, which are specifically for symmetric, positive definite coefficient matrices; both compute accurate solutions. F04ASF computes a solution to a problem having only a single right-hand side, whilst F04ABF is designed to handle the case of multiple right-hand sides.

A typical call to F04ASF takes the form

```
CALL F04ASF(A,IA,B,N,X,WK1,WK2,IFAIL)
```

where A, dimensioned (IA, N) with IA $\geq$ N, should contain, on entry, the upper triangular part of A only (including the principal diagonal). It is not necessary to supply the lower triangular part since A is symmetric. The vector B, dimensioned at least N, should contain the right-hand side vector and WK1 and WK2, both declared to be of length at least N, are used for workspace. Since the routine attempts to compute an accurate solution it may terminate with IFAIL = 2 due to the refinement process failing to improve the solution. It will also halt, with IFAIL = 1, if it detects that the coefficient matrix is not positive definite. This condition is apparent if the argument of a square root, used to compute the diagonal element of L in (4.14), is not positive. This may be caused either by an improper use of the routine, or by the build up of rounding errors. On successful exit, X (of length at least N) will contain the solution vector. The upper triangular components of the coefficient matrix and the right-hand side vector are not destroyed by the routine, but any values placed in the lower triangular part of A will be overwritten by those elements of L below the diagonal. WK1 is used to store the reciprocals of the diagonal elements of L. Note that B and X must be distinct arrays.

A call of F04ABF takes the form

```
CALL F04ABF(A,IA,B,IB,N,M,X,IX,WKSPCE,BB,IBB,IFAIL)
```

with B dimensioned (IB($\geq$ N), $p(\geq$ M)), containing the M right-hand sides on entry. On exit, X, dimensioned (IX($\geq$ N), $p(\geq$ M)), contains the solution vectors, and BB, dimensioned (IBB($\geq$ N), $p(\geq$ M)), contains the corresponding residuals. The three arrays B, BB and X must be distinct. Finally, WKSPCE, of length at least N, is used for workspace.

Being able to guarantee a numerically stable solution to symmetric, positive definite systems is especially helpful when dealing with sparse systems. For these problems we wish to economize the storage overheads as far as possible, and both the equations and the unknowns may be reordered prior to the factorization to minimize the amount of *fill-in* (entries which were zero in the original coefficient matrix but become non-zero during the factorization process). Details of reordering algorithms may be found in George and Liu [35].

EXERCISE 4.12 Write down the stages of the Crout (LU) factorization for a symmetric, positive definite matrix. Show that U may be written as DL^T and hence derive an algorithm equivalent to Cholesky decomposition which requires the same amount of storage but avoids the need to compute square roots.

4.12 Banded matrices

Banded matrices occur frequently in practice in the analysis of complex problems, for example, the numerical solution of partial differential equations. In some cases it may be necessary to reorder the equations and/or the unknowns to generate the coefficient matrix of the required form. This effort is especially worthwhile when the systems are large and a substantial gain in efficiency may be obtained by forming a matrix with a narrow band width.

We recall from Section 4.11 that for a symmetric, positive definite matrix a factorization of the form $A = LL^T$ exists. If A is banded with band width k, and there is no pivoting, L will only have non-zero entries on the leading diagonal and the k subdiagonals. Thus, when forming the Cholesky decomposition using (4.14), the number of terms which need to be computed in L is considerably reduced. Furthermore, within the computation of each l_{ij}, the number of terms which we have to sum over is bounded by the band width. In view of these potential savings, it is clear that special purpose routines should be used to solve systems of equations in which the coefficient matrix is banded. We consider in detail the case of a tridiagonal matrix ($k = 1$); the extension to higher band width matrices will be clear.

For convenience we express our symmetric, positive definite tridiagonal matrix as

$$A = \begin{pmatrix} a_1 & c_1 & & & & & 0 \\ c_1 & a_2 & c_2 & & & & \\ & c_2 & a_3 & c_3 & & & \\ & & \ddots & \ddots & \ddots & & \\ & & & c_{n-2} & a_{n-1} & c_{n-1} \\ 0 & & & & c_{n-1} & a_n \end{pmatrix}. \tag{4.15}$$

Then we need to compute the entries in

$$
L = \begin{pmatrix}
p_1 & & & & & & 0 \\
q_1 & p_2 & & & & & \\
 & q_2 & p_3 & & & & \\
 & & & \ddots & \ddots & & \\
 & & & & q_{n-2} & p_{n-1} & \\
0 & & & & & q_{n-1} & p_n
\end{pmatrix},
$$

such that $A = LL^T$. Since no pivoting is required we may equate coefficients (or use (4.14)) to give

$$
\begin{aligned}
p_1 &= \sqrt{a_1}, \\
q_{i-1} = c_{i-1}/p_{i-1}, \qquad p_i &= \sqrt{a_i - q_{i-1}^2}, \qquad i = 2, 3, \ldots, n.
\end{aligned} \tag{4.16}
$$

Returning to (4.14), we recall that for a general positive definite, symmetric matrix Cholesky factorization is an $O\left(n^3\right)$ process. L has $(n^2+n)/2$ entries to be determined, and the computation of each will involve the evaluation of $O(n)$ summations. If (4.16) is used to form L, we compute only $2n-1$ terms, each involving an $O(1)$ sum; hence we have reduced the operation count from $O\left(n^3\right)$ to $O(n)$ and this represents a considerable saving for large n. For matrices with band width k, Cholesky decomposition is reduced to an $O(n)$ process for small values of k; as k approaches n the method, which has an operation count of $O\left(nk^2\right)$, reverts to $O\left(n^3\right)$.

Having formed the LL^T factorization of A the system $Ax = b$ is solved in the normal manner; i.e., solve $Ly = b$ for the intermediate vector $\mathbf{y}$ using forward substitution and then solve $L^T\mathbf{x} = \mathbf{y}$ for $\mathbf{x}$ using backward substitution. For a tridiagonal system

$$
\begin{aligned}
y_1 &= b_1/p_1, \\
y_i &= (b_i - q_{i-1}y_{i-1})/p_i, \qquad i = 2, 3, \ldots, n,
\end{aligned} \tag{4.17}
$$

and

$$
\begin{aligned}
x_n &= y_n/p_n, \\
x_i &= (y_i - q_i x_{i+1})/p_i, \qquad i = n-1, n-2, \ldots, 1.
\end{aligned} \tag{4.18}
$$

We have reduced an inherently $O\left(n^2\right)$ process to one which is $O(n)$ (since the sum in (4.4) is now $O(1)$). Again, if k is small compared to n the argument holds for matrices with band width k; for larger values of k the scheme, which is $O(kn)$, reverts to $O\left(n^2\right)$.

For banded matrices we can do more than reduce the operation count and hence the execution time required to solve the problem. For example, a tridiagonal symmetric matrix may be stored in an array of size $n \times 2$ in the form

$$A = \begin{pmatrix} * & a_1 \\ c_1 & a_2 \\ \vdots & \vdots \\ c_{n-2} & a_{n-1} \\ c_{n-1} & a_n \end{pmatrix}, \qquad (4.19)$$

a saving of $n^2 - 2n$ elements. The $(1,1)$ entry in this array representation is arbitrary. In addition, the matrix L in the LL^T factorization of A may also be stored in a matrix of this size. These techniques have been incorporated into the subroutine TRIDIA, Code 4.4, which solves a system of equations in which the coefficient matrix is symmetric, positive definite and tridiagonal. Note that it is possible for the intermediate vector, y, and the solution vector, x, to use the same array.

```
      SUBROUTINE TRIDIA(A,NROWS,B,X,LLT,N,IFAIL)
*
*..GIVEN A TRIDIAGONAL MATRIX A AND A RIGHT-HAND
*..SIDE VECTOR B THIS ROUTINE
*..COMPUTES AN APPROXIMATE SOLUTION, X, TO THE
*..EQUATION A*X = B USING THE
*..CHOLESKY FACTORIZATION ( L L(TRANSPOSE) ) OF A
*..THE LOWER TRIANGULAR PART OF A ONLY NEED BE SUPPLIED
      DOUBLE PRECISION ZERO
      PARAMETER (ZERO = 0.0D0)
      INTEGER IFAIL,N,NROWS
      DOUBLE PRECISION A(NROWS,2),B(NROWS),X(NROWS),
     +                 LLT(NROWS,2)
      DOUBLE PRECISION ARG
      INTEGER I
      INTRINSIC SQRT
*
*..FORM THE CHOLESKY DECOMPOSITION OF A
      ARG = A(1,2)
      IF (ARG.LE.ZERO) THEN
        IFAIL = 1
        RETURN
      ENDIF
      LLT(1,2) = SQRT(ARG)
      DO 10 I = 2,N
        LLT(I,1) = A(I,1)/LLT(I-1,2)
        ARG = A(I,2)-LLT(I,1)**2
        IF (ARG.LE.ZERO) THEN
```

```
            IFAIL = 1
            RETURN
         ELSE
            LLT(I,2) = SQRT(ARG)
         ENDIF
10    CONTINUE
*
*..FORM THE INTERMEDIATE VECTOR Y USING FORWARD SUBSTITUTION
      X(1) = B(1)/LLT(1,2)
      DO 20 I = 2,N
         X(I) = (B(I)-LLT(I,1)*X(I-1))/LLT(I,2)
20    CONTINUE
*
*..FORM THE SOLUTION VECTOR USING BACKWARD SUBSTITUTION
      X(N) = X(N)/LLT(N,2)
      DO 30 I = N-1,1,-1
         X(I) = (X(I)-LLT(I+1,1)*X(I+1))/LLT(I,2)
30    CONTINUE
      END
```

Code 4.4   Tridiagonal equation solver.

The NAG equivalent of TRIDIA is F04FAF for which a typical call takes the form

```
CALL F04FAF(JOB,N,D,E,B,IFAIL)
```

This routine may be used in two modes depending on whether the input value of the parameter JOB is zero or one. When JOB is set to zero the tridiagonal coefficient matrix of order N is defined not by a matrix, but by the two vectors D and E. The diagonal elements, a_1, a_2, ..., a_n are stored in the vector D and the elements $E(2)$, $E(3)$, ..., $E(N)$ contain the values c_1, c_2, ..., c_{n-1} in (4.15). The right-hand side vector is specified in B and on successful exit these values are overwritten by the solution vector. On exit the arrays D and E will have been overwritten by a factorization of the original matrix, this form being slightly different to the Cholesky decomposition (see Dongarra et al. [26] for more details). F04FAF will halt with IFAIL $= 1$ if JOB or N are given invalid values. Note that in the latter case the run-time system may abort the program before a check is made that $N > 1$, since N is used to dimension the arrays D and E (see Exercise 1.14). When JOB $= 0$ a failure with IFAIL $= 2$ may be detected if the decomposition breaks down, either because the coefficient matrix is not positive definite, or because it is computationally singular.

For a multiple right-hand side problem the factorization need only be performed once. The first right-hand side is solved as described above.

Further calls to F04FAF are then made with JOB set to one, and the arrays
D and E containing the factorized form returned by the first call.

For the case of a non-symmetric tridiagonal matrix A, F04EAF is available. In the call sequence

```
CALL F04EAF(N,D,DU,DL,B,IFAIL)
```

the diagonal, principal super-diagonal and principal sub-diagonal of A
should be supplied in D, DU and DL, respectively, with DU(1) and DL(1)
unused. The routine does not allow reuse of the factorization for further
right-hand sides. Note that pivoting may be necessary here in forming the
LU factors. This may increase from one to two the number of non-zero
elements to the right of the diagonal of U.

For systems of the form $Ax = b$ where A is symmetric, positive definite and of band width $M(> 1)$ F04ACF should be used. The call sequence
is of the form

```
CALL F04ACF(A,IA,B,IB,N,M,IR,X,IX,RL,IRL,M1,IFAIL)
```

where A should, on entry, contain the entries in the coefficient matrix stored
in the form

$$
A = \begin{pmatrix}
* & \cdots & * & * & a_{11} \\
* & \cdots & * & a_{21} & a_{22} \\
* & \cdots & a_{31} & a_{32} & a_{33} \\
\vdots & & \vdots & \vdots & \vdots \\
a_{n-1,n-m-1} & \cdots & a_{n-1,n-3} & a_{n-1,n-2} & a_{n-1,n-1} \\
a_{n,n-m} & \cdots & a_{n,n-2} & a_{n,n-1} & a_{nn}
\end{pmatrix}.
$$

A should be of dimension $\big(\text{IA}(\geq \text{N}), p(\geq \text{M1} = \text{M}+1)\big)$. The routine uses the
array RL, of dimension $\big(\text{IRL}(\geq \text{N}), p(\geq \text{M1})\big)$, to store the lower triangular
matrix L in the LL^T decomposition of A; note that the reciprocals of the
diagonal elements are stored (equivalent to $1/p_i$ in (4.16)). The routine
will solve IR right-hand sides and these should be stored in the matrix
B dimensioned $\big(\text{IB}(\geq \text{N}), p(\geq \text{IR})\big)$, and the solutions are returned in the
matrix X of dimension $\big(\text{IX}(\geq \text{N}), p(\geq \text{IR})\big)$. The parameters B and X may
be the same, in which case the solution vectors overwrite the original right-
hand sides.

The three routines mentioned in this section compute an approximate
solution only. There is no special purpose black-box routine for solving a
system in which the coefficient matrix is banded but non-symmetric. A
general purpose solver may be used for such a system. Alternatively, use
the pair of routines F03LBF/F04LDF (see Section 4.13).

EXERCISE 4.13 Generalize the equations (4.16), (4.17) and (4.18) to the
case of a symmetric, positive definite system of band width k.

EXERCISE 4.14   Consider the relative merits of storing a banded matrix A with large band width in a two-dimensional array A whose columns

- correspond to the columns of A,
- correspond to the diagonals of A.

Derive an expression involving the order of the matrix and the band width which defines the cross-over point at which the second representation ceases to be the better storage scheme.

EXERCISE 4.15   For a general banded matrix A what are the implications of pivoting with respect to the minimum amount of storage required for the LU factors if the second strategy of Exercise 4.14 is used?

EXERCISE 4.16   A data file contains all n^2 elements of a banded, symmetric positive definite matrix of band width k stored by rows. Write a subroutine to read this data into an $n \times k + 1$ array suitable for passing to F04ACF.

EXERCISE 4.17   The subroutine TRIDIA (Code 4.4) was written to match the notation used in (4.19); if you were to design a general purpose code suitable for a subroutine library what change would you make to the interface?

4.13   Black-box vs. general purpose routines

The NAG routines we have looked at so far are classified as black-box, designed to be used in a single call for a given coefficient matrix within a program run. These may be divided into two subclasses

- routines which handle a single right-hand side vector,
- routines which accept several right-hand sides.

It may be that the number and/or form of the right-hand sides may not be known in advance. For example, an approximation to the smallest eigenvalue, λ_1, and its associated eigenvector of a positive definite matrix A (whose eigenvalues are all positive) using the inverse power method requires the solution of a sequence of linear systems of the form $A\mathbf{x}_n = \mathbf{b}_n$ where $\mathbf{b}_n$ is dependent on $\mathbf{x}_{n-1}$. The iteration converges provided that λ_1 is unique. This is a multiple right-hand side problem since the coefficient matrix remains unchanged. However $\mathbf{b}_n$ cannot be formed until $\mathbf{x}_{n-1}$ itself has been calculated. To call a black-box, single right-hand side routine at each iteration would be very wasteful. A better approach would be to call one routine to form the LU, or LL^T, factorization of A (or, more commonly, PA) and then another, general purpose, routine which uses this factorization and performs the backward and forward substitutions for each new right-hand side. Routines which perform just a matrix factorization are available in chapters F01 and F03. For example, to solve a general real linear system of order N we first use F03AFF to form the Crout factorization of the coefficient matrix. A call is of the form

CALL FO3AFF(N,EPS,AA,IAA,D1,ID,P,IFAIL)

where on entry EPS should be set to the machine epsilon returned via
X02AJF, and AA, dimensioned $(IAA(\geq N), r(\geq N))$, contains the matrix
which, on successful exit, is overwritten with the LU factors. Also on
exit the real/double-precision vector P, of length at least N, contains the
permutation information. The purpose of D1 and ID is described in Section 4.14. The routine will fail with IFAIL = 1 if the matrix is found to be
computationally singular.

Having obtained the LU factors of A, FO4AHF may be called to solve
the system $Ax = b$. A typical call to this latter routine takes the form

CALL FO4AHF(N,IR,A,IA,AA,IAA,P,B,IB,EPS,X,
+   IX,BB,IBB,K,IFAIL)

where AA and P contain the factorized form and permutation vector respectively, obtained from the call to FO3AFF, and EPS is again the machine
epsilon. FO4AHF can solve for several right-hand sides at one call; these
should be stored in B which should be dimensioned $(IB(\geq N), s(\geq IR))$
and the corresponding solution vectors are returned in X of size $(IX(\geq N),$
$s(\geq IR))$, where IR is the number of right-hand sides. The routine computes an accurate solution (i.e., one obtained using iterative refinement,
see Section 4.9) and thus requires the coefficient matrix to be supplied via
A, dimensioned $(IA(\geq N), s(\geq N))$, in order to compute the residuals. On
exit K gives the number of iterations employed by the refinement process,
and BB, of size $(IBB(\geq N), s(\geq IR))$, contains the final residuals, that is,
the amount by which the values returned in X fail to satisfy the original
system.

The essence of the general purpose routines is, therefore, that an appropriate factorization of the coefficient matrix must have been performed
before they may be called. In the following list of available routines the
factorization routine is given first followed by the respective solver:

FO3AFF/FO4AJF: for an approximate solution to a general real system,

FO3AFF/FO4AHF: for an accurate solution to a general real system,

FO3AEF/FO4AGF: for an approximate solution to a real, symmetric, positive definite system,

FO3AEF/FO4AFF: for an accurate solution to a real, symmetric, positive
definite system,

FO3AGF/FO4ALF: for an approximate solution to a real, symmetric, positive definite banded system,

FO1BTF/FO4AYF: as FO3AFF/FO4AJF but designed for efficiency in a paged
environment (see Section 1.6),

F01BXF/F04AZF: as F03AEF/F04AGF but designed for efficiency in a paged environment (see Section 1.6),

F01BFF/F04AQF: as F01BXF/F04AZF but the lower triangular factor L is supplied via a one-dimensional array in order to economize on storage,

F01MCF/F04MCF: for an efficient approximate solution to a system whose coefficient matrix is real, symmetric, positive definite and has a variable band width,

F01LBF/F04LDF: for an approximate solution to a general real banded system,

F01LEF/F04LEF: for the solution of a general real tridiagonal system.

It should be noted that certain F01 routines (e.g., F01BXF) accumulate inner products in normal precision, and that certain F04 routines (e.g. F04AQF) accept a single right-hand side only.

4.14   Determinant evaluation and matrix inversion

It is rare that the explicit evaluation of the determinant of a matrix is required. However for those instances which do require such computation we remark that a direct approach based on the mathematical definition (see Strang [79], p. 162ff) requires $O(n!)$ arithmetic operations and is thus, obviously, computationally useless for any but the smallest values of n. The preferred approach is to evaluate $\det(A)$ as a by-product of the factorization of A. For example, if Crout's method is used to factor A into $P^{-1}LU$ then $\det(A) = \pm \det(L)\det(U)$. Now, the determinant of a lower or upper triangular matrix is equal to the product of the elements on the diagonal, but since U has a unit diagonal, $|\det(A)|$ is simply the product of the diagonal elements of L. The sign is the determinant of P^{-1} and is positive/negative depending on whether an even/odd number of row interchanges have taken place. The operation count for this part alone is $O(n)$ and so the overall cost is dominated by the $O(n^3)$ factorization process (see Section 4.6).

Chapter F03 of the NAG library contains several routines for determinant evaluation based on matrix factorization. 'Black-box' routines are available for certain types of real matrix and these return the determinant as a single real value. 'General purpose' routines return the value in the form $d1 \times r^{d2}$ where $\frac{1}{16} \le |d_1| < 1$ and the radix, r, is 2. Some routines allow a radix of 10 to be specified in which case $\frac{1}{16} \le |d_1| < 10$. Using this representation helps to guard against the possibility of underflow or overflow. The routines may be further divided into those which return the appropriate factorization of A, and those which require the factors to be supplied as input by a prior call to a suitable F01 routine.

In Section 4.1 we remarked that the system $Ax = b$ is never solved numerically for x by forming the inverse of A, followed by the matrix–vector multiply $A^{-1}b$. One reason is efficiency; we first consider Cramer's rule, a well-known theoretical technique for matrix inversion (see Strang [79], p.171 for details). An analysis of the method shows that, even if we use decomposition to evaluate the determinants, the method is still a very expensive $O\left(n^4\right)$.

Consider now the n systems $Ax_i = e_i$, for $i = 1, 2, \ldots, n$, where e_i is the ith column of the unit matrix, I_n. The solution vectors, x_i, then form the columns of A^{-1}. The cost of the method is made up of

- $O\left(n^3\right)$ for the factorization of A,

- n forward and backward substitutions at $O\left(n^2\right)$ each; $O\left(n^3\right)$ in all,

giving a grand total of $O\left(n^3\right)$ operations to compute the inverse. Compared to the cost of factorization plus one forward and one backward substitution required to solve a linear system by the methods discussed in earlier sections, we see that forming the explicit inverse of A cannot yield a competitive method.

Despite what we have said above, there are instances (for example, in regression analysis) when the explicit inverse of a matrix is required. Clearly, any linear system solver capable of dealing with multiple right-hand sides may be used. In NAG terminology this means using the appropriate routine from the F04 chapter, although the F01 chapter does contain a small number of routines which compute A^{-1} without requiring the user to set up the right-hand side vectors, e_i.

For the general real matrix, A, of order N, a typical call sequence might consist of the following:

(1) Read in the values of the elements of A.

(2) Copy A into the real/double-precision matrix LU using

```
CALL F01CMF(A,N,LU,N,N,N)
```

(3) Set up an identity matrix, IDENT, of order N using

```
CALL F01CBF(IDENT,N,N,IFAIL)
```

(4) Form the Crout factorization of A in LU using

```
EPS = X02AJF(DUMMY)
CALL F03AFF(N,EPS,LU,N,D1,ID,P,IFAIL)
```

This routine also returns the determinant of A in the form D1 $\times$ 2^{ID}, but this value is not required.

(5)   Form `AINV`, a real/double-precision accurate inverse, by the call

```
      CALL F04AHF(N,N,A,N,LU,N,P,IDENT,N,EPS,AINV,
     +            N,BB,N,K,IFAIL)
```

If, at a later stage, the matrix $C = A^{-1}B$ is required, this may be determined either by calling `F04AHF` with B as the set of right-hand sides or, if this is not possible, as follows:

(6)   Read the elements of B into array B of size (N, N).

(7)   Form in C, an array of size (N, N), the required product using

```
      OPT = 1
      CALL F01CKF(C,AINV,B,N,N,N,Z,IZ,OPT,IFAIL)
```

Of course stages (2) and (3) of this process are easily programmed and calls to the NAG routines `F01CBF` and `F01CMF` are not really necessary. It might be thought that the same could be said of forming the matrix–matrix product $A^{-1}B$. However, in view of the need to minimize the effects of rounding errors by using multiple length inner product routines, the use of `F01CKF` in step (7) is advised.

EXERCISE 4.18   Show that if the LU factors of a matrix A are available the cost of forming $A^{-1} = U^{-1}L^{-1}$ is still $O\left(n^3\right)$.

4.15   Complex problems

As we have already indicated, the methods outlined in this chapter for the solution of real problems may be readily extended to deal with complex systems. The general system (Section 4.1) may be solved by forming the Crout (LU) factors of PA, where P is an appropriate permutation matrix, followed by the application of forward and backward substitution. The black-box (`F04ADF`) and general purpose (`F04AKF`) routines, based on this approach, provide for the approximate solution of complex systems.

Cholesky factorization may be applied to systems for which the co-efficient matrix is positive definite and Hermitian (see Section 4.2). We may form $PA = LL^H$, where L^H is the complex conjugate transpose of L. `F04AWF` is a general purpose routine for determining an approximate solution to such a system. Perversely, it is L^H (denoted by U in the routine documentation) which needs to be supplied using a prior call to `F01BNF`. All three F04 routines mentioned in this section are capable of dealing with several right-hand sides.

EXERCISE 4.19   Let A be a positive definite Hermitian matrix, $A^{(k)}$ be A after $k-1$ stages of Gaussian elimination have been applied, and A_k, of size $(n-k+1) \times (n-k+1)$, be $A^{(k)}$ with the first $k-1$ rows and columns removed. Show that A_k is Hermitian positive definite and $\left|a_{ij}^{(k)}\right| \leq \max_{i,j}\left|a_{ij}\right|$.

4.16   Summary

We concentrate here on the case in which the number of equations equals the number of unknowns. Methods based on the solution of over-determined problems (that is, more equations than unknowns) can be applied to this particular case, but are likely to be less efficient than the special purpose routines. The first decision that the user has to make is whether to use a black-box or a general purpose routine. All methods implemented in the F04 chapter use a decomposition of the coefficient matrix. If several problems are to be solved in which the coefficient matrix is the same in each case, it makes sense to perform this decomposition once only, and this is achieved using an appropriate routine from the F03 or F01 chapter. The system is then solved by a call of an appropriate general purpose F04 routine. If, on the other hand, only a single call is expected, or several calls are expected but the coefficient matrix changes, a black-box F04 routine will be found easier to use.

We consider the black-box routines first. If the coefficient matrix is real, and the system involves only a single right-hand side vector, F04ARF will determine a solution (Section 4.8). It is possible to improve such an 'approximate' solution using a process known as iterative refinement, and if greater accuracy is required, F04ATF should be used to compute an 'accurate' solution (Section 4.9). For the case of several right-hand sides, the corresponding routines are F04AAF and F04AEF (Section 4.10). If it is known that the coefficient matrix is symmetric and positive definite, F04ASF and F04ABF compute an accurate solution in the case of a single and multiple right-hand sides respectively (Section 4.11). For a symmetric positive definite banded coefficient matrix and several right-hand sides, F04ACF computes an approximate solution (Section 4.12). F04EAF deals with the case of a general tridiagonal matrix, whilst F04FAF should be used if, in addition, the matrix is symmetric and positive definite; both accept a single right-hand side only, and compute an approximate solution (Section 4.12). If the matrix is complex and of general form, F04ADF is available. Several right-hand sides are permitted, and an approximate solution is computed (Section 4.15).

If the use of a general purpose routine seems appropriate, and the coefficient matrix contains real components only, F04AJF computes an approximate solution (Section 4.13), whilst F04AHF computes an accurate solution (Section 4.13). However, in the case of a paged environment F04AYF may be used in preference to F04AJF. For a real symmetric positive definite matrix use F04AGF to determine an approximate solution (F04AZF in a paged environment), and F04AFF if an accurate solution is required. A further alternative is F04AQF which expects the matrix decomposition to be supplied in a condensed form, thus saving on storage costs (Section 4.13). If, in addition, the matrix has a fixed band width, F04ALF computes an approximate solution (Section 4.13), whilst F04MCF deals with the case of a variable band width (Section 4.13). For a general banded matrix F04LDF is available, whilst F04LEF deals with the special case of a tridiagonal matrix (Section 4.13). If the coefficient matrix is complex, F04AKF is available, whilst F04AWF deals with the case in which, in addition, the matrix is Hermitian and positive definite (Section 4.15). For a complex matrix of fixed band width, F04NAF may be used. All routines for complex coefficient matrices return an approximate solution only. With the exception of F04AQF and F04LEF, general purpose routines accept several right-hand sides.

Clearly, categorization of the coefficient matrix is an important, probably *the* most important, factor in routine selection. If the matrix is tridiagonal then use of, say F04ARF, is likely to prove considerably less efficient than the use of a special purpose routine. This property should be easy to detect, and so should symmetry, but classification of a matrix as being positive definite is far from straightforward and we can only give a few guidelines of real use in this respect (Section 4.2). A further consideration in routine selection is whether an approximate or accurate solution is required. For certain classes of problems, routines are available which determine either. For others (for example, real sparse coefficient matrix) only a routine for determining an approximate solution is available. If an accurate solution is required, it is possible for the user to code the iterative refinement process himself, and thus improve the initial estimate. The alternative is to use a less specialized routine. If a problem falls into a particular classification, and needs to be solved for a number of right-hand sides at one go, but such a routine is not available, then the specialized routine should be called several times; this is likely to prove more efficient than a single call of a more general routine which does accept multiple right-hand sides. In contrast, a routine which can accept several right-hand sides, but is called with just one, is likely to prove less efficient than an equivalent routine designed to take a single right-hand side. Moreover, problem set-up is likely to prove less amenable.

The use of the sparse linear system solvers is beyond the scope of the present volume and we will content ourselves with just a mention of the available routines. For a general sparse matrix, F04AXF computes an approximate solution once the coefficient matrix has been decomposed using F01BRF or F01BSF. An alternative routine is F04QAF which employs an iterative technique and allows the user to define the desired accuracy of the solution. If the matrix is symmetric, or symmetric and positive definite, the routines F04MBF or F04MAF (precede with F01MAF) should be more efficient than the general solvers. The F04 sparse matrix routines admit a single right-hand side only.

Chapter 5
Curve Fitting

In this chapter we are concerned with fitting a curve to a set of data points. Initially we restrict the discussion to the one-dimensional case and look for a close fit. We

- introduce the concept of least squares approximation;

- describe the least squares process in terms of approximation by Chebyshev polynomials;

- develop the idea of piecewise polynomial (spline) approximation;

- show how splines can be used in least squares approximation.

These ideas are then extended to the problem of fitting a surface to a two-dimensional set of data. Returning to the one-dimensional case, we next consider the problem of fitting a curve which exactly passes through the data (interpolation). We consider methods

- when function values only are available;

- when both function and derivative values are available;

- using splines.

5.1 Introduction

We interpret *curve fitting* to mean the process by which some function f, with values $\{ f_k \mid k = 1, 2, \ldots, m \}$, known at a discrete set of points $\{ x_k \mid k = 1, 2, \ldots, m \}$, is approximated with the aim of producing a continuous function which reflects the behaviour of f and is easy to manipulate. This is a somewhat wider definition than that used in the NAG documentation. It includes the process of interpolation, where we attempt to pass a curve exactly through the given points. We shall restrict the discussion to approximation by a polynomial p, although facilities exist within the NAG library for the user to choose other types of approximating function. It should be noted at the outset that, in this chapter, the notation used to represent a polynomial is different to that employed in Chapter 2. Here,

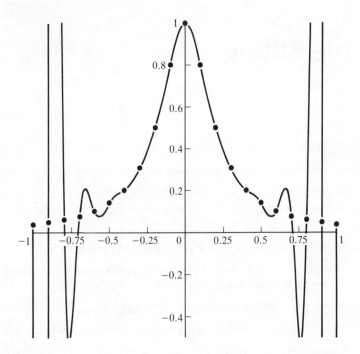

Figure 5.1   A wildly oscillating interpolating polynomial.

a_0 will be used to denote the constant term, a_1 the coefficient of x, a_2 that of x^2, and so on. This inconsistency is inconvenient, but reflects the different representations used in the respective sections of the NAG library. Techniques for curve fitting may readily be extended to the surface fitting problem where we attempt to represent a two-dimensional grid of data values using a polynomial in both x and y, and these modifications to the basic algorithms will also be considered.

It is well known that the process of evaluating a high-order interpolating polynomial in order to obtain an estimate of f at some point not in the given set is far from satisfactory. The approximation is likely to exhibit unwanted fluctuations between data points. (See Figure 5.1 in which a polynomial of degree 20 has been used to interpolate the function $f(x) = 1/(x^2 + 25)$ at 21 equally spaced data points in the range $[-1, 1]$.) To avoid this problem we could determine a low degree polynomial which may not pass through any of the coordinates $\{(x_k, f_k) \mid k = 1, 2, \ldots, m\}$, but which ensures that the *residuals*, $r_k = f_k - p(x_k)$, are small. A further approach is to group the points into subsets and seek a different low degree approximation for each subset, whilst maintaining suitable continuity conditions throughout the range. We shall consider each idea in turn, and attempt, at the end of the chapter, to summarize their respective benefits/drawbacks. As a deliberate policy we shall consider non-interpolatory approximation first; in general, this process is likely to give the best results.

As with most topics in numerical analysis, some preliminary investigation can prove of immense benefit. It may be immediately apparent that the function being approximated has a number of phases. For certain values of x it might behave like a straight line; elsewhere it could have sharp peaks. In such circumstances it often proves beneficial to seek a different type of approximation in different ranges. Sometimes it is immediately obvious that particular data values are completely erroneous. Whilst certain of the techniques we describe can be tailored so that they do not give undue weight to these 'wild points', it is best if they are removed at the outset. It may be possible to improve matters using an appropriate transformation. It is not difficult to approximate points on the curve $y = e^x$ by a polynomial; it is even easier if we write the equation as $\ln y = x$. Curve and surface fitting subroutines should be used only after some initial analysis of the data has been performed.

Generally speaking, as indicated above we reserve m to denote the number of data values available, with n (or, $n-1$) being used for the order of some polynomial approximation. However, we shall deviate from this, particularly in the sections on interpolation. This inconsistency arises from the fact that for interpolation, n and m are directly related, and the NAG routines do not adhere to any fixed notation. The reader should take care to interpret n and m appropriately, and to relate their meaning to the N and M of any NAG routines. This is occasionally the opposite way round to that expected.

5.2 Least squares approximation

We begin our discussion of numerical methods for approximation by considering the problem of fitting a straight line (polynomial of degree 1) to a set of coordinates. This process is referred to as *linear regression* in statistical parlance. What we are attempting to do is to solve the system of equations

$$a_0 + a_1 x_k = f_k, \qquad k = 1, 2, \ldots, m.$$

Assuming $m > 2$, this means that we have more equations than unknowns, and we refer to an *overdetermined* system. Unless the coordinates happen to lie on a straight line there will be no exact solution to the system, and we attempt to find values of the slope, a_1, and the y intercept, a_0, such that each equation is approximately satisfied. One way of determining a 'solution' is to find that which gives the minimum sum of the squares of the residuals

$$S = \sum_{k=1}^{m} r_k^2 = \sum_{k=1}^{m} (f_k - a_0 - a_1 x_k)^2. \qquad (5.1)$$

Using partial differentiation with respect to a_0 and a_1, it can be shown (Exercise 5.1) that these optimal values are solutions to the symmetric system

$$a_0 m + a_1 \sum_{k=1}^{m} x_k = \sum_{k=1}^{m} f_k,$$

$$a_0 \sum_{k=1}^{m} x_k + a_1 \sum_{k=1}^{m} x_k^2 = \sum_{k=1}^{m} f_k x_k,$$

(5.2)

known as the *normal equations*. The statistical properties of the solution to this system are well known (see Press et al. [66], p. 504) and special routines exist in the G02 chapter of the NAG library for determining confidence intervals, correlation coefficients, etc.

Here, we extend the idea to approximation by $p(x) = \sum_{j=0}^{n} a_j x^j$ a polynomial of degree n ($\leq m$) and associate a weight, $w_k > 0$, with each x_k. The residuals are now defined as $r_k = w_k(f_k - p(x_k))$. By using a large weight at certain points we can emphasize the corresponding residuals and hence ensure that the approximating polynomial will fit closer to these values than if an even weighting were used. Conversely, by choosing a small weight we can reduce the influence of a data point on the computed solution and hence limit the effect of values which we consider to be suspect or inaccurate. We define the *inner product* $< \cdot , \cdot >$ by

$$<g,h> = \sum_{k=1}^{m} w_k g_k h_k$$

(5.3)

for any two functions g and h, where $g_k = g(x_k)$ and $h_k = h(x_k)$. We again start with an overdetermined system, which here takes the form

$$w_k p(x_k) = w_k f_k, \qquad k = 1, 2, \ldots, m.$$

This may be written

$$X \mathbf{a} = \mathbf{F},$$

where X has elements

$$X_{ij} = w_i x_i^j, \qquad i = 1, 2, \ldots, m, \quad j = 0, 1, \ldots, n,$$

$\mathbf{F}$ has elements

$$F_i = w_i f_i, \qquad i = 1, 2, \ldots, m,$$

and $\mathbf{a}^T = (a_0, a_1, \ldots, a_n)$. Forming partial derivatives with respect to the free parameters a_0, a_1, ..., a_n, the normal equations become

$$A \mathbf{a} = \mathbf{b},$$

where

$$A_{ij} = X^T X = <x^i, x^j>, \qquad i,j = 0,1,\ldots,n,$$
$$b_i \;= X^T \mathbf{F} = <x^i, f>, \qquad i = 0,1,\ldots,n.$$

The problem has been reduced to the solution of a symmetric system of $n+1$ equations in $n+1$ unknowns. We could simply halt at this stage and use one of the routines discussed in Chapter 4 for solving such a system. However, a reformulation of the problem can be used to provide a solution directly.

Let p_0, p_1, ... be a sequence of polynomials, with p_i of degree i, defined by the three-term recurrence relation

$$p_i(x) = \lambda_{i-1}(x - \alpha_i)p_{i-1}(x) - \beta_{i-1}p_{i-2}(x), \qquad (5.4)$$

where

$$\alpha_i \;= \; <xp_{i-1}, p_{i-1}> \,/\, <p_{i-1}, p_{i-1}>,$$
$$\beta_i \;= \; \lambda_i <p_i, p_i> \,/\, (\lambda_{i-1} <p_{i-1}, p_{i-1}>),$$

and $\beta_0 = 0$ (q.v. Section 3.4). The λ_is are present as normalizing factors; in particular, the choice $\lambda_{i-1} = 1$ ensures that the coefficient of x^i in p_i is 1 as long as we choose $p_0(x) = 1$ and $p_1(x) = x$. Now

$$<p_1, p_0> \;=\; \lambda_0 <(x - \alpha_1)p_0, p_0>$$
$$=\; \lambda_0(<xp_0, p_0> - \alpha_1 <p_0, p_0>).$$

By definition $\alpha_1 = <xp_0, p_0> / <p_0, p_0>$, and it follows that $<p_1, p_0> = 0$. We say that p_1 and p_0 are *orthogonal* with respect to the inner product (5.3). In fact we can prove by induction that the polynomials p_i defined by (5.4) are mutually orthogonal, that is, $<p_i, p_j> = 0$ for $i \neq j$, and we say that they form an *orthogonal set* (see Exercise 5.4).

How can orthogonality help in least squares approximation? First we observe that any polynomial p of degree n may be expressed as a linear combination of $p_0, p_1, \ldots, p_n$, that is, we may write $p(x) = \sum_{i=0}^{n} c_i p_i(x)$ for some c_i. If we now redefine the residuals in terms of this new representation and form the partial derivatives of the sum of the squares with respect to the c_is, the entries in the coefficient matrix and right-hand side of the normal equations are now given by $A_{ij} = <p_i, p_j>$ and $b_i = <f, p_i>$. The orthogonality property means that A is diagonal, and the solution to the problem may immediately be determined as

$$c_i = b_i/A_{ii} = <f, p_i> / <p_i, p_i>. \qquad (5.5)$$

This solution is now expressed in terms of the derived orthogonal polynomials, although it is possible to express it in standard, power series, form by using the inverse transformation. An alternative is to rewrite it as a sum of Chebyshev polynomials, as we shall see in Section 5.3.

EXERCISE 5.1 By considering the sum of the squares of the residuals (5.1) for a straight line approximation, derive the normal equations (5.2) defining the least squares approximation, and show that the solution to this system is

$$a_0 = \frac{\sum_{k=1}^{m} x_k^2 \sum_{k=1}^{m} f_k - \sum_{k=1}^{m} x_k \sum_{k=1}^{m} f_k x_k}{m \sum_{k=1}^{m} x_k^2 - (\sum_{k=1}^{m} x_k)^2},$$

$$a_1 = \frac{m \sum_{k=1}^{m} f_k x_k - \sum_{k=1}^{m} f_k \sum_{k=1}^{m} x_k}{m \sum_{k=1}^{m} x_k^2 - (\sum_{k=1}^{m} x_k)^2}.$$

EXERCISE 5.2 The pressure of a gas (p) and its volume (v) are related by an equation of the form $pv^\gamma = c$. In an experiment the following values were obtained

p	0.50	1.00	1.50	2.00	2.50	3.00
v	1.62	1.00	0.75	0.62	0.52	0.46

Find γ and c by the method of least squares.

EXERCISE 5.3 Derive the normal equations for the least squares weighted approximation of m coordinates by a polynomial of degree n written in power series form.

EXERCISE 5.4 Prove by induction that the polynomials defined by (5.4) form an orthogonal set.

EXERCISE 5.5 Show that the sum of the squares of the residuals $\sum_{k=1}^{m} (f_k - p(x_k))^2$, where $p(x) = \sum_{i=0}^{n} p_i(x)$, with the p_is orthogonal, may be written

$$\sum_{k=1}^{m} \left(f_k^2 - \sum_{i=0}^{n} c_i^2 p_i^2(x_k) \right).$$

5.3 Chebyshev expansions

Many routines in the NAG library compute, or work with, polynomials expressed in *Chebyshev form*, that is, as a linear combination of the Chebyshev polynomials T_i (q.v. Section 3.8). (The T comes from an alternative spelling Tschebysheff. For an interesting discourse on the spelling of the name of this Russian mathematician, see Clenshaw [16], p. 17.) These polynomials possess a number of important properties when the interval of approximation is $[-1, 1]$ (again, see Clenshaw [16]), and when working with them it is normal to transform to this standard interval. The T_is may be defined via the three-term recurrence relation

$$T_i(x) = 2x T_{i-1}(x) - T_{i-2}(x), \tag{5.6}$$

with $T_0(x) = 1$ and $T_1(x) = x$. Suppose

$$p(x) = \tfrac{1}{2} a_0 T_0(x) + a_1 T_1(x) + \cdots + a_n T_n(x). \tag{5.7}$$

Note the half associated with the first term in this representation. It is a standard convention to include this factor; its presence simplifies the mathematics later on. Now, if we let $b_{n+1} = b_{n+2} = 0$ and define

$$b_i = 2xb_{i+1} - b_{i+2} + a_i, \qquad i = n, n-1, \ldots, 0, \qquad (5.8)$$

the value of p at any point x may be determined as $p(x) = (b_0 - b_2)/2$. This is simply nested multiplication (see Section 2.7).

Relation (5.8) forms the basis of E02AEF, whose call sequence is

CALL E02AEF(NPLUS1,A,XCAP,P,IFAIL)

The NPLUS1 coefficients in (5.7) should be in A(1), A(2), ... (with A(1) = a_0, A(2) = a_1, etc.) where A is an array of size at least NPLUS1, and, on exit, P contains the value $p($XCAP$)$. (The Chebyshev coefficients may have been computed by E02ADF, a routine which determines a least squares approximation in Chebyshev form; see Section 5.4.) Assuming the points x_k to be ordered so that $x_k < x_{k+1}$, the range of interest will be $x_1 \le x \le x_m$, and before E02AEF can be used to evaluate the polynomial a transformation to $[-1, 1]$ must be made. Thus, if x is the required point of evaluation, XCAP must be given the value $\hat{x}$, where

$$\hat{x} = \big((x - x_1) - (x_m - x)\big)/(x_m - x_1), \qquad (5.9)$$

and the routine will fail, with IFAIL = 1, if XCAP lies outside this interval. The coefficients supplied to E02AEF must therefore correspond to approximation on $[-1, 1]$.

A routine which achieves a similar effect is E02AKF for which the call sequence is

CALL E02AKF(NP1,XMIN,XMAX,A,IA1,LA,X,RESULT,IFAIL)

Here X represents the actual value (between XMIN and XMAX) that is to be used, the transformation (5.9) being made within the routine. The polynomial coefficients are again provided via A, of size LA, but need not be in consecutive locations. IA1 defines a constant increment indicating where the a_is are to be found starting at A(1). With this facility, the routine may be used to evaluate a row, or column, of a two-dimensional Chebyshev polynomial expansion, that is, a polynomial of the form

$$p(x, y) = \sum_{i=0}^{n_1}{}' \sum_{j=0}^{n_2}{}' a_{ij} T_i(x) T_j(y).$$

Note that the primes on the summation signs mean that the zeroth row and column of the matrix of coefficients are each multiplied by $\frac{1}{2}$, and hence that a_{00} is multiplied by $\frac{1}{4}$. If the evaluation of such an expansion at some point (x, y) is required, E02CBF may be used. Its call sequence takes the form

```
CALL EO2CBF(MFIRST,MLAST,K,L,X,XMIN,XMAX,Y,YMIN,
+           YMAX,FF,A,NA,WORK,NWORK,IFAIL)
```

The polynomial approximation is defined by the coefficients contained in A. This is a one-dimensional array of size NA. The first $L + 1$ locations should hold the first column of the matrix of coefficients, the next $L + 1$ locations should hold the second column, and so on. K defines the order of the expansion in the x direction. The polynomial is evaluated for a given value of Y, and at x values in positions MFIRST to MLAST of the array X. The f values are returned in positions MFIRST to MLAST of FF. The parameters XMIN, XMAX, YMIN and YMAX define any prior transformation that has to be made. WORK is a workspace array of size NWORK, which must be at least $K + 1$.

Now, if $q(x) = \frac{1}{2}b_0 T_0(x) + b_1 T_1(x) + \cdots + b_{n+1} T_{n+1}(x)$ is the polynomial of degree $n + 1$ corresponding to the indefinite integral of p, and we define $a_{n+1} = a_{n+2} = 0$, then the coefficients in q are defined by

$$b_i = \frac{a_{i-1} - a_{i+1}}{2i}, \qquad i = 1, 2, \ldots, n, \tag{5.10}$$

and b_0 is arbitrary. This result follows from the identity

$$\int T_i(x)dx = \frac{1}{2}\left(\frac{T_{i+1}(x)}{i+1} - \frac{T_{i-1}(x)}{i-1}\right). \tag{5.11}$$

EO2AJF uses (5.10) to integrate a Chebyshev polynomial expansion. In the call sequence

```
CALL EO2AJF(NP1,XMIN,XMAX,A,IA1,LA,QATM1,QUAD,IQUAD1,
+           LQUAD,IFAIL)
```

the NP1 coefficients of p must be supplied via A, of size LA, and the coefficients of q are returned via QUAD of size LQUAD. These coefficients need not necessarily be stored in consecutive positions. The user is allowed to specify, via IA1 and IQUAD1, a constant increment indicating the spacing of the coefficients within the arrays A and QUAD respectively. Like EO2AKF, the routine may be used to integrate a row or column of a two-dimensional Chebyshev expansion. The two parameters XMIN and XMAX are used to define any transformation that may have taken place. The derivative of (5.9) is $2/(x_m - x_1)$ and so the b_is are multiplied by $(\text{XMAX} - \text{XMIN})/2$ to ensure that the integration is performed correctly. In terms of the standard interval, EO2AKF determines $q(x) = \int_{-1}^{x} p(z)\,dz$, and the value of b_0 is fixed by requiring that $q(-1) = \text{QATM1}$. Since $T_i(-1) = (-1)^i$ (see Exercise 5.6), this means that $\frac{1}{2}b_0 = b_1 - b_2 + b_3 - \cdots - (-1)^{n+1}b_{n+1} + \text{QATM1}$.

Now, let $r(x) = \frac{1}{2}c_0 T_0(x) + c_1 T_1(x) + \cdots + c_{n-1} T_{n-1}(x)$ be the polynomial expansion for the derivative of p. Since differentiation is just the inverse of integration, we immediately have from (5.10)

$$c_{i-1} = c_{i+1} + 2ia_i, \qquad i = n, n-1, \ldots, 1,$$

with $c_{n+1} = c_n = 0$. This relation forms the basis of E02AHF whose call sequence takes the form

```
      CALL E02AHF(NP1,XMIN,XMAX,A,IA1,LA,PATM1,ADIF,IADIF1,
    +              LADIF,IFAIL)
```

and this is similar to the call sequence for E02AJF. The parameter PATM1 gives, on exit, the value $r(-1)$. Again, the routine allows for any transformation that may have taken place.

EXERCISE 5.6   Prove that the polynomials defined by

$$T_i(x) = \cos\bigl(i\cos^{-1}(x)\bigr)$$

satisfy the three-term recurrence relation (5.6). Determine the polynomials T_0, T_1 and T_2, and confirm that these are the first three Chebyshev polynomials. Verify that the maximum absolute value of T_i on $[-1, 1]$ is 1, and that $T_i(-1) = (-1)^i$ and $T_i(1) = 1$.

EXERCISE 5.7   Derive expressions for the indefinite integrals of T_0 and T_1. Using the representation of Exercise 5.6, prove the identity (5.11). Determine an expression for the definite integral over $[-1, 1]$ of a polynomial written in Chebyshev form.

EXERCISE 5.8   Write and test subroutines which take as input the coefficients for p, a Chebyshev polynomial expansion, and produce as output the series for (i) the integral of p and (ii) the derivative of p.

EXERCISE 5.9   Prove that the Chebyshev polynomials satisfy

$$T_i T_j = \tfrac{1}{2}(T_{i+j} + T_{i-j}),$$

where $T_{-i} = T_i$. Use this identity to formulate an algorithm which determines the Chebyshev coefficients of a polynomial which is the product of two polynomials expressed in Chebyshev form.

5.4 Chebyshev approximation

E02ADF computes a weighted least squares polynomial approximation to a given set of coordinates. Forsythe's algorithm (Forsythe [30]) is used to generate orthogonal polynomials using (5.4). The result is returned not in terms of the orthogonal polynomials, nor in standard form, but as a Chebyshev series, and we now consider the advantages of expressing the approximation in such a way. Our discussion relates to the interval $[-1, 1]$ only, but an appropriate transformation can be used if the range of interest is other than this.

The first thing to note is that, compared to a standard power series form, the coefficients in a Chebyshev series expansion tend to converge more rapidly. Indeed, if we form a Chebyshev series directly from a power series, it is often possible to remove the highest-order term with little, or no, loss in accuracy (see Exercise 5.10). Moreover, since the maximum absolute value of each T_i is 1 we should experience few problems with differencing.

Now, in forming a least squares approximation using orthogonal polynomials we need to keep account of the p_is as they are formed. In fact it is sufficient to store just their values at the points x_k. Further, the generation of these polynomials by a three-term recurrence relation means that we only need to store two such polynomials at any given time. The drawback with this scheme is that 4 arrays, each of size m, are required; one each for the x_ks and f_ks, and two for the polynomial values. If m is large (and, typically, this will be the case) this represents a considerable amount of storage. Clenshaw [15] suggests that the orthogonal polynomials should be stored as coefficients in some series, and, in view of what we have just said, this should be in Chebyshev form. Now two arrays, each of size $n+1$, are required for these coefficients, and for $m \gg n$ this represents a significant saving. Whenever a polynomial evaluation is required, we simply use the technique outlined at the beginning of Section 5.3.

Let the ith orthogonal polynomial have Chebyshev expansion

$$p_i(x) = \sum_{j=0}^{i}{}' a_j^{(i)} T_j(x).$$

Set $\lambda_i = 2$ in (5.4) and define $p_0(x) = \frac{1}{2}$ (so that $a_0^{(0)} = 1$). Substituting for p_{i-1} and p_{i-2} in (5.4) we have

$$p_i(x) = 2(x - \alpha_i) \sum_{j=0}^{i-1}{}' a_j^{(i-1)} T_j(x) - \beta_{i-1} \sum_{j=0}^{i-2}{}' a_j^{(i-2)} T_j(x).$$

But $x = T_1(x)$, and $T_1 T_j = \frac{1}{2}(T_{j+1} + T_{j-1})$ (see Exercise 5.9), and so,

$$p_i(x) = \sum_{j=0}^{i-1}{}' a_j^{(i-1)} (T_{j+1}(x) - 2\alpha_i T_j(x) + T_{j-1}(x)) - \beta_{i-1} \sum_{j=0}^{i-2}{}' a_j^{(i-2)} T_j(x),$$

which means that the coefficients $a_j^{(i)}$ may be determined from

$$a_j^{(i)} = a_{j+1}^{(i-1)} + a_{j-1}^{(i-1)} - 2\alpha_i a_j^{(i-1)} - \beta_{i-1} a_j^{(i-2)}.$$

Note that $a_i^{(i)} = 1$. If we define

$$q_i(x) = \sum_{j=0}^{i}{}' A_j^{(i)} T_j(x)$$

to be the Chebyshev representation of the least squares polynomial approximation after i terms have been computed, then $q_i(x) = q_{i-1}(x) + c_i p_i(x)$, where c_i is the newly determined least squares coefficient defined by (5.5). Matching coefficients of T_i, we immediately have $A_j^{(i)} = A_j^{(i-1)} + c_i a_j^{(i)}$.

There is a further, unrelated, advantage to expressing an approximating polynomial in Chebyshev form which stems from an orthogonality property possessed by the Chebyshev polynomials. If, in the inner product (5.3), the weights are all chosen to be unity, except for w_1 and w_m which are given the value $\frac{1}{2}$, and the points x_k are chosen to be $x_k = \cos(\pi(k-1)/(m-1))$, then the T_is form an orthogonal set. If we can ensure that the x_ks are of this special form then the approximation process is greatly simplified. There is no need to generate orthogonal polynomials, we already have them; and there is no need to perform any conversions, either in the evaluation of each p_i, or in expressing the final approximation in Chebyshev form. We immediately have that the coefficients, a_j, in the Chebyshev series least squares approximation are given by $a_j = <T_j, f> / <T_j, T_j>$, and the identity

$$<T_j, T_j> = \begin{cases} (m-1)/2, & j \neq 0 \text{ or } m-1; \\ m-1, & j = 0 \text{ or } m-1; \end{cases} \qquad (5.12)$$

simplifies the computation further.

An extension to the basic method is to produce a solution which satisfies the least squares criterion, and additionally takes on prescribed function and derivative values at specified points. Let $\{ z_l \mid l = 1, 2, \ldots, q \}$ be a set of points, not necessarily coincident with the x_ks, at which we wish to impose the constraints

$$p^{(r)}(z_l) = f_l^{(r)}, \qquad r = 0, 1, \ldots, r_l - 1,$$

that is, at z_l we wish to ensure that the first $r_l - 1$ derivatives of p take on prescribed values. We begin by forming u, a polynomial of degree $\bar{n} - 1$, where $\bar{n} = \sum_{l=1}^{q} r_l$, which satisfies the given constraints. This is an interpolation problem; we have the same number of equations as unknowns. We also form v, a polynomial of degree $\bar{n}$, such that it, and its appropriate derivatives, vanish at the z_ls. This may be written

$$v(x) = (x - z_1)^{r_1}(x - z_2)^{r_2} \cdots (x - z_q)^{r_q},$$

and we convert it to Chebyshev form. We now aim to determine the coefficients in the polynomial $\bar{p}$, of degree $s - 1$ with $s = n - \bar{n}$, such that $v\bar{p}$ is the least squares approximation to the data $(x_k, f_k - u(x_k))$. Finally, we construct $p = v\bar{p} + u$, which is the solution to the constrained least squares problem.

EXERCISE 5.10 The first few terms in the Taylor series expansion of e^x about $x = 0$ can be used to form a polynomial approximation. Let

$$p(x) = 1 + x + \frac{x^2}{2!} + \frac{x^3}{3!} + \frac{x^4}{4!}$$

be such an approximation. Express p as a linear combination of Chebyshev polynomials and compare the convergence of the coefficients in each expansion. If the approximation is to be used to determine an estimate of e^x for $-1 \leq x \leq 1$, what would be the effect of ignoring the x^4 term in the power series form, and the T_4 term in the Chebyshev form?

5.5   The NAG least squares approximation routines

We choose to distinguish between those routines in the NAG library which return a least squares approximation in the form of a spline, and those which return a Chebyshev expansion. Spline approximation will be considered in the next section, and for the moment we concentrate on E02ADF, E02AGF, and E02AFF.

Clenshaw's modification to Forsythe's original algorithm has been incorporated into E02ADF, the call sequence for this routine being

```
        CALL E02ADF(M,KPLUS1,NROWS,X,Y,W,WORK1,WORK2,
    +                A,S,IFAIL)
```

The M coordinates should be supplied via X and Y, with the weights stored in W. The x_ks must be supplied in non-decreasing order. The computed coefficients are returned through the two-dimensional array A, which should have NROWS rows, and at least KPLUS1 columns. Since the polynomial approximation is built up term by term, E02ADF is able to return the intermediate results. On exit the first row of A will contain the Chebyshev series for the least squares constant function approximation, in the second row will be the least squares straight line approximation, and so on. The corresponding square roots of the mean square residuals are returned in S; formally, we have that $S(i)$ is the square root of the weighted sum of the squares of the residuals divided by $M - i$. The maximum degree that the routine goes up to is $k =$ KPLUS1 $- 1$. Note that if M = KPLUS1 the resulting approximation will interpolate the data; this must be true since m coefficients can certainly be chosen so that the interpolation conditions $p(x_k) = f_k$ for $k = 1, 2, \ldots, m$ are satisfied, and this must give the minimum sum of the squares of the residuals (namely zero). The workspace array WORK2, with 2 rows and at least KPLUS1 columns, is used to store the coefficients of two consecutive orthogonal polynomials written in Chebyshev form. The other workspace array, WORK1, must have 3 rows and at least M columns. The first row is used to store the weighted residuals, the second to store the x_ks, transformed to $[-1, 1]$, and the third to store the values at these points of the most recently computed orthogonal polynomial. This ensures efficient evaluation of the inner products.

E02AGF computes a least squares approximation to function values and allows constraints to be imposed. The call sequence for this routine, namely

```
    CALL E02AGF(M,KPLUS1,NROWS,XMIN,XMAX,X,Y,W,
  +             MF,XF,YF,LYF,IP,A,S,NP1,WRK,LWRK,
  +             IWRK,LIWRK,IFAIL)
```

is long, and care must be exercised if the routine is to be used successfully. M, KPLUS1, NROWS, X, Y, W, A and S have the same role as in E02ADF. The solution is again returned in Chebyshev form, XMIN and XMAX defining the transformation to $[-1,1]$. The routine allows constraints to be imposed at MF points supplied in XF. Normally XMIN and XMAX should be given the minimum and maximum values held in the two arrays X and XF. The number of derivative constraints to be imposed at each point is supplied in IP, and the values defining the constraints in the vector YF, of size LYF. The first $IP(1) + 1$ positions of YF should contain specified values of the function, its first derivative, second derivative, ..., $IP(1)^{th}$ derivative at XF(1). The next block of $IP(2) + 1$ values should contain information at XF(2), and so on.

Although included in the NAG curve fitting chapter, the prime purpose of E02AFF is to produce an interpolating polynomial approximation. It assumes that the f_ks, supplied via the array F, of size at least NPLUS1, in the call

```
    CALL E02AFF(NPLUS1,F,A,IFAIL)
```

are function values at the points

$$(x_m - x_1)\cos\big(\pi(k - 1)/(m - 1)\big)/2 + (x_m + x_1)/2,$$

where x_1 and x_m define the user's range of interest, and $m = $ NPLUS1. The choice of points ensures that, when appropriate weights are taken, the Chebyshev polynomials form an orthogonal set. Hence, the coefficients returned via A, also of size NPLUS1, can be computed immediately. To obtain a least squares approximation of degree n, say, with $n < m - 1$, we simply take the first $n + 1$ coefficients of A.

The simplicity of the call sequence of E02AFF, and the ease with which the polynomial approximation is computed, make this a very attractive routine to use. However, the choice of points at which function values are supplied is crucial. If f is known as a continuous function this presents no problem, but the user may have no choice over the selection of the coordinates to which an approximation is required. Moreover he has no opportunity to choose his own weights to reflect confidence, or lack of it, in certain of the values. E02ADF offers greater flexibility, but this is offset by the cost of the generation of orthogonal polynomials, and the conversion of these, and the final solution, to Chebyshev form. E02AGF offers yet further flexibility, but it should only be used in those, relatively rare, cases which it is designed to cater for, namely where derivative, as well as function, values are available.

Having produced a least squares approximation in Chebyshev form, we may now evaluate, integrate, or differentiate it using the routines discussed in Section 5.3.

EXERCISE 5.11   Show that

$$\sum_{k=2}^{m} \cos\big((k-1)x\big) \;=\; \frac{1}{2}\left(\frac{\sin\big((m-\tfrac{1}{2})x\big)}{\sin(\tfrac{x}{2})} - 1 \right),$$

$$\sum_{k=1}^{m}{}'' \cos\big((k-1)x\big) \;=\; \frac{\sin\big((m-1)x\big)\cos(\tfrac{x}{2})}{2\sin(\tfrac{x}{2})},$$

where the double prime on the second summation sign means that the first and last terms are to be halved. Use these results to prove the discrete orthogonality property of the Chebyshev polynomials.

5.6   Spline approximation

We next consider a rather different way of representing an approximation to a given set of coordinates. Instead of looking for an overall polynomial approximation possessing continuity of all derivatives, we restrict the order to 3 and seek a *piecewise polynomial approximation*. The range of interest $[x_1, x_m]$ is divided into a number of subranges by the introduction of the points $\{ z_i \mid i = 1, 2, \ldots, n \}$ (which we assume to be ordered), with $z_1 = x_1$ and $z_n = x_m$. The remaining points z_2, z_3, $\ldots$, z_{n-1}, interior to $[x_1, x_m]$, are called (interior) *knots*, and each may, or may not, correspond to an x_k. Within the subinterval $[z_i, z_{i+1}]$ we seek a cubic polynomial approximation, but ensure continuity of adjacent cubics, and their first two derivatives, at the knots. The advantage of such a piecewise approximation is that if we know in advance that there are certain areas where f is rapidly varying, and others where f is more smooth, we can concentrate the knots accordingly. A disadvantage is that the approximating cubic does not possess continuity of its third derivative. The coefficients defining our piecewise cubic will have been obtained, for example, from the solution of an appropriate least squares problem. For the moment we simply assume the existence of such a solution and consider its manipulation. The calculation of the solution we leave to the next section.

Within each subinterval the cubic may be expressed in standard, power series, form. An alternative is to represent it in *B-spline* form. To ease the mathematics we assume that the z_is are equally spaced, distance

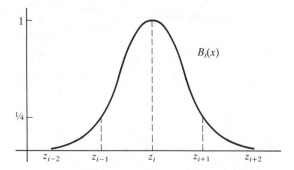

Figure 5.2   Cubic B-spline.

h apart. We define

$$B_i(x) = \frac{1}{4h^3} \begin{cases} (x - z_{i-2})^3, & z_{i-2} \leq x \leq z_{i-1}; \\[2mm] \begin{aligned} &-3(x - z_{i-1})^3 + 3h(x - z_{i-1})^2 \\ &\quad +3h^2(x - z_{i-1}) + h^3, \end{aligned} & z_{i-1} \leq x \leq z_i; \\[2mm] \begin{aligned} &-3(z_{i+1} - x)^3 + 3h(z_{i+1} - x)^2 \\ &\quad +3h^2(z_{i+1} - x) + h^3, \end{aligned} & z_i \leq x \leq z_{i+1}; \\[2mm] (z_{i+2} - x)^3, & z_{i+1} \leq x \leq z_{i+2}; \\[2mm] 0, & x < z_{i-2} \text{ or } x > z_{i+2}, \end{cases}$$

$$(5.13)$$

and note that B_i is a cubic polynomial in $[z_{i-2}, z_{i+2}]$ with continuous derivatives up to second order, and values $B_i(z_{i-2}) = B_i(z_{i+2}) = 0$, $B_i(z_{i-1}) = B_i(z_{i+1}) = \frac{1}{4}$, and $B_i(z_i) = 1$. We obtain the bell-shaped function shown in Figure 5.2. Since B_i is zero outside the interval $[z_{i-2}, z_{i+2}]$, we say that it has *local support*.

Now, the B_is form a basis for the space of cubic splines (see Powell [64], p. 232). By this we mean that there exist constants c_i such that, if p is our spline approximation,

$$p(x) = \sum_{i=0}^{n+1} c_i B_i(x).$$

Note that this representation requires the introduction of the additional points $z_{-2}, z_{-1}, z_0, z_{n+1}, z_{n+2}$ and z_{n+3}, and that $B_i(x) = 0$ for $x \in [z_1, z_n]$ if $i < 0$ or $i > n+1$.

To evaluate p for some chosen value of x we first need to locate the interval $[z_i, z_{i+1}]$ within which the point lies. The obvious way of doing this is to perform the tests $x \geq z_1$ (which must be made to ensure that the evaluation can be performed), $x \geq z_2$, and so on. On average x will be found in the middle of the range $[z_1, z_n]$, implying that the operation count

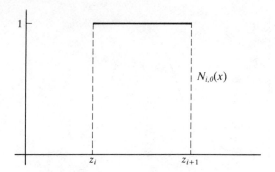

Figure 5.3   Constant B-spline.

of this *direct search* technique is approximately $\frac{1}{2}n$ comparisons. Suppose, instead, we first test x to see whether it lies to the left or right of the middle value z_{mid}. The original sequence has now been split into two halves, one of which may be discarded from further consideration. We now test x against the middle value of the half in which it is known to lie (assuming a preliminary check to ensure that it is within $[z_1, z_n]$). Further z_is are now discarded, and the process continues until the required subinterval is located. If we assume that $n = 2^l$, then the number of comparisons made will be $l = \log_2 n$, and so the operation count of this (the *binary search*) method is $O(\ln n)$. This can represent a significant saving compared with the $O(n)$ count of the direct search method. Having located the correct interval, we next observe that only the B-splines B_{i-1}, B_i, B_{i+1}, and B_{i+2} are non-zero in this range, so that

$$p(x) = c_{i-1}B_{i-1}(x) + c_i B_i(x) + c_{i+1}B_{i+1}(x) + c_{i+2}B_{i+2}(x).$$

We now consider ways of computing the values of the B-splines in this expression efficiently.

The notation we have used so far means that B_i is centred on z_i. We now change this and generate cubic B-splines, $N_{i,3}$, which are non-zero on $[z_i, z_{i+3}]$. We relax the constraint that the z_is be equally spaced, and define $h_i = z_{i+1} - z_i$. We introduce the *constant B-spline*

$$N_{i,0}(x) = \begin{cases} 1, & z_i \le x < z_{i+1}; \\ 0, & \text{otherwise} \end{cases}$$

(see Figure 5.3), and the *linear B-spline*

$$N_{i,1}(x) = \begin{cases} (x - z_i)/h_i, & z_i \le x \le z_{i+1}; \\ (z_{i+2} - x)/h_{i+1}, & z_{i+1} \le x \le z_{i+2}; \\ 0, & x \le z_i \text{ or } x \ge z_{i+2} \end{cases}$$

(see Figure 5.4).

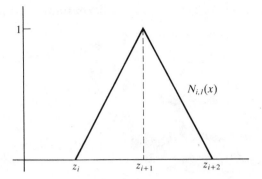

Figure 5.4   Linear B-spline.

It can be readily seen that

$$N_{i,1}(x) = \frac{(x - z_i)}{h_i} N_{i,0}(x) + \frac{(z_{i+2} - x)}{h_{i+1}} N_{i+1,0}(x). \tag{5.14}$$

In fact if $N_{i,j}$ denotes the spline of order j, non-zero on $[z_i, z_{i+j+1}]$, then (5.14) generalizes to

$$N_{i,j}(x) = \frac{(x - z_i)}{z_{i+j} - z_i} N_{i,j-1}(x) + \frac{(z_{i+j+1} - x)}{z_{i+j+1} - z_{i+1}} N_{i+1,j-1}(x), \tag{5.15}$$

(see de Boor [24]), and this recurrence relation can be used to evaluate the $N_{i,j}$s. In particular, we observe that, in the case of equally spaced knots, $N_{i,3}$ has the values 0, $\frac{1}{6}$, $\frac{2}{3}$, $\frac{1}{6}$ and 0 at the points z_i, z_{i+1}, z_{i+2}, z_{i+3} and z_{i+4} respectively, in each case two-thirds of that of B_{i+2}. Further, regardless of the relative spacing of the knots, if, for fixed j, we sum over i the values of $N_{i,j}$ at some point x, we obtain 1 (Powell [64], p. 242), and the $N_{i,j}$s are therefore referred to as *normalized splines*.

An alternative, and equivalent, scheme for evaluating a spline approximation is derived as follows. We consider

$$p(x) = \sum_{i=-2}^{n-1} d_i N_{i,3}(x)$$

and assume that the evaluation point, x, lies in $[z_k, z_{k+1}]$. From (5.15) we have

$$\begin{aligned}
p(x) &= \sum_{i=-2}^{n-1} d_i \left(\frac{x - z_i}{z_{i+3} - z_i} N_{i,2}(x) + \frac{z_{i+4} - x}{z_{i+4} - z_{i+1}} N_{i+1,2}(x) \right) \\
&= \sum_{i=-1}^{n-1} d_i^{(1)} N_{i,2}(x),
\end{aligned}$$

(noting that $N_{-2,2}(x) = 0$ on $[z_1, z_n]$) and, continuing the process,

$$p(x) = \sum_{i=0}^{n-1} d_i^{(2)} N_{i,1}(x) = \sum_{i=1}^{n-1} d_i^{(3)} N_{i,0}(x),$$

where

$$d_i^{(j)} = \begin{cases} d_i, & j = 0; \\ (x - z_i)d_i^{(j-1)}/(z_{i+4-j} - z_i) & \\ \quad +(z_{i+4-j} - x)d_{i-1}^{(j-1)}/(z_{i+4-j} - z_i), & j > 0. \end{cases} \tag{5.16}$$

Now, for $x \in [z_k, z_{k+1}]$, $N_{k,0}(x) = 1$ and $N_{i,0}(x) = 0$ for $i \neq k$. Hence, to evaluate p we determine sufficient of the $d_i^{(j)}$ s using the recurrence relation (5.16) to enable $d_k^{(3)}$ to be computed.

By simple manipulation (de Boor [24]) it may be shown that

$$N_{i,3}'(x) = -3\left(\frac{1}{z_{i+4} - z_{i+1}} N_{i+1,2}(x) - \frac{1}{z_{i+3} - z_i} N_{i,2}(x)\right). \tag{5.17}$$

Hence

$$p'(x) = \sum_{i=-1}^{n-1} D_i N_{i,2}(x),$$

where

$$D_i = 3\frac{d_i - d_{i-1}}{z_{i+3} - z_i}, \tag{5.18}$$

and so $p'(x)$ may be evaluated for any value of x using the techniques outlined above. The evaluation of second and third derivatives follows in a similar manner.

The recurrence relation (5.16) is used in E02BBF to evaluate a cubic spline approximation at a given point. In the call sequence

CALL E02BBF(NCAP7,K,C,X,S,IFAIL)

the NCAP7 coefficients in a spline expansion are supplied via C and the corresponding knots via the real/double-precision array K. In terms of the notation used elsewhere in this section, NCAP7 must be equal to $n + 6$ (i.e., $n - 1 + 7$, where $n - 1$ is the number of subintervals in the range $[z_1, z_n]$). Consequently X(1) should be equal to the introduced point z_{-2}, X(2) should correspond to z_{-1}, and so on. The arrangement for the coefficients is that C(1) should contain the coefficient of the spline centred on z_0 ($N_{-2,3}$), C(2) the coefficient for $N_{-1,3}$, and so on. X is used as the evaluation point and, on successful exit, S contains the value $p(X)$.

E02BCF evaluates a cubic spline and its first three derivatives at a point. Its call sequence is

CALL E02BCF(NCAP7,K,C,X,LEFT,S,IFAIL)

We observe that the derivative of a linear spline, and hence the third derivative of a cubic spline, is discontinuous at the knots. Further, it is possible to use a number of coincident knots, and at such points lower-order derivatives (including zeroth order) may be discontinuous. In such circumstances we have to talk about the limit, as x tends to z_k, of the appropriate derivative. Since we will obtain a different result if we approach from the left or from the right, we must indicate which we mean. The integer parameter LEFT is used to indicate this. A value of 1 means that discontinuous derivatives are determined from the left; any other value will result in them being determined as the limit from the right. On exit the array S, of length at least 4, gives the computed derivative values. $S(1)$ contains $p(X)$, $S(2)$ contains $p'(X)$, and so on.

Cox [21] has shown that

$$\int_{z_1}^{z_n} p(x)\, dx = \tfrac{1}{4} \sum_{i=-2}^{n-1} (z_i - z_{i+4}) d_i$$

and this formula is used in E02BDF, for which the call sequence is

CALL E02BDF(NCAP7,K,C,DEFINT,IFAIL)

On successful exit, DEFINT contains the required result.

In Section 5.8 we consider the fitting of a bicubic spline to two-dimensional data. It is natural to express this approximation as a product of cubic B-splines, and hence to evaluate it we simply have to evaluate the individual B-splines in the manner indicated above. E02DBF performs this operation. Its call sequence is

CALL E02DBF(M,PX,X,Y,FF,LAMDA,MU,POINT,NPOINT,
+   C,NC,IFAIL)

The routine returns M values of $f(x, y)$ in FF at the points contained in X and Y. The ordering of these values requires careful description, and we leave this until Section 5.8 where the construction of a two-dimensional spline approximation using E02DAF is discussed in detail.

EXERCISE 5.12 Draw a normalized quadratic B-spline for the case of equally spaced points and determine its equation.

EXERCISE 5.13 Derive a recurrence relation for the jth derivative of a kth degree normalized B-spline.

EXERCISE 5.14 Show that $\sum_i N_{i,1}(x) = 1$.

EXERCISE 5.15 Show that the third derivative of the cubic B-spline (5.13) is discontinuous.

5.7   Least squares spline approximation

We return to the problem of weighted least squares approximation and express the approximating polynomial as a linear combination of cubic B-splines. For convenience we shall express this as

$$p(x) = \sum_{i=0}^{n+1} a_i B_i(x),$$

where the B_is are defined by (5.13), but observe that in practice we would work with normalized splines. The aim is to determine those values of the a_is which minimize the sum of the squares of the weighted residuals r_k, where

$$r_k = w_k\big(f_k - p(x_k)\big), \qquad k = 1, 2, \ldots, m.$$

From this overdetermined system we form the normal equations

$$X^T X \mathbf{a} = X^T \mathbf{F},$$

where X now has elements

$$X_{ij} = w_i B_j(x_i), \qquad i = 1, 2, \ldots, m, \quad j = 0, 1, \ldots, n+1 \qquad (5.19)$$

and $\mathbf{F}$ has elements

$$F_i = w_i f_i, \qquad i = 1, 2, \ldots, m.$$

We observe that X is sparse; if we look at the ith row of X and assume that x_i is in $[z_k, z_{k+1}]$, then the four values $B_{k-1}(x_i)$, $B_k(x_i)$, $B_{k+1}(x_i)$, and $B_{k+2}(x_i)$ only are non-zero, which means that X has at most 4 adjacent non-zeros in any row. If x_i corresponds to a knot then only three entries will be non-zero.

Suppose that we can factor X as $X = QR$, where Q has rows 1 to m and columns 0 to $n+1$, and R is square with rows and columns 0 to $n+1$. Then,

$$X^T X = (QR)^T (QR) = R^T Q^T Q R.$$

If we can arrange that the columns of Q are orthonormal, that is, $Q^T Q = I_m$, where I_m is the identity matrix of size $m \times m$, then the normal equations become

$$R^T R \mathbf{a} = R^T Q^T \mathbf{F},$$

which means that we must have

$$R \mathbf{a} = Q^T \mathbf{F}. \qquad (5.20)$$

We now try to determine a sequence of orthonormal transformations to X which produces a matrix R of upper triangular form. If, at the same time, we modify the right-hand side vector and form $\tilde{\mathbf{F}} = Q^T \mathbf{F}$, the coefficients defining the least squares approximation can be computed directly using backward substitution, which here takes the form

$$a_{n+1} = \tilde{F}_{n+1}/R_{n+1,n+1},$$

$$a_i = \left(\tilde{F}_i - \sum_{j=i+1}^{n+1} a_j R_{ij}\right) \Bigg/ R_{ii}, \qquad i = n, n-1, \ldots, 0.$$

In fact, we shall see that the summation is over at most 3 elements.

The starting point is the use of *Givens plane rotations*. For convenience we regard Q as being of size $m \times m$, and R as being of size $m \times 0 : (n+1)$. Suppose we construct the matrix $Q^{(pq)}$ defined by

$$Q_{ij}^{(pq)} = \begin{cases} 1, & i = j \text{ and } i \neq p \text{ or } q; \\ \sin(\theta), & i = j = p \text{ or } q; \\ \cos(\theta), & i = p, \quad j = q; \\ -\cos(\theta), & i = q, \quad j = p; \\ 0, & \text{otherwise.} \end{cases} \qquad (5.21)$$

It can be readily seen that $Q^{(pq)}$ is orthonormal. Now, if we set $\theta = \tan^{-1}(X_{pp}/X_{qp})$, and form the product $\tilde{X} = Q^{(pq)}X$, we find that $\tilde{X}_{qp} = 0$ (see Exercise 5.17).

We can avoid evaluating the arctangent, sine, and cosine functions by observing that $\sin(\theta) = X_{pp}\Big/\sqrt{X_{pp}^2 + X_{qp}^2}$ and $\cos(\theta) = X_{qp}\Big/\sqrt{X_{pp}^2 + X_{qp}^2}$. However, we now have to form square roots, and this itself can be a relatively expensive process. The so-called *square root free Givens* process of Gentleman [34] can be used to reduce the operation count further. See also Lawson and Hanson [53].

We generate R in the following manner:

- The first row of R is formed as the first row of X. There are at most three non-zero entries only (see below); each involves the evaluation of a B-spline at x_1.

- For each subsequent row of R we form the corresponding row of X in the same manner. Each row will have at most 4 non-zero entries. Conceptually, we view this as being added into R, and then Givens plane rotations are used to zero elements in this row below the diagonal (and, in the later stages of the algorithm, this means all of them). All rotations made to the coefficient matrix must be made to the vector of right-hand sides as well.

Note that we avoid the need to store X explicitly since we generate it a row at a time. This row may be stored in a one-dimensional array of size 4. The only two-dimensional array we require is R. An array of size $0 : (n+1) \times 4$ will suffice. The following illustrates the way the method works.

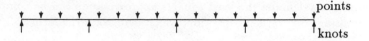

Figure 5.5   Example of knots and data points.

Suppose we have 16 data points and place knots between points 4 and 5, 12 and 13, and at 9 (see Figure 5.5). In our overdetermined system of equations we will have 7 unknowns, namely a_0, a_1, ..., a_6. X will be an array with 16 rows and 7 columns. In an attempt to avoid confusion, we regard rows and columns as being numbered from 1. The pattern of non-zero elements in X is as shown in Figure 5.6. The first row corresponds to $B_0(x_1)$, $B_1(x_1)$ and $B_2(x_1)$. All other B-splines are zero at this point. In particular, since x_1 corresponds to an 'external' knot, $B_3(x_1)$ is zero. This row forms the first row of R. We now form the second row of X, (consisting of B_0, B_1, B_2 and B_3 evaluated at x_2) and, using a rotation involving rows 1 and 2, zero entry X_{21}. X_{14} will become non-zero, but all other zeros will remain unchanged. The third row of X is now formed, and X_{31} zeroed using a rotation involving rows 1 and 3. X_{32} is then zeroed using row 3 and the new row 2. Row 4 is dealt with in a similar manner. Row 5 already has a leading zero and so we proceed immediately to the second element. Forcing this to be zero will make X_{25} non-zero. The third and fourth elements of the new row are then made zero. We continue in this manner with rows 6 and 7, after which the pattern of non-zeros is as shown in Figure 5.7. The first 7 rows of the matrix are of upper triangular form, and can be used to determine a solution using the first 7 data points only. This is, of course, an interpolation process. To obtain the required least squares solution it is necessary to add the rest of the points in. We do this by including the remaining rows of X and performing rotations with the first 7 rows which ensure that all entries in each new row are zeroed. This will necessarily alter the values held in the first 7 rows. At the end of this process we solve for the B-spline coefficients using backward substitution as indicated above. Note, however, that in any row at most three entries to the right of the diagonal will be non-zero.

The process of B-spline approximation using Givens reductions is implemented in E02BAF, for which a typical call takes the form

```
CALL E02BAF(M,NCAP7,X,Y,W,K,WORK1,WORK2,C,SS,IFAIL)
```

The M data points and weights should be stored, before entry, in the arrays X, Y, and W. NCAP7 is used to define the number of terms in the expansion. If $n - 2$ interior knots are to be used then NCAP7 should be given the value $n + 6$ (i.e., $n - 1 + 7$, where $n - 1$ is the number of panels that the data is split into by the knots). The interior knots should be specified via the real/double-precision array K, starting at position 5. In terms of the

$$
\begin{array}{ccccccc}
x & x & x & 0 & 0 & 0 & 0 \\
x & x & x & x & 0 & 0 & 0 \\
x & x & x & x & 0 & 0 & 0 \\
x & x & x & x & 0 & 0 & 0 \\
0 & x & x & x & x & 0 & 0 \\
0 & x & x & x & x & 0 & 0 \\
0 & x & x & x & x & 0 & 0 \\
0 & 0 & x & x & x & 0 & 0 \\
0 & 0 & x & x & x & x & 0 \\
0 & 0 & x & x & x & x & 0 \\
0 & 0 & x & x & x & x & 0 \\
0 & 0 & 0 & x & x & x & x \\
0 & 0 & 0 & x & x & x & x \\
0 & 0 & 0 & x & x & x & x \\
0 & 0 & 0 & 0 & x & x & x \\
\end{array}
$$

Figure 5.6   Initial pattern of zeros.

notation used earlier, z_2 should be placed in K(5), z_3 in K(6), and so on. The two exterior knots, x_1 and x_m, will, on exit, be found in K(4) and K($n+3$), respectively. The introduced knots are, on exit, in the first three and last three positions of K (assuming its length to be NCAP7). Note that the routine uses coincident knots at x_1 and x_m with obvious repercussions regarding the continuity of the solution and its derivatives at these points. This choice also means that the off-diagonal elements in the first and last rows of the matrix X defined by (5.19) are zero. The solution coefficients are returned via C, the coefficient of $N_{-2,3}$ being in C(1), that of $N_{-1,3}$ in C(2), and so on. SS will contain the sum of the squares of the residuals. The two workspace arrays, WORK1 and WORK2, dimensioned at least M and $(4, p(\geq \text{NCAP7}))$ respectively, are used to store a newly introduced row of X, and R. Note that the routine uses an array with 4 rows, as opposed to columns, for the latter.

EXERCISE 5.16   A cubic B-spline approximation to seven data points is sought. If a knot is placed between x_3 and x_4, indicate the structure of the coefficient matrix in the corresponding overdetermined system of equations.

EXERCISE 5.17   Show that $Q^{(pq)}$ as defined by (5.21) is orthonormal, that premultiplication of X by $Q^{(pq)}$ leaves all the elements unchanged except those in rows p and q, and that the $(q,p)^{th}$ element of the product is zero.

5.8 Surface fitting

The ideas of previous sections may be readily extended to provide surface fitting algorithms. Here we have two independent variables, x and y, and for each coordinate pair (x_k, y_l) a function value, f_{kl}, is known. If a polynomial least squares approximation is sought, we look for coefficients in a

```
x  x  x  x  0  0  0
0  x  x  x  x  0  0
0  0  x  x  x  x  0
0  0  0  x  x  x  x
0  0  0  0  x  x  x
0  0  0  0  0  x  x
0  0  0  0  0  0  x
0  x  x  x  x  0  0
0  0  x  x  x  0  0
0  0  x  x  x  x  0
0  0  x  x  x  x  0
0  0  x  x  x  x  0
0  0  x  x  x  x  0
0  0  0  x  x  x  x
0  0  0  x  x  x  x
0  0  0  x  x  x  x
0  0  0  0  x  x  x
```

Figure 5.7   Pattern of zeros after first stage of transformation.

bi-variate polynomial expansion $p(x, y)$ which minimizes a weighted sum of the squares of the residuals

$$S = \sum_{k=1}^{m_1} \sum_{l=1}^{m_2} w_{kl} r_{kl}^2 = \sum_{k=1}^{m_1} \sum_{l=1}^{m_2} w_{kl} \left(f_{kl} - p(x_k, y_l)\right)^2. \tag{5.22}$$

We will again find it convenient to express the approximating polynomial, and orthogonal polynomials used in its determination, in Chebyshev form, which means that appropriate transformations will need to be made to ensure that the domain of definition in each direction is $[-1, 1]$.

The process of determining the optimal coefficients is straightforward if the data points lie on a regular rectangular mesh. We consider the unweighted case and begin by writing

$$p(x, y) = \sum_{i=0}^{n_1} \sum_{j=0}^{n_2} a_{ij} p_i(x) q_j(y),$$

where the p_is and q_js are orthogonal on the respective sets of grid points. Now, we wish to minimize

$$S = \sum_{k=1}^{m_1} \sum_{l=1}^{m_2} \left(f_{kl} - \sum_{i=0}^{n_1} \sum_{j=0}^{n_2} a_{ij} p_i(x_k) q_j(y_l)\right)^2$$

with respect to the a_{ij}s. This problem has a solution when

$$0 = \frac{\partial S}{\partial a_{rs}} = -2 \sum_{k=1}^{m_1} \sum_{l=1}^{m_2} \left(f_{kl} - \sum_{i=0}^{n_1} \sum_{j=0}^{n_2} a_{ij} p_i(x_k) q_j(y_l)\right) p_r(x_k) q_s(y_l).$$

Re-ordering the summation signs and making use of the orthogonality of the p_is and q_js, we immediately have

$$a_{ij} = \sum_{k=1}^{m_1}\sum_{l=1}^{m_2} f_{kl}p_i(x_k)q_j(y_l) \bigg/ \left(\sum_{k=1}^{m_1} p_i^2(x_k) \sum_{l=1}^{m_2} q_j^2(y_l) \right).$$

However, there is another way of approaching the problem. Suppose that for each line of constant y we find coefficients $c_i^{(l)}$ which minimize

$$S_l = \sum_{k=1}^{m_1} \left(f_{kl} - \sum_{i=0}^{n_1} c_i^{(l)}p_i(x_k) \right)^2$$

and then find coefficients $\tilde{a}_{ij}$ which minimize

$$\tilde{S} = \sum_{l=1}^{m_1} \left(c_i^{(l)} - \sum_{j=0}^{n_2} \tilde{a}_{ij}q_j(y_l) \right)^2.$$

Then the minimum of S_l occurs when

$$c_i^{(l)} = \sum_{k=1}^{m_1} f_{kl}p_i(x_k) \bigg/ \sum_{k=1}^{m_1} p_i^2(x_k),$$

whilst the minimum of $\tilde{S}$ occurs when

$$\tilde{a}_{ij} = \sum_{l=1}^{m_2} c_i^{(l)}q_j(y_l) \bigg/ \sum_{l=1}^{m_2} q_j^2(y_l) = a_{ij}.$$

If the x ordinates at which function values are known lie on lines of constant y, but are not the same, either in number or position, along each line, then the alternative approach outlined above may still be used. We introduce functions which approximate the x boundaries of the region (smooth curves which approximately fit the smallest and largest x values along each line), and these are used to define the transformation of each line to the standard interval $[-1, 1]$ (see Figure 5.8). Note that no data value should lie outside these limits. The solution we obtain will, in general, be only an approximation to the required solution, namely that which gives the minimum sum of the squares of the residuals. In principle, this should not worry us unduly, since the least squares criterion for defining a solution to a curve or surface fitting problem is but one from a multitude that can be used. However, there are dangers arising from this (see Clenshaw and Hayes [17]).

An extension to the basic method is to incorporate into the approximation known polynomial factors in x and y. Now we seek an approximation

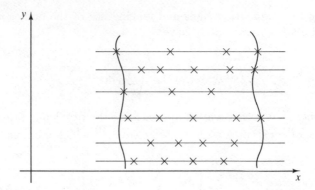

Figure 5.8   Grid of observations and smooth boundaries.

of the form $q(x,y) = u(x)v(y)p(x,y)$, where u and v are known polynomials. The motivation for including these factors is that the least squares approximation can be forced to take on specified values on lines of constant x or y. The process is equivalent to the extension of the one-dimensional least squares process outlined at the end of Section 5.4. As a simple example suppose that $y_1 = 0$ and that $f_{k1} = 0$ for all k. Whilst the least squares solution can be expected to approximate these values closely (and we can use weights to influence this), the fit will not be exact. If, however, we choose as factors $u(x) = 1$ and $v(y) = y$, we must have $q(x,0) = 0$, and the constraint is satisfied. If $f_{k1} = c$ we use the same factors but fit the curve to the values $f_{kl} - c$; when evaluating the polynomial approximation we must then remember to add c back on. For further details of this technique, including the specification of derivatives, see Clenshaw and Hayes [17].

The above surface fitting algorithm is implemented in E02CAF, for which a typical call takes the form

```
CALL E02CAF(M,N,K,L,X,Y,F,W,NX,A,NA,XMIN,XMAX,
+             NUX,INUXP1,NUY,INUYP1,WORK,NWORK,IFAIL)
```

It can be seen that this routine has a large number of parameters, and its use requires care. Further, although we are dealing with a two-dimensional problem, all data must be packed into one-dimensional arrays. N indicates the number of lines of constant y at which data values are known, and these lines are defined by the array Y, of length N. The number of values along each line is indicated by the array parameter M, and the corresponding x values must be supplied in X, of size $NX(\geq M(1) + M(2) + \cdots + M(N))$. The $M(1)$ x values on $y = Y(1)$ should be located in positions 1 to $M(1)$ of X, the $M(2)$ x values on $y = Y(2)$ in positions $M(1) + 1$ to $M(1) + M(2)$, and so on. The known function values, and associated weights, should be supplied via the one-dimensional array parameters F and W respectively, in the same order as the x values. The smooth curves which define region borders corresponding to minimum and maximum values of x along each line are defined pointwise via the array parameters XMIN and XMAX. These values may be determined by eye, or by using a one-dimensional curve fitting routine, adding a suitable small amount so that all observations lie within these curves, and then evaluating them at the specified y values. K and L give the expansion degree in x and y required by the user. The polynomial coefficients are returned via A, of size NA which must be at least $(K + 1) \times (L + 1)$. The first $L + 1$ terms give the first row of coefficients, the next $L + 1$ terms the second row, and so on. The array NUX, of size INUXP1, defines u, the polynomial factor in x which the solution is required to have. Note that this needs to be specified in Chebyshev form. NUY, of size INUYP1, is similarly used to represent the polynomial factor v. Despite their names, these are real/double-precision arrays, as would be expected. The workspace array WORK should be of size NWORK; this must be at least the sum of

(1) the sum of the number of x values along each line of constant y;

(2) twice the maximum of the maximum number of x values along a line and the number of lines;

(3) twice the product of the number of lines of constant y and $K + 2$ (the expansion size in x plus 2);

(4) five times the sum of (1) and the maximum of the expansion sizes in x and y.

If the user has control over where to place the points, clustering them towards the ends of the lines of constant y is likely to give improved results. If possible, they should be chosen to be the zeros or extrema of Chebyshev polynomials, transformed appropriately.

Fitting a spline to a two-dimensional set of data values is also a straightforward extension of the one-dimensional case. A rectangular grid is defined using knots in the x and y directions and then we seek an approximation of the form

$$p(x, y) = \sum_{i=0}^{n_1+1} \sum_{j=0}^{n_2+1} a_{ij} B_i(x) B_j(y),$$

(where the B_i and B_j are cubic B-splines) which minimizes the sum of the squares of the residuals (5.22). The tricky part is ensuring that the data points are grouped so that the sparsity pattern existing in the original matrix X in the overdetermined system is not destroyed. At any data point there are at most four x and four y B-splines which are non-zero. Since we are forming the product, this means that at most 16 terms in any row of the matrix are non-zero. The matrix is to be reduced to upper triangular form using Givens plane rotations, and if the data is ordered

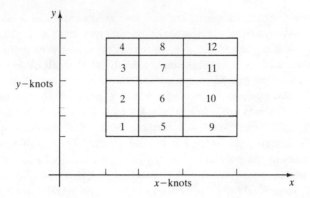

Figure 5.9   Panels defining the ordering of data points.

haphazardly many of the entries above the diagonal will become non-zero. However, if we arrange that X consists of blocks, each block of which has the sparsity pattern exhibited in the one-dimensional case, and the pattern of non-zero blocks has the same pattern, then we can ensure that, even after plane rotations have been applied, each row will have no more than 16 non-zero entries, and these will be grouped into 4 identifiable blocks of 4. For this pattern to be achieved we regard the data as being split into panels by the knots (see Figure 5.9). We order the panels in y ordering, that is, the first y column, then the second y column, and so on. The ordering within each panel is unimportant.

E02DAF forms a least squares bicubic spline approximation to a given set of data points. Its call sequence is

```
CALL E02DAF(M,PX,PY,X,Y,F,W,LAMDA,MU,POINT,NPOINT,DL,
+           C,NC,WS,NWS,EPS,SIGMA,RANK,IFAIL)
```

Again we have a large number of parameters and any call requires care in setting up the input values. M gives the total number of data points, defined in any consistent ordering by the arrays X, Y and F; the associated weights are supplied via W. Note that we have said that any ordering may be used; the ordering required to ensure the required sparsity pattern of the coefficient matrix is defined, on input, via the integer array POINT. Essentially, this array contains index pointers to the X, Y, F, and W arrays. Any software which expected the user to define this ordering would not be very appealing; fortunately E02ZAF exists to sort the pointers into the required order, and we return to it shortly. The size, NPOINT, of POINT must be at least $M + (PX - 7) \times (PY - 7)$, where PX and PY are integer parameters giving the total number of knots in the x and y directions. Observe the use of the word 'total'. If we recall the situation in the one-dimensional case, the total number of knots is equal to the sum of the number of internal knots, the two knots corresponding to the end-points x_1 and x_m, plus the two groups of three knots introduced at either end. Hence, the total number of knots is $n - 2 + 8$, where $n - 2$ is the number of internal knots. Here PX should be equal to the number of internal knots in the x direction, plus 8. PY is defined in an analogous manner. The actual values used for the interior knots should be supplied via the real/double-precision arrays LAMDA and MU. The first interior x knot should be in position 5 of LAMDA, the second in position 6, and so on. This leaves space for the remaining knots to be inserted. The same argument holds for MU.

Once the coefficient matrix X has been reduced to upper triangular form, the squares of the diagonal elements, divided by the mean squared weight, are formed in DL, of size at least $NC = (PX - 4) \times (PY - 4)$. These values are examined for significance by comparison with EPS. If any entry is found to be small the diagonal entry of X is set to zero, and the remaining non-zero entries on that row are zeroed using plane rotations. This reduces the rank of the coefficient matrix, and the final rank determined by the routine is returned in the integer parameter RANK. The solution coefficients are then computed using backward substitution and returned in C (of size at least NC). The sum of the squares of the residuals corresponding to this solution is returned via SIGMA. The workspace array WS, of size NWS, should contain at least $2 \times NC \times (p + 2) + p$ elements, where $p = 3 \times (PY - 4) + 4$.

As we have already indicated, the data for E02DAF can be sorted using a prior call of E02ZAF, for which the call sequence is

```
    CALL E02ZAF(PX,PY,LAMDA,MU,M,X,Y,POINT,NPOINT,
  +             ADRES,NADRES,IFAIL)
```

Here, ADRES, of size NADRES, is an integer workspace array. NADRES must be at least $(PX - 7) \times (PY - 7)$.

5.9   Interpolation – Aitken's method

So far we have looked at the problem of fitting a curve to data from the viewpoint that we are quite happy if the curve does not actually pass through any of the points, as long as it forms a good overall approximation to the function represented by the data values. It should be stressed that it is frequently the case that a 'best' (for example, least squares) approximation yields a better solution for subsequent manipulation than one which interpolates, that is, passes through, the data. However, there are circumstances where interpolation can be of use, particularly if we are interested in producing just one interpolated value, that is, a function value at a point not in the data set. Interpolating polynomials can be produced using the least squares approach simply by letting the number of unknowns equal the number of data values. However, more efficient, direct, schemes exist, and we now look at some of these in detail.

$$
\begin{array}{cccccc}
d & 1 & 2 & 3 & \cdots & n \\
x_1 & P_{11} \\
x_2 & P_{21} & P_{22} \\
x_3 & P_{31} & P_{32} & P_{33} \\
\vdots & \vdots & \vdots & \vdots & \ddots \\
x_n & P_{n1} & P_{n2} & P_{n3} & \cdots & P_{nn}
\end{array}
$$

Figure 5.10   Aitken's method.

The first method we consider is *Aitken's method* of successive linear interpolations. Let $\{\, x_k \mid k = 1, 2, \ldots, n \,\}$ be a set of points at which the function values $\{\, f_k \mid k = 1, 2, \ldots, n \,\}$ are known, and define the polynomials

$$
\begin{aligned}
P_{k1}(x) &= f_k, \\
P_{k,d+1}(x) &= \frac{(x_k - x)P_{dd}(x) - (x_d - x)P_{kd}(x)}{x_k - x_d}.
\end{aligned}
$$

Figure 5.10 indicates the way these polynomials are constructed. We begin by forming the second column from the known data values. Each entry in the next column is determined as a simple linear combination of the element in the same row in the previous column, and at the head of the column. All subsequent columns are formed in an analogous manner.

We observe that each P_{k1} is a constant function passing through (x_k, f_k). We refer to it as the *constant interpolant* based on x_k. By substitution, it is easily verified that P_{k2} passes through the coordinates (x_1, f_1) and (x_k, f_k); it is therefore denoted the *straight line interpolant* based on the points x_1 and x_k. In fact it can be shown (see Henrici [43], p. 207) that P_{kd} is the interpolating polynomial of degree $d - 1$ for the points x_1, x_2, $\ldots$, x_{d-1}, x_k, and, hence, that P_{nn} is of degree $n - 1$ and passes through all the given coordinates. To evaluate this polynomial at some point $\overline{x}$, say, we simply construct the table with $x = \overline{x}$, and then $P_{nn}(\overline{x})$ will give the required value.

Subroutine AITKEN in Code 5.1 implements the algorithm described above. We obviate the need to use a two-dimensional array in which only half the elements are used by performing the computation in a one-dimensional array. When the P_{k2}s are computed they are stored in the first $n - 1$ positions of C, the P_{k3}s are stored in the next $n - 2$ positions, and so on. The final required value will occupy the final position of the array when the process terminates. Note that the value returned in C(M − 2) will be a lower-order approximation and can be used to test the accuracy of the process, and that the original values supplied in X are destroyed.

E01AAF implements Aitken's method. A typical call takes the form

CALL E01AAF(A,B,C,N1,N2,N,X)

and the argument list is similar to that of AITKEN. The parameters A, B, and C (arrays of size N1, N1 and N2 respectively) correspond to X, Y and C in Code 5.1; X is the interpolation point. The NAG documentation stresses the fact that interpolation can yield very poor results, particularly if the interpolation point is near one end of the interval spanned by the x_is. In order to improve the convergence of the estimates $P_{d,d}(\overline{x})$ it is recommended that, if possible, the x_is be arranged in increasing order of distance from $\overline{x}$, with $\overline{x}$ somewhere in the middle of the range.

```
      SUBROUTINE AITKEN(X,Y,N,C,M,XBAR)
*
*.. GIVEN N FUNCTION VALUES Y(I) AT THE N POINTS X(I),
*.. THIS ROUTINE PRODUCES AN INTERPOLATED VALUE AT XBAR
*.. THE METHOD USED IS AITKEN'S METHOD OF SUCCESSIVE
*.. LINEAR INTERPOLATION AND THE ESTIMATES ARE STORED
*.. IN THE ARRAY C, THE FINAL INTERPOLATED VALUE IS
*.. RETURNED IN C(M) WHERE M = N*(N-1)/2
*.. FOR AN ACCURACY TEST THIS SHOULD BE COMPARED
*.. WITH C(M-2)
      DOUBLE PRECISION XBAR
      INTEGER M,N
      DOUBLE PRECISION C(M),X(N),Y(N)
*
      INTEGER D,DD,DP1,K,KD,KDP1
      DOUBLE PRECISION PDD,PKD
*
      KDP1 = 0
      DO 10 K = 1,N
        X(K) = X(K)-XBAR
10    CONTINUE
      DD = -N+1
      DO 30 DP1 = 2,N
        D = DP1-1
        IF(DP1.EQ.2)THEN
          PDD = Y(1)
        ELSE
          PDD = C(DD)
        ENDIF
        KD = DD
        DO 20 K = DP1,N
          KD = KD+1
          KDP1 = KDP1+1
```

```
        IF(DP1.EQ.2)THEN
          PKD = Y(K)
        ELSE
          PKD = C(KD)
        ENDIF
        C(KDP1) = (X(K)*PDD-X(D)*PKD)/(X(K)-X(D))
20    CONTINUE
      DD = DD+N-DP1+2
30  CONTINUE
    END
```

Code 5.1   Subroutine for Aitken's method.

EXERCISE 5.18 Values of the function $f(x) = e^{-x}\sin(4x)$ are available at $x = 0.1$, 0.6 and 2.0. An estimate of $f(1.25)$ is required. Use subroutine AITKEN to compute such a value. Generate the quadratic polynomial approximation and, using a sketch, indicate why the estimate of $f(1.25)$ is so poor. Evaluate f at 11, 21, 31, 41 and 51 equally spaced points in $[0,2]$ and obtain estimates of $f(0.0125)$, $f(0.5)$, $f(1.25)$, and $f(1.985)$. Analyse your results carefully.

5.10 Interpolation – Everett's formula

Here we again assume the existence of n data points, but now insist that the x_ks be equally spaced, say distance h apart. Further, we require that there be an even number of them, so that they may be relabelled as $\{\,\overline{x}_k \mid k = -(\overline{n}-1), -\overline{n}, \ldots, -1, 0, 1, \ldots, \overline{n}-1, \overline{n}\,\}$, where $\overline{n} = n/2$, and $\overline{x}_k = \overline{x}_0 + kh$. Accordingly, we let $\overline{f}_k$ denote the known function value at $\overline{x}_k$. We introduce the *central difference operator*, δ, which we define as

$$\delta f(x) = f\left(x + \frac{h}{2}\right) - f\left(x - \frac{h}{2}\right).$$

Note that, for example, $\delta^2 f(x) = f(x+h) - 2f(x) + f(x-h)$, and it follows that if x is one of the points $\overline{x}_k$, then $\delta^{2j}f(x)$ does not involve function values at points mid-way between those available. We also introduce the *binomial coefficient*

$$\binom{r}{m} = \frac{r(r-1)(r-2)\cdots(r-m+1)}{m!},$$

where r is any real number greater than the integer value m. Then by substitution or otherwise (see Henrici [43], p. 226), it may be verified that the interpolating polynomial for the coordinates $(\overline{x}_k, \overline{f}_k)$ may be written

$$p_{n-1}(\overline{x}_0 + ph) = \sum_{m=0}^{\overline{n}-1} \binom{1-p-m}{2m+1}\delta^{2m}\overline{f}_0 + \sum_{m=0}^{\overline{n}-1} \binom{p+m}{2m+1}\delta^{2m}\overline{f}_1,$$

(*Everett's formula*) for $-1 < p < 1$.

E01ABF uses Everett's formula to produce an interpolated value. A typical call takes the form

```
CALL E01ABF(N,P,A,G,N1,N2,IFAIL)
```

On entry, the $2N$ known function values should be stored, in order of increasing x, in A of size $N1(\geq 2N)$. P defines the interval fraction at which an interpolated value is required; this is restricted to lie within the range $-1 < P < 1$. The even-order central differences are stored by the routine in G, of size at least $N2(= 2N + 1)$; G(1) and G(2) hold $\delta^0 \overline{f}_0 (= \overline{f}_0)$, and $\delta^0 \overline{f}_1 (= \overline{f}_1)$ respectively, G(3) and G(4) hold $\delta^2 \overline{f}_0$ and $\delta^2 \overline{f}_1$, and so on. The interpolated value is returned via G(N2), and an upper bound on the contribution of the highest-order differences (and hence the error in the interpolated value) can, for $0 < P < 1$, be determined by forming $c \times |G(N2 - 1)| + |G(N2 - 2)|)$. The value to be used for c depends on N; for $N = 1, 2, 3, 4$ and 5 the recommended choice is 0.1, 0.02, 0.005, 0.001 and 0.0002, respectively. For each higher value of N, c should be reduced by a factor of about 4.

5.11 Function and derivative interpolation

We now widen the discussion to permit the possibility that derivative, as well as function, values are available at the points $\{ x_k \mid k = 1, 2, \ldots, n \}$. We assume that at each x_k the first q_k derivative values of f are known. The aim is to construct a polynomial p_{m-1}, of degree $m - 1$, where $m = \sum_{k=1}^{n}(q_k + 1)$, such that

$$p_{m-1}^{(q)}(x_k) = f_k^{(q)}, \qquad q = 0, 1, \ldots, q_k, \quad k = 1, 2, \ldots, n.$$

We begin by assuming that $q_k = 0$ for $k = 1, 2, \ldots, n$ (so that we are reverting to the case of function interpolation) and recall (Section 3.9) that the jth divided difference of f with respect to $x_k, x_{k+1}, \ldots, x_{k+j}$ is defined by

$$f[x_k x_{k+1} \cdots x_{k+j}] = \frac{f[x_{k+1} x_{k+2} \cdots x_{k+j}] - f[x_k x_{k+1} \cdots x_{k+j-1}]}{x_{k+j} - x_k}.$$

Let $\Pi_j = (x - x_1)(x - x_2) \cdots (x - x_j)$. Then by substitution it can be verified that the polynomial

$$\begin{aligned} p_{n-1}(x) = {} & f_1 + f[x_1 x_2]\Pi_1 + f[x_1 x_2 x_3]\Pi_2 \\ & + \cdots + f[x_1 x_2 \cdots x_n]\Pi_{n-1} \end{aligned} \qquad (5.23)$$

passes through the coordinates $\{(x_k, f_k) \mid k = 1, 2, \ldots, n\}$. (See Johnson and Riess [46], p. 212 for a formal proof of this result.) This is known as the *Newton form* of the interpolating polynomial.

We now extend the Newton form to the case that derivative values are available. From the original set of x values we construct a new set, $\{ \overline{x}_k \mid k = 1, 2, \ldots, m \}$, as follows:

$$
\overline{x}_k = \begin{cases}
x_1, & k = 1, 2, \ldots, q_1 + 1; \\
x_2, & k = q_1 + 2, q_1 + 3, \ldots, q_1 + q_2 + 2; \\
x_3, & k = q_1 + q_2 + 3, q_1 + q_2 + 4, \ldots, q_1 + q_2 + q_3 + 3; \\
\cdots
\end{cases}
$$

that is, if derivative values of order up to and including q_k are known at the point x_k, then $q_k + 1$ copies of x_k are included in the new set. We introduce the function values $\overline{f}_k$ in an exactly equivalent manner and define

$$
f[\overbrace{x_k x_k \cdots x_k}^{s \text{ copies}}] = f_k^{(s)} / s!
$$

This may seem rather complicated, but a tabular representation reveals that it is, in fact, quite straightforward. In Figure 5.11 we list the difference table for the case $n = 4$, $q_1 = 2$, $q_2 = 1$, $q_3 = 0$, $q_4 = 2$. When constructing first differences we proceed as normal, except that if two x values are identical we insert the corresponding known first derivative value. Hence, for example, for the entry $f[\overline{x}_4 \overline{x}_5] = f[x_2 x_2]$, we insert the value $f_2^{(1)}$. Similarly, second differences are computed as normal, except in the case of equal x values, where we insert known second derivative values (divided by 2). From this stage onwards, the table is constructed in the usual manner.

Using the above definitions the Newton form (5.23) is now valid (with overbars added where appropriate) for the case when both function and derivative values are available. See Steffenson [77] for a formal proof, but note that, for example, in Figure 5.11

$$
f[\overline{x}_1 \overline{x}_2] = f[x_1 x_1] = \lim_{\epsilon \to 0} \frac{f(x_1 + \epsilon) - f(x_1)}{\epsilon} = f_1^{(1)}.
$$

Note further that only the components f_1, $f[\overline{x}_1 \overline{x}_2]$, $f[\overline{x}_1 \overline{x}_2 \overline{x}_3]$, ... of the divided difference table (that is, those on the top diagonal line) are used in the construction of p_{m-1}. Having calculated the first few coefficients of this polynomial from l, say, function and derivative values, and associated differences, the next term may be determined by introducing a further function value and computing the corresponding reverse diagonal of the difference table.

Using a method due to Krogh [51], E01AEF constructs the Newton form of the interpolating polynomial from the divided difference table. Note that this routine returns the polynomial coefficients themselves, not an interpolated value. The Newton form is converted before exit into Chebyshev form, which means that the transformation $z = (2x - a - b)/(b - a)$ is used to map $[a, b]$ onto $[1, 1]$, with a and b chosen so that $a \le x_k \le b$, for all k.

$$
\begin{array}{lll}
\bar{x}_1 & x_1 & f_1 \\
 & & \quad\quad f_1^{(1)} \\
\bar{x}_2 & x_1 & f_1 \quad\quad\quad\quad\quad f_1^{(2)}/2 \\
 & & \quad\quad f_1^{(1)} \\
\bar{x}_3 & x_1 & f_1 \quad\quad\quad\quad\quad f[\bar{x}_2\bar{x}_3\bar{x}_4] \\
 & & \quad\quad f[\bar{x}_3\bar{x}_4] \\
\bar{x}_4 & x_2 & f_2 \quad\quad\quad\quad\quad f[\bar{x}_3\bar{x}_4\bar{x}_5] \\
 & & \quad\quad f_2^{(1)} \\
\bar{x}_5 & x_2 & f_2 \quad\quad\quad\quad\quad f[\bar{x}_4\bar{x}_5\bar{x}_6] \\
 & & \quad\quad f[\bar{x}_5\bar{x}_6] \\
\bar{x}_6 & x_3 & f_3 \quad\quad\quad\quad\quad f[\bar{x}_5\bar{x}_6\bar{x}_7] \\
 & & \quad\quad f[\bar{x}_6\bar{x}_7] \\
\bar{x}_7 & x_4 & f_4 \quad\quad\quad\quad\quad f[\bar{x}_6\bar{x}_7\bar{x}_8] \\
 & & \quad\quad f_4^{(1)} \\
\bar{x}_8 & x_4 & f_4 \quad\quad\quad\quad\quad f_4^{(2)}/2 \\
 & & \quad\quad f_4^{(1)} \\
\bar{x}_9 & x_4 & f_4 \\
\end{array}
$$

Figure 5.11   Derivative difference table.

To evaluate the polynomial, E02AKF should be called with the coefficients produced by E01AEF passed in the parameter list. The polynomial may be differentiated and integrated by calling E02AHF and E02AJF respectively. See Section 5.3 for further details.

A typical call of E01AEF takes the form

```
    CALL E01AEF(M,XMIN,XMAX,X,Y,IP,N,ITMIN,ITMAX,A,
   +            WRK,LWRK,IWRK,LIWRK,IFAIL)
```

On entry, the M distinct x values are supplied via X, and the N function and derivative values via Y (that is, M and N correspond to n and m above). These latter values should appear in the order

$$
\begin{array}{lll}
Y(k) & = f_1^{(k-1)}, & k = 1, 2, \dots, q_1 + 1, \\
Y(k + q_1 + 1) & = f_2^{(k-1)}, & k = 1, 2, \dots, q_2 + 1, \\
Y(k + q_1 + q_2 + 2) & = f_3^{(k-1)}, & k = 1, 2, \dots, q_3 + 1, \\
\quad\vdots & \quad\vdots \quad\vdots & \quad\vdots
\end{array}
$$

The M q_ks are supplied via the integer array IP, and XMIN and XMAX correspond to a and b above. ITMIN and ITMAX are used to control an outer iteration which successively refines the current estimate of the interpolating polynomial $p_{m-1}^{[l]}(x)$, say. The process is similar to that met in Chapter 4 where an iterative refinement technique was used to produce an accurate solution to a linear system of equations. After calculating the residuals at the x_ks, that is, the discrepancies between the known function and derivative values, and those given by the polynomial, each iteration consists of forming the interpolating polynomial for these residuals, and adding the result to $p_{m-1}^{[l]}(x)$ to form $p_{m-1}^{[l+1]}(x)$. The arrays WRK and IWRK, of size LWRK ($\geq 7N + 5 \times q_{max} + M + 7$), where $q_{max} = \max_k q_k$, and LIWRK($\geq 2M + 2$), respectively, are essentially work arrays, but do return some information to the calling routine. In particular, the entries WRK($q_{max} + 1 + j$) for $j = 1, 2,$ ..., N are the final residuals, and inspection of these quantities will reveal how well, or otherwise, the routine has worked.

5.12   Interpolation using splines

When discussing curve fitting using the least squares criterion we commented that it is sometimes inadvisable to look for an overall continuous approximation. Frequently the use of piecewise polynomials will yield superior results, and we dealt with the case of cubic spline approximation in some detail. The same arguments apply here, and so we now consider ways of constructing cubic (and, later, bicubic) spline interpolating polynomials. Naturally we shall wish to express such an approximation in terms of B-splines. If we write

$$p(x) = \sum_{i=0}^{n+1} a_i B_i(x),$$

where B_i is the cubic B-spline, defined by (5.13), then the aim is to find values for the a_is such that $p(x_k) = f_k$. Given the form of the B-splines, and, in particular, their local support, and choosing the knots to be the x_ks, it follows immediately that we must have

$$a_{i-1}B_{i-1}(x_i) + a_i B_i(x_i) + a_{i+1}B_{i+1}(x_i) = f_i, \qquad (5.24)$$

which gives n equations in the $n+2$ unknowns $\{ a_i \mid i = 0, 1, \ldots, n+1 \}$. In order to define a solution to the interpolation problem it would appear that we need to find a further two equations. Schemes based on interpolating known derivatives at x_1 and x_n have been proposed, and an alternative idea is to force the second derivative of p to be zero at the two end-points. A different approach, proposed by Cox [21], is to remove x_2 and x_{n-1} from the set of knots, reducing the number of unknowns to n. Since we are still attempting to satisfy the n equations (5.24), the required interpolatory property of p is maintained.

Having reduced the number of knots in the manner outlined above, the problem of cubic spline interpolation may be viewed as a special case of least squares cubic spline approximation and E01BAF, which determines an interpolating spline in terms of normalized cubic B-splines, simply sets up appropriate information for a call to E02BAF (Section 5.7). In the call

   CALL E01BAF(M,X,Y,K,C,LCK,WRK,LWRK,IFAIL)

the M coordinates that the spline is being fitted to should be supplied via X and Y. On exit the B-spline coefficients are contained in C, and the knots for this solution in K. It will be found that the interior knots $x_3, x_4, \ldots, x_{n-2}$ are in the real/double-precision array positions 5, 6, ..., M of K. Positions K(1), K(2), K(3) and K(4) will all contain x_1. The last of these corresponds to the use of x_1 as the external knot, and the first three result from the repeated use of x_1 as an introduced knot. Similarly, the array positions K(M + 1), K(M + 2), K(M + 3) and K(M + 4) will each contain x_n. LCK must, on entry, indicate the size of C, or K, whichever is the smaller; its value must be at least M + 4. The size of the work array WRK is specified by LWRK, and its value must be at least 6M + 16.

Fitting a bicubic spline to data in two-dimensional space is, in principle, straightforward. We simply derive the appropriate interpolating equations, make modifications to these, or the knots, to ensure that a solution exists, and then solve. If we are simply interested in obtaining a single interpolated value, and the data is available in the required form, then the following approach is likely to prove more efficient.

Suppose we start with a sequence of values $\{ x_k \mid k = 1, 2, \ldots, n \}$, and $\{ y_l \mid l = 1, 2, \ldots, m \}$, which define a regular grid on which function values f_{kl} are known. (Note that this is a different requirement to that for E02DAF in Section 5.7.) Let $x = \overline{x}$ and $y = \overline{y}$ define the point at which an interpolated value is required. Then for each y_l we can compute an estimate of $f(\overline{x}, y_l)$ by fitting a one-dimensional cubic spline to the data $\{ (x_k, f_{kl}) \mid k = 1, 2, \ldots, n \}$ and evaluating it at $x = \overline{x}$. This will give a sequence of m values, $\{ (y_l, \overline{f}_l) \mid l = 1, 2, \ldots, m \}$ to which we can fit another cubic spline, and obtain an interpolated value at $y = \overline{y}$. This final value will be our approximation to $f(\overline{x}, \overline{y})$. The process may, of course, be applied in the reverse direction; fix the x_ks first, fit splines in the y direction, and then fit a spline in the x direction. However, it should be noted that the two approaches will not necessarily produce the same interpolated value, so perhaps we should produce both and inspect them carefully. If there is reasonable agreement then we can accept either with some confidence; if there is some considerable discrepancy then further analysis of the function values is required.

E01ACF implements the above scheme to produce an interpolated value from a set of function values specified on a regular grid. In the call

   CALL E01ACF(A,B,X,Y,F,VAL,VALL,IFAIL,XX,WORK,AM,
   +   D,IG1,M1,N1)

the N1 x values and the M1 y values should be supplied via X and Y, respectively, in strictly ascending order. The corresponding function values should be placed in the two-dimensional array F, which must have N1 rows and at least M1 columns. A and B define the x and y coordinates at which an interpolated value is required, and two approximations to this are returned in VAL and VALL. The variable VAL corresponds to approximating along lines of constant y first; x is then fixed, and the final approximation takes place. Similarly, VALL is obtained by first approximating along lines of constant x. The routine uses 4 workspace arrays, XX, WORK, AM, and D; each should be of size at least IG1, whose value must be the larger of N1 and M1.

5.13 Summary

In this chapter of the book we have considered a number of routines from the NAG E01 and E02 chapters of the library. Many share a common purpose, although the form in which a solution is returned, and the manner in which it is determined, will be different. It is not possible to lay down hard and fast rules which should be strictly adhered to when the user wishes to select a routine. Rather we attempt to provide some general guidelines which we hope will be of use. However, the user is still expected to employ a certain amount of intuition in order to make a final judgement.

The first thing that should be done is to attempt to make an assessment of the quality of the data. If some of the given function values are likely to be in error by a significant amount, then an interpolating approximation will follow these errors. Even if an equal weighting is used, a least squares approximation is unlikely to be too distracted by such rogue values. Further, if we have a large number of points then an interpolating polynomial is likely to exhibit unwanted fluctuations. This suggests that, for general curve-fitting to data produced by an experiment, from the digitization of a graph, etc., E02ADF should be used with a relatively low-order approximation, and E02AGF can be used to impose constraints (Section 5.5). If we can ensure that the x_ks are of a special form, then E02AFF will obtain a solution more efficiently. The solutions returned by these routines are in Chebyshev form, so that evaluation, differentiation and integration of the result can easily be obtained using E02AEF or E02AKF, E02AHF and E02AJF (Section 5.3). A least squares surface approximation, with constraints, can be formed using E02CAF (Section 5.8), provided the data points lie along lines of constant y. An interpolated value may subsequently be obtained from E02CBF (Section 5.3)

If the user is not too worried about the continuity properties of the solution, then subdivision of the range using knots followed by least squares cubic spline approximation may prove an attractive alternative. In the one-dimensional case the relevant routine for fitting to function values is E02BAF (Section 5.7). Again, a polynomial approximation is returned, this time as a linear combination of normalized B-splines. E02BBF and E02BCF can subsequently be used to evaluate the approximation and its derivatives at a specified point. The definite integral of the approximation over the range of interest may be obtained from E02BDF (Section 5.6). In the two-dimensional case, a bicubic spline can be fitted to a rectangular grid of values using E02DAF. However the points have to be ordered in an appropriate fashion using a prior call of E02ZAF (Section 5.8). Evaluation of the approximation is achieved using E02DBF (Section 5.6).

If the data is exact and/or only a few values are available, interpolation may be of use. If the data is unequally spaced the routine to use is E01AAF (Section 5.9). If equally spaced, use E01ABF (Section 5.10). If function and derivative values are available, use E01AEF (Section 5.11). The first two return a single interpolated value. The last computes a polynomial solution in Chebyshev form which may be evaluated, etc., as indicated above. If a large number of exact values are available, spline interpolation is likely to yield the best results. The relevant routine is E01BAF. For two-dimensional data on a rectangular grid, E01ACF returns two estimates of an interpolated value (Section 5.12).

Finally we give brief mention of those routines in chapters E01 and E02 which have not been discussed in the foregoing discussion of curve and surface fitting and interpolation. Space limitations preclude a full discussion of the way these routines work and, in any case, they are of relatively limited use. E01RAF produces an interpolating rational polynomial approximation, that is, one which is expressed as the quotient of two polynomials in continued fraction form. The solution may be evaluated using E01RBF. In general it is unlikely that this approximation will be any better than those we have already discussed. It has the disadvantage that, if centred on the origin, interpolated values some distance from the origin are likely to be quite poor. E02RAF computes a rational approximation, expressed in quotient form, from a power series expansion. E02RBF may be used to evaluate this approximation at a point.

Throughout this chapter we have relied on the least squares criterion for determining a best approximation. Other criteria are possible. The *least first power* (or l_1) approximation minimizes the sum of the absolute values of the residuals, and its properties suggest that it may be of use when fitting to inexact data. The use of this criterion forms the basis of E02GAF, for which the user is expected to supply the coefficient matrix in the overdetermined system of equations to be solved; this means that approximations expressed as linear combinations of functions other than polynomials may be used. The method of solution allows constraints to be imposed readily, and E02GBF is available to deal with this case. In contrast, the *minimax* (or l_∞) approximation minimizes the maximum absolute residual, and should only be used when the data is known to be exact. For the general minimax fit E02GCF should be used; for polynomial approximation use E02ACF.

Chapter 6
Ordinary Differential Equations

In this chapter we consider the solution of a system of initial value problems in ordinary differential equations. We

- introduce the idea of step-by-step methods using Euler's method, discuss the errors associated with the method, and outline the problems associated with stability;

- consider the Runge–Kutta class of methods and step-control strategies;

- develop the 'linear multistep' class of methods;

- introduce the concept of a 'stiff' system and describe special methods for their solution.

6.1 Introduction

The mathematical modelling of a wide range of physical processes results in a system of ordinary differential equations (ODEs). Such models include chemical kinetics, atmospheric pollution, and trajectory and ballistic problems.

Some examples of ODEs are

(1) $$\frac{dy}{dt} = g - \frac{k}{m}y^2, \qquad t > 0, \quad \text{with} \quad y(0) = 0.$$

This equation models a falling body of mass m; y represents the velocity of the body whose acceleration, dy/dt, is equal to that due to gravitational force, less a component which represents the retardation effect of air resistance,

(2) $$\frac{d^2y}{dt^2} = y^{3/2}t^{-1/2}, \quad 0 < t < \infty, \quad \text{with} \quad y(0) = 1 \text{ and } y(\infty) = 0.$$

This is the Thomas–Fermi equation, used to model the charge density at high atomic numbers,

(3) $c^2 \dfrac{d^2 y}{dt^2} + \lambda y = 0,$ $0 < t < 1,$ with $y(0) = y(1) = 0.$

This equation results from the use of the separation of variables technique on the hyperbolic wave equation.

During the 17th and 18th centuries it was hoped to find solutions to all ODEs in terms of a finite number of functions and their integrals. We now know this to be unachievable and there are very simple equations for which no such analytic solution exists. Even when analytic solutions are known it is often cheaper and sometimes more accurate to solve the original equation numerically than to evaluate the analytic solution. Whenever we apply numerical techniques we shall assume that the equation being solved is well-defined and admits a unique solution. (For details of existence and uniqueness of solutions, see Henrici [42] and Keller [47].)

ODEs come in many diverse forms and some initial categorization is called for. At the same time we introduce some terminology which will be used throughout this chapter.

The *order* of an ODE is defined to be the order of the highest derivative appearing in the equation. Thus, the first of our examples is a first-order equation, whilst the others are both second order. The solution, $y(t)$, to each of these problems is termed the dependent variable, and t the independent variable. (Because y is a function of one variable only the problems are ODEs; problems involving partial derivatives of dependent functions of more than one independent variable are termed partial differential equations (PDEs).)

Returning to our example equations, we observe that each is accompanied by a number of additional constraints, or conditions. (For example, in (1) we have that the initial velocity is zero.) These are present to ensure that a unique solution exists. Consider the first-order equation

$$y'(t) = -y(t), \qquad t > 0, \tag{6.1}$$

which has the analytic solution $y(t) = ae^{-t}$. The differential equation defines a family of possible solutions (see Figure 6.1) since the value of a is arbitrary. We need a condition of the form

$$y(0) = y_0, \tag{6.2}$$

where y_0 is some known constant, to indicate which particular member of this family is the required solution. (It is easy to see that this condition ensures that $a = y_0$.) For an nth-order equation we require n constraints of this form to 'tie down' the solution.

For many practical applications we are interested not in a single equation, but in a 'coupled' system of ODEs. The *Lotka–Volterra*, or predator-prey, equations for $t > 0$,

$$\begin{aligned} f' &= \alpha f + \beta f r, \\ r' &= \gamma r + \delta f r, \end{aligned} \tag{6.3}$$

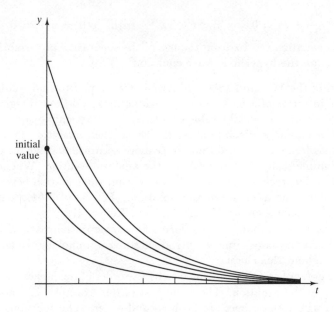

Figure 6.1   A family of solutions.

with conditions

$$f(0) = f_0, \qquad r(0) = r_0, \tag{6.4}$$

describe a very simple ecological system consisting of a predator population of foxes whose size is given by f, and a prey population of rabbits, size r. It is assumed that the rabbits have an inexhaustible supply of food, and that the foxes feed solely on the rabbits. Clearly the rate of change of the fox population f' will depend on the number of foxes and on the number of rabbits, and vice versa. This coupling is represented by the appearance of both f and r in each of the right-hand sides of the system (6.3). The conditions (6.4) define some initial distribution of foxes and rabbits at time $t = 0$. Note that although we only have a first-order system, two conditions are required, one for each equation. In fact, for a system of n equations in which the ith equation is of order i_j, a total of $N = \sum_{i=1}^{n} i_j$ constraints will need to be imposed.

Systems of ODEs occur naturally, but can also result from the application of approximation techniques to more complex problems. For example, the parabolic PDE

$$\frac{\partial u(x,t)}{\partial t} = \frac{\partial^2 u(x,t)}{\partial^2 x}, \qquad 0 < x < 1, \quad t > 0,$$

along with appropriate initial and boundary conditions which ensure a unique solution, may be approximated using the method of lines (Sincovec and Madsen [75]) to yield the system

$$\mathbf{u}'(t) = A\mathbf{u}(t) + \mathbf{b}(t),$$

where A is a tridiagonal matrix, $\mathbf{u}^T = \big(u_1(t), u_2(t), \ldots, u_n(t)\big)$, and $u_i(t)$ represents an approximation to $u(x_i, t)$ for some $x_i \in [0, 1]$ and $t > 0$.

It is natural, and convenient, to classify ODEs further along the following lines:

- *Initial value problems* (IVPs) are systems for which all conditions (initial conditions) are given at the same value of the independent variable. Each condition will typically involve a constraint on y, y', y'', etc., or some linear combination of these. Initial value problems are essentially evolutionary in nature and represent some dynamic system. Consequently, the independent variable usually represents time.

- *Two-point boundary value problems* (BVPs) are systems for which the conditions (*boundary conditions*) are given at two extreme values of the independent variable. Examples (2) and (3) above are of this form. BVPs are used to model steady state situations and the independent variable often represents a space dimension.

- *Sturm–Liouville problems* are BVPs for which non-trivial (i.e., not identically zero) solutions only exist for certain values of a model parameter (λ in Example (3) above). These special values are known as *eigenvalues* and their associated solutions are termed *eigenfunctions*.

In this chapter we are concerned with systems of first-order IVPs only. However, if a system of higher-order IVPs may be written so that the highest-order derivative appears explicitly on the left-hand side of an equation and nowhere else, then it may also be reformulated as a first-order system. For example, the second-order IVP

$$
\begin{aligned}
y'' &= f(t, y, y'), \\
y(0) &= \alpha, \\
y'(0) &= \beta,
\end{aligned}
$$

may be written

$$
\begin{aligned}
\mathbf{y}' &= \mathbf{g}(t, \mathbf{y}), \\
\mathbf{y}(0) &= \boldsymbol{\eta},
\end{aligned}
$$

where $\mathbf{y}^T(t) = \big(y_1, y_2\big)$, with $y_1(t) = y(t)$ and $y_2(t) = y'(t)$, $\mathbf{g}^T = (g_1, g_2)$, with $g_1(t, y_1, y_2) = y_2$ and $g_2(t, y_1, y_2) = f(t, y_1, y_2)$, and $\boldsymbol{\eta}^T = (\alpha, \beta)$. In addition, *shooting techniques* for certain BVPs also require the problem to be recast as an IVP (see Exercise 6.12), and hence the methods we describe here have wider application than might at first be thought. Although

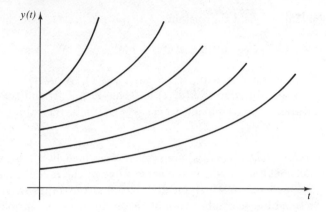

Figure 6.2   Unstable equation.

initially we shall describe these methods in the context of a single equation, the extension to a system is essentially trivial.

Precisely what is required as a 'solution' will vary from problem to problem. For example, for the predator–prey equations we may wish to determine

- when the prey population is zero,

- what the predator/prey populations are at some specified time,

- what is the maximum attainable population of the predator/prey,

- if the population cycles, what is the cycle time,

and so on. The fundamental techniques for the numerical solution of each problem will be the same, although the actual algorithms used will vary slightly, and we shall see that the proliferation of routines in the NAG library for solving IVPs reflects this diversity in the problem specification.

Before considering numerical methods in detail, we consider the effect of an error in the initial conditions. For equation (6.1) an error in the initial condition (6.2) would result in the wrong trajectory being selected, but as time increases the distance between this curve and the correct one decreases. At time $t = b > 0$ the distance between the two curves is just e^{-b} times the initial error, and this will be very small for large b. The equation is termed *stable* . However, for the equation $y'(t) = y(t)$ a small error in the initial condition will cause the difference between the selected and true trajectories to increase exponentially as time increases (see Figure 6.2). Here we have an *unstable* equation.

We may generalize this idea of stability by considering equations of the form

$$y'(t) = \lambda y(t), \tag{6.5}$$

where λ is a given constant. An error in the initial value will lead to an error growth at the rate of $e^{\lambda t}$. For $\lambda \leq 0$ the initial error is not amplified as time increases and the equation is stable. If $\lambda > 0$, any initial error grows unboundedly with t and the equation is unstable. Note that the concept of stability introduced here is a property of the problem itself; the numerical method chosen for its solution may itself exhibit unstable characteristics even when applied to an inherently stable equation.

EXERCISE 6.1   Rewrite the fourth-order IVP $y'''' - 3601y'' + 3600y = 1800t^2$, subject to $y(0) = y'(0) = y''(0) = y'''(0) = 0$, as a coupled system of four first-order IVPs.

EXERCISE 6.2

(a)   Express the following pair of high-order equations as a system of first-order equations

$$\frac{d^4x}{dt^4}\frac{dx}{dt} + \sin(x)\frac{d^2y}{dt^2} + \frac{d^2x}{dt^2} + \frac{d^2x}{dt^2}\frac{dy}{dt} = e^{txy},$$
$$\frac{d^2x}{dt^2} + \frac{d^3y}{dt^3}\sin(t) = \frac{dx}{dt}\frac{dy}{dt}.$$

What extra conditions would make this an IVP?

(b)   Prove that a system of ordinary differential equations can be rewritten in the form $\mathbf{y}' = \mathbf{f}(x, \mathbf{y})$ if, and only if, the system can also be rewritten with the highest-order derivative in each dependent variable appearing as the left-hand side of one equation and nowhere else.

EXERCISE 6.3   Consider the second-order differential equation

$$\frac{d^2y}{dt^2} = 4y,$$

with initial conditions $y(0) = \frac{1}{10}$ and $y'(0) = -\frac{1}{5}$. Verify that the solution is $y(t) = e^{-2t}/10$. Show that if the initial conditions are changed to $y(0) = \frac{1}{10} + \epsilon$ and $y'(0) = -\frac{1}{5}$, the solution becomes $y(t) = \frac{1}{2}\epsilon e^{2t} + \left(\frac{1}{10} + \frac{1}{2}\epsilon\right)e^{-2t}$. What difficulties would you expect if you were to attempt to solve this problem numerically?

6.2   Introduction to numerical methods

We consider the single first-order IVP

$$\begin{aligned} y'(t) &= f(t, y), & t > a, \\ y(a) &= \eta, \end{aligned} \tag{6.6}$$

where f is some known function and η defines an initial condition. Each method we describe for the numerical solution of (6.6) attempts to find an approximation y_i to the true solution $y(t_i)$ at a set of discrete points $\{t_i \mid i = 0, 1, \ldots\}$ where $a = t_0 < t_1 < \cdots$. This is done in a step-by-step fashion; that is starting with the initial value $y_0 = \eta$, we compute y_1, then y_2, and so on until some appropriate criterion is satisfied (say, we have obtained an approximation at $t = T$). For the moment we shall assume that the distance between successive discretization points is a constant, h; that is $t_{i+1} - t_i = h$ for $i = 0, 1, \ldots$.

The Taylor series expansion for $y(t_i + h)$ about $t = t_i$ may be written

$$y(t_i + h) = y(t_i) + hy'(t_i) + \frac{h^2}{2!}y''(t_i) + \cdots + \frac{h^r}{r!}y^{(r)}(t_i) + R_r, \qquad (6.7)$$

where

$$R_r = \frac{h^{r+1}}{(r+1)!}y^{(r+1)}(\xi), \qquad \xi \in (t_i, t_{i+1}),$$

(see Section 1.15). If we have available approximations to y and its first r derivatives (assuming they exist) at $t = t_i$, the Taylor series (6.7) may be truncated before the remainder term R_r. If we denote these approximations by y_i, y_i', y_i'', etc., we have the scheme

$$y_{i+1} = y_i + hy_i' + \frac{h^2}{2!}y_i'' + \cdots + \frac{h^r}{r!}y_i^{(r)}. \qquad (6.8)$$

The derivatives of y may not have a simple analytic form but can in principle be generated from

$$\begin{aligned} y' &= f, \\ y'' &= f_t + f_y f, \\ y''' &= f_{tt} + 2f f_{ty} + f^2 f_{yy} + f_y(f_t + f f_y), \end{aligned} \qquad (6.9)$$

etc., where the subscripts denote partial derivatives.

The *local truncation error* (LTE) of any numerical scheme is defined to be the difference between $y(t_{i+1})$ and y_{i+1} under the assumption that all values used in the calculation of y_{i+1} are exact. (This definition assumes the absence of rounding error.) Further, the scheme is said to be of *order* r if the LTE is $O\left(h^{r+1}\right)$. For the Taylor series scheme just outlined this means that if $y_i' = y'(t_i)$, etc., the LTE is just R_r, and the method is of rth-order.

The simplest form of a Taylor series method is derived by setting $r = 1$ in (6.8). We obtain

$$y_{i+1} = y_i + hy_i' = y_i + hf_i, \qquad (6.10)$$

where $f_i = f(t_i, y_i)$. Assuming that an analytic form for f is available, we may easily compute y_1, y_2, $\ldots$ in turn. This first-order method, *Euler's method*, is rarely used in practice, but a close examination of its properties will prove instructive.

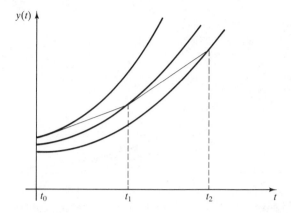

Figure 6.3   Euler's method on an unstable problem.

Geometrically, the first step of the method consists of extrapolating from the point (t_0, y_0) along the tangent to the true solution (see Figures 6.3 and 6.4). The effect of this error is to cross over onto another member of the family of solutions which corresponds to a different initial value. The second step transfers to yet another member, and so on. We observe that for an unstable problem (Figure 6.3) the error is being amplified at each step and our approximation is, in an absolute sense, moving rapidly away from the true solution. For a stable problem any errors introduced into the calculations will ultimately be damped out. (In Figure 6.4 both approximate and true solutions tend to zero as t increases.)

For the simple equation (6.5) with initial condition $y(0) = 1.0$ we may solve the Euler recurrence relation $y_{i+1} = y_i + \lambda h y_i$ exactly to give $y_{i+1} = (1 + \lambda h)^{i+1} y_0$. We note that y_{i+1} will decrease in magnitude with i only if $-1 < 1 + \lambda h < 1$, that is if $-2 < \lambda h < 0$. Assuming a positive value for h, we compare the behaviour of the approximate solution with that of $e^{\lambda t}$, the analytic solution. If λ is positive we have the correct qualitative behaviour. However, when λ is negative, choices of $h \geq 2/|\lambda|$ will result in the approximation increasing, rather than decreasing, in value. Further, the y_is will alternate in sign; this cannot be correct since $e^{\lambda t}$ is everywhere positive. For such values of h the method is said to be unstable (on what is an inherently stable problem). We say that $(-2, 0)$ is the *interval of stability* for the method.

We are normally interested in obtaining approximate solutions to within some user specified accuracy, and the quantity we would therefore like to control is the *global truncation error* (GTE), defined to be the difference between the computed y_{i+1} and $y(t_{i+1})$. This differs from the local truncation error in that it recognizes the existence of errors in all previously computed approximations. From Figures 6.3 and 6.4 we can see

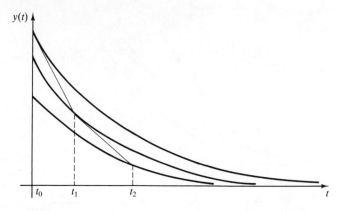

Figure 6.4   Euler's method on a stable problem.

that the GTE for Euler's method is caused by the accumulation of LTEs, and at $t = t_{i+1}$ is equal to the LTE for the $i + 1^{st}$ step, plus the errors which arise from the fact that we are using y_i, not $y(t_i)$, in (6.10). This error build up results in the global error for the method being of $O(h)$ (as opposed to $O(h^2)$ for the LTE). For higher-order methods we have an analogous situation, and this explains why, earlier in this section, we said that a method with LTE of $O(h^{r+1})$ is classified as being of order r.

Subroutine EULER (Code 6.1) is a simple routine for integrating the system of first-order initial value problems

$$\begin{aligned} \mathbf{y}' &= \mathbf{f}(t, \mathbf{y}), \\ \mathbf{y}(a) &= \boldsymbol{\eta}, \end{aligned}$$

where $\mathbf{y}^T(t) = (y_1(t), y_2(t), \ldots, y_n(t))$, $\mathbf{f}^T(t, \mathbf{y}) = (f_1(t, \mathbf{y}), f_2(t, \mathbf{y}), \ldots, f_n(t, \mathbf{y}))$, and $\boldsymbol{\eta}^T = (\eta_1, \eta_2, \ldots, \eta_n)$. The user must supply the number of equations in the system (N), the starting position (A), an upper limit on the integration interval (B) and the number of equal steps to be used (NTIME). Each step will be of length $(B - A)/NTIME$. The initial values of $\mathbf{y}$ at $t = A$ should be stored in the array Y prior to a call of EULER. The user supplied routine

```
SUBROUTINE FUNCT(T,Y,F,N)
INTEGER N
DOUBLE PRECISION T,Y(N),F(N)
```

must return in F the value of $\mathbf{f}(t, \mathbf{y})$ for given values of t and $\mathbf{y}$. MONIT is also user supplied, with specification

```
SUBROUTINE MONIT(T,Y,N)
INTEGER N
DOUBLE PRECISION T,Y(N)
```

MONIT allows the progress of the integration to be monitored at the end of each step. For example, suppose that we wish to print all intermediate approximate solutions. Then the inclusion within MONIT of the statement

```
WRITE(6,*)T,(Y(I),I = 1,N)
```

will achieve the desired result. If no such information is required the body of MONIT should be left empty.

```
      SUBROUTINE EULER(A,B,NTIME,N,Y,F,FUNCT,MONIT)
*
*..ROUTINE EULER IMPLEMENTS THE SIMPLE ONE-STEP
*..EULER METHOD FOR A SYSTEM OF N FIRST ORDER
*..INITIAL VALUE PROBLEMS.
*
*..THE RIGHT-HAND SIDES OF THE SYSTEM ARE DEFINED
*..BY THE USER SUPPLIED FUNCTION FUNCT.
      INTEGER N,NTIME
      DOUBLE PRECISION A,B,F(N),Y(N)
      EXTERNAL FUNCT,MONIT
*
      DOUBLE PRECISION H,T
      INTEGER I,J
*
      H = (B-A)/NTIME
      DO 20 I = 0,NTIME-1
        T = A+I*H
        CALL FUNCT(T,Y,F,N)
        DO 10 J = 1,N
          Y(J) = Y(J)+H*F(J)
  10    CONTINUE
        CALL MONIT(T+H,Y,N)
  20 CONTINUE
      END
```

Code 6.1   Subroutine EULER.

As an example of the performance of EULER we use it to solve the pair of linear IVPs

$$
\begin{aligned}
y' &= -y + z, \\
z' &= y - z,
\end{aligned}
\qquad (6.11)
$$

NTIME	$\left|y_{t=2}^{true} - y_{t=2}^{approx}\right|$
16	0.8293×10^{-2}
64	0.2240×10^{-2}
256	0.5694×10^{-3}
1024	0.1429×10^{-3}
4096	0.3576×10^{-4}
16384	0.8942×10^{-5}
65536	0.2236×10^{-5}

Figure 6.5   Solution using Euler's method with double-precision arithmetic.

for $t > 0$ with initial conditions $y(0) = 3$, $z(0) = 1$. The exact solution is $y(t) = 2 + e^{-2t}$, $z(t) = 2 - e^{-2t}$. For this problem the routine FUNCT takes the form

```
SUBROUTINE FUNCT(T,Y,F,N)
INTEGER N
DOUBLE PRECISION  Y(N),F(N),T
F(1) = -Y(1)+Y(2)
F(2) =  Y(1)-Y(2)
END
```

Choosing B $= 2.0$ and NTIME $= 16, 64, 256, \ldots, 65536$ we list in Figure 6.5 the errors in the approximations to $y(2.0)$ returned by EULER. The error in $z(2.0)$ is very similar due to the symmetry of the two solutions. The linear nature of the global error is clearly illustrated; a reduction of the step length by a certain factor reduces the error by approximately the same amount. It is also clear that the use of Euler's method involves a considerable amount of work if an accurate solution is to be obtained even for a relatively simple problem. Further, if we repeat the calculations using single-precision arithmetic (24 bit mantissa), the use of very small steps may be counter-productive due to the accumulation of rounding errors (see Figure 6.6). Hence we must seriously consider using methods which have higher order.

Retaining an extra term in the truncated Taylor series yields the scheme

$$y_{i+1} = y_i + hf_i + \tfrac{1}{2}h^2 \left(\frac{\partial f_i}{\partial t} + f_i \frac{\partial f_i}{\partial y} \right)$$

which has an $O\left(h^3\right)$ LTE and hence an $O\left(h^2\right)$ GTE. With this method we would anticipate that at each halving of the step size we would get a four-fold increase in accuracy. However, there is a price to pay for this quadratic rate of convergence, namely the need to determine, and subsequently evaluate, the derivative terms. For a system this additional effort

NTIME	$\lvert y_{t=2}^{true} - y_{t=2}^{approx} \rvert$
16	0.8293×10^{-2}
64	0.2240×10^{-2}
256	0.5684×10^{-3}
1024	0.1421×10^{-3}
4096	0.3195×10^{-4}
16384	0.4530×10^{-5}
65536	0.1049×10^{-4}

Figure 6.6   Solution using Euler's method with single-precision arithmetic.

can be considerable, since the approximation to each equation will involve partial differentiation of f with respect to t and each dependent variable. For example, for a system of just two equations we have

$$y_{1,i+1} = y_{1i} + hf_{1i} + \tfrac{1}{2}h^2 \left(\frac{\partial f_{1i}}{\partial t} + f_i \frac{\partial f_{1i}}{\partial y_1} + f_i \frac{\partial f_{1i}}{\partial y_2} \right),$$

$$y_{2,i+1} = y_{2i} + hf_{2i} + \tfrac{1}{2}h^2 \left(\frac{\partial f_{2i}}{\partial t} + f_i \frac{\partial f_{2i}}{\partial y_1} + f_i \frac{\partial f_{2i}}{\partial y_2} \right).$$

It is a simple matter to modify subroutine EULER in order to implement this second-order method, but the user must supply expressions for the partial derivatives. For example, for the system (6.11) we need to form $\partial f_1/\partial t = \partial f_2/\partial t = 0$, $-\partial f_1/\partial y_1 = \partial f_1/\partial y_2 = \partial f_2/\partial y_1 = -\partial f_2/\partial y_2 = 1$. Results obtained from such a modification are summarized in Figure 6.7 and the improvement in convergence rate is apparent.

Our subroutine EULER, or its modification, gives little indication of what factors must be borne in mind when developing robust, efficient software for the numerical solution of ODEs. Since there is a close relationship between ODEs and quadrature (differentiation and integration being opposite operations) there would appear to be some possibility of overlap when

NTIME	$\lvert y_{t=2}^{true} - y_{t=2}^{approx} \rvert$
16	0.7888×10^{-2}
64	0.4922×10^{-3}
256	0.3072×10^{-4}
1024	0.1919×10^{-5}
4096	0.1199×10^{-6}
16384	0.7496×10^{-8}
65536	0.4685×10^{-9}

Figure 6.7   Solution using a second-order Taylor series method.

it comes to algorithm design. Recalling the important features of quadrature routines, in the present context we observe the need to minimize the number of function evaluations necessary to obtain approximations within some user specified tolerance. Some adaptive technique for varying the step size would therefore appear to be appropriate. It is usual to compute an LTE estimate and hope that by controlling this we can keep the global error within bounds. Since we sample the solution at a finite set of points only, care must be exercised to ensure that we do not miss important features, such as sharp peaks.

In the next three sections we consider methods which are of higher order than Euler's method but require evaluations of f only. Further, we begin to examine ways of controlling the LTE in an adaptive fashion.

EXERCISE 6.4 By considering the Euler solution of $y' = y$ with $y(0) = 1$, show that $\lim_{i \to \infty} (1 + x/i)^i = e^x$.

EXERCISE 6.5 Write a subroutine TAYLOR which solves a system of N initial value problems using the second-order Taylor series approximation. Attempt to make the interface to the user supplied function as simple as possible.

EXERCISE 6.6 Use EULER (Code 6.1) and the second-order Taylor series routine from Exercise 6.5 to approximate the solution to

$$y'(t) = \frac{-y^4 \cos(t) + t^2 y}{t^3}, \qquad (6.12)$$

$$y(\pi/2) = \frac{\pi 3^{-\frac{3}{2}}}{2},$$

from $t = \pi/2$ to $t = \pi$. [True solution: $y(t) = t/(3 \sin t)^{\frac{3}{2}}$.] Use a monitor routine to compute the average absolute error, A_e, defined by

$$A_e = \frac{1}{\text{NTIME}} \sum_{i=1}^{\text{NTIME}} |y(t_i) - y_i|,$$

and the position at which the maximum absolute error occurs. How does A_e vary with h?

EXERCISE 6.7 Extend the solution of (6.11) to higher-order Taylor series approximations.

EXERCISE 6.8 Use Euler's method to solve $y'(t) = \cos(\pi t y)$ for the sequence of starting values $y(0) = 0, 0.2, 0.4, \ldots, 2.8$. For a given value of h plot the resultant solutions on one frame. Repeat for different values of h. How many different asymptotic solutions are there?

EXERCISE 6.9 The oscillations (angle to vertical) in a vacuum of a pendulum of length l released at time $t = 0$ at an angle θ from the vertical satisfy the second-order ODE

$$\frac{d^2\theta}{dt^2} + gl\sin(\theta) = 0,$$

where $g = 9.81$ m/sec^2 is the acceleration due to gravity. Use the differential equation to obtain expressions for the third, fourth and fifth derivatives of θ, and use the Taylor series expansions

$$\theta(t) \approx \theta(0) + t\theta'(0) + \frac{t^2}{2!}\theta''(0) + \frac{t^3}{3!}\theta'''(0) + \frac{t^4}{4!}\theta^{(4)}(0) + \frac{t^5}{5!}\theta^{(5)}(0),$$

$$\theta'(t) \approx \theta'(0) + t\theta''(0) + \frac{t^2}{2!}\theta'''(0) + \frac{t^3}{3!}\theta^{(4)}(0) + \frac{t^4}{4!}\theta^{(5)}(0),$$

to find approximations to θ and θ' at $t = 0.1, 0.2, 0.3$ when $\theta_0 = 2\pi/3$, $\theta'(0) = 0$ and $l = 0.981$ m.

6.3   Introduction to Runge–Kutta methods

We begin our discussion of *Runge-Kutta* methods by considering again the geometrical interpretation of Euler's method (Figures 6.3 and 6.4). At the $(i + 1)^{st}$ step we follow the tangential direction of the trajectory passing through (t_i, y_i). Thus, Euler's method generates a polygonal arc approximation to the required solution. In a bid to obtain a more accurate approximation we start with a half step along the Euler direction. We then use the value of the tangent at this new point to adjust the slope to be used at t_i, that is we compute

$$y_{i+\frac{1}{2}} = y_i + \tfrac{1}{2}hf_i, \tag{6.13}$$

followed by

$$y_{i+1} = y_i + hf\left(t_{i+\frac{1}{2}}, y_{i+\frac{1}{2}}\right). \tag{6.14}$$

We recall that when applied to the simple problem (6.5) with initial condition $y(0) = 1$, Euler's method gives $y_{i+1} = (1 + h\lambda)^i$. Using the binomial expansion, and remembering that $t_i = ih$, we may rewrite this as

$$y_{i+1} = 1 + \lambda t_i + \tfrac{1}{2}i(i - 1)\lambda^2 t_i^2 + O\left(h^3\right).$$

The first two terms here may be identified with the Taylor series expansion of $e^{\lambda t_i}$ about the point $t = 0$ and hence we have $y_{i+1} = e^{\lambda t_i} + O\left(h^2\right)$. Applying the same technique to the new method, (6.13), (6.14), we have

$$\begin{aligned}
y_{i+\frac{1}{2}} &= (1 + \tfrac{1}{2}h\lambda)y_i, \\
y_{i+1} &= (1 + h\lambda + \tfrac{1}{2}h^2\lambda^2)y_i \\
&= (1 + h\lambda + \tfrac{1}{2}h^2\lambda^2)^i \\
&= 1 + \lambda t_i + \tfrac{1}{2}\lambda^2 t_i^2 + \tfrac{1}{6}i(i - 1)(i + 1)\lambda^3 t_i^3 + O\left(h^4\right) \\
&= e^{\lambda t_i} + O\left(h^3\right),
\end{aligned}$$

which is one order of magnitude more accurate than Euler's method.

To determine the order of the new method for the more general single equation (6.6) we require the two-variable Taylor series expansion

$$f(t+h, y+k) = f(t,y) + \left(h\frac{\partial}{\partial t} + k\frac{\partial}{\partial y}\right)f(t,y)$$
$$+ \left(h\frac{\partial}{\partial t} + k\frac{\partial}{\partial y}\right)^2 \frac{f(t,y)}{2!} + \cdots .$$

Applying this to the right-hand side of (6.14), we have

$$f(t_i + \tfrac{1}{2}h, y_i + \tfrac{1}{2}hf_i) = f_i + \tfrac{1}{2}h\frac{\partial f_i}{\partial t} + \tfrac{1}{2}hf_i\frac{\partial f_i}{\partial y}$$
$$+ \frac{1}{8}\left(h^2\frac{\partial^2 f_i}{\partial t^2} + 2h^2 f_i\frac{\partial^2 f_i}{\partial t\partial y} + h^2 f_i^2\frac{\partial^2 f_i}{\partial y^2}\right)$$
$$+ O\left(h^3\right).$$

Under the usual assumptions, the LTE may now be determined from

$$\begin{aligned}
y(t_{i+1}) - y_{i+1} &= y(t_i) + hf\left(t_i, y(t_i)\right) + \tfrac{1}{2}h^2 F\left(t_i, y(t_i)\right) \\
&\quad + \tfrac{1}{6}h^3\left(G\left(t_i, y(t_i)\right) + f_y\left(t_i, y(t_i)\right)F\left(t_i, y(t_i)\right)\right) \\
&\quad + O\left(h^4\right) \\
&\quad - y_i - hf_i - \tfrac{1}{2}h^2 F(t_i, y_i) + \tfrac{1}{8}h^3 G(t_i, y_i) + O\left(h^4\right) \\
&= O\left(h^3\right)
\end{aligned}$$

where $F(t,y) = f_t + ff_y$ and $G(t,y) = f_{tt} + 2ff_{ty} + f^2 f_{yy}$, showing that the method is second order.

We may use this idea of 'scouting ahead', sampling the slope and adjusting the direction from t_i repeatedly. This leads to the general explicit Runge–Kutta formulation

$$y_{i+1} = y_i + h\sum_{r=1}^{R} c_r s_r,$$

where

$$s_r = f\left(t_i + a_r h, y_i + h\sum_{j=1}^{r-1} b_{rj} s_j\right), \qquad r = 1, 2, \ldots, R \qquad \textbf{(6.15)}$$

and $a_1 = 0$. R represents the number of stages (forward sampling forays used), and the a_r, c_r and b_{rj} are free parameters chosen to make the method as high order as possible. Note that the b_{rj} are only defined for $j < r$. As a result the method is said to be *explicit*; by this we mean that at each stage the s_rs and y_{i+1} may be computed immediately since we have explicit expressions for them. It is possible to derive *implicit* Runge–Kutta methods in which the upper limit on the summation sign in (6.15) is R. Now, at each step we have a nonlinear system to solve for the s_rs and finally y_{i+1}. Implicit methods possess some very desirable properties, but the need to use an iterative process at each stage to solve the nonlinear system is a serious drawback. Only explicit Runge–Kutta methods are implemented in the NAG library and we restrict ourselves accordingly. For details of implicit methods see Hall and Watt [41], p. 136.

The algebra involved in deriving high-order Runge–Kutta formulae is considerable and tedious. Here we confine our attention to the family of two-stage methods; higher-order methods can be derived in a completely analogous manner, as shown by Ralston and Rabinowitz [67], p. 208. The two-stage family takes the form

$$y_{i+1} = y_i + h(c_1 s_1 + c_2 s_2), \tag{6.16}$$

with $s_1 = f(t_i, y_i)$ and $s_2 = f(t_i + a_2 h, y_i + h b_{21} s_1)$. The four free parameters at our disposal are c_1, c_2, a_2 and b_{21}. We begin by expanding s_2 about $t = t_i$ and $y = y_i$. Expanding first about $t = t_i$ we have

$$s_2 = f(t_i, y_i + h b_{21} s_1) + a_2 h f_t(t_i, y_i + h b_{21} s_1) + O\left(h^2\right). \tag{6.17}$$

Expanding each function value in (6.17) about y_i then gives

$$s_2 = f(t_i, y_i) + h b_{21} s_1 f(t_i, y_i) f_y(t_i, y_i) + a_2 h f_t(t_i, y_i) + O\left(h^2\right),$$

and (6.16) may thus be written

$$y_{i+1} = y_i + h c_1 f_i + h c_2 f_i + h^2 c_2 b_{21} f_i(f_y)_i + h^2 c_2 a_2 (f_t)_i + O\left(h^3\right). \tag{6.18}$$

Now, the Taylor series expansion for $y(t_{i+1})$ about $t = t_i$ is

$$y(t_{i+1}) = y(t_i) + h y'(t_i) + \tfrac{1}{2} h^2 y''(t_i) + O\left(h^3\right),$$

and using the results (6.9) this may be written

$$y(t_{i+1}) = \left[y + h f + \tfrac{1}{2} h^2 (f_t + f f_y)\right]\big|_{t=t_i,\, y=y(t_i)} + O\left(h^3\right). \tag{6.19}$$

The final stage is to match coefficients in the two expansions (6.18) and (6.19) under the assumption that y_i is exact. This gives the equations

$$
\begin{array}{rclcl}
y & : & 1 & = & 1, \\
hf & : & c_1 + c_2 & = & 1, \\
h^2 f_t & : & c_2 \times a_2 & = & \tfrac{1}{2}, \\
h^2 f f_y & : & c_2 \times b_{21} & = & \tfrac{1}{2},
\end{array}
\tag{6.20}
$$

and the final three may be used to determine values for the four unknowns. Since the number of unknowns is one more than the number of equations (that is, we have an overdetermined system) one value may be chosen arbitrarily and the rest then determined. It is natural, though not necessary, to constrain a_2 so that it lies in $(0, 1]$, since f is then evaluated at a value for t in the range $(t_i, t_{i+1}]$. The resultant method is at least second-order since the LTE is at least $O\left(h^3\right)$. To show that all members of this family are exactly second-order we need to take the Taylor expansions one stage further and show that the $O\left(h^3\right)$ term does not vanish (see Exercise 6.10). The choice of a_2 cannot therefore influence the order of the method, although we can hope to minimize the coefficient in the LTE by making a judicious choice. We might also hope to influence the stability characteristics of the method by choosing a_2 appropriately. Unfortunately this is not possible. If we apply the method to the simple problem (6.5) and proceed in the same way as for Euler's method, it transpires that the interval of stability for the method defined by (6.16) is again $(-2, 0)$, regardless of the choice of a_2 (see Atkinson and Harley [6], p. 278).

For higher-order Runge–Kutta methods we will again have an overdetermined system to solve, and hence at least one degree of freedom available to minimize the LTE, or even to increase the order of the method. It can be shown (see Ralston and Rabinowitz [67], p. 222) that for an R-stage R^{th}-order method the interval of stability is independent of the choice of these parameters. However, for $R > 4$ the maximum attainable order is less than R and the parameters can be varied to adjust the interval of stability, as well as reduce the size of the LTE.

EXERCISE 6.10 Determine the two-stage, second-order Runge–Kutta methods corresponding to the choices $a_2 = 1$, $\frac{1}{2}$, $\frac{1}{3}$ and $\frac{1}{4}$. By expanding (6.18) and (6.19) up to $O\left(h^4\right)$ terms, obtain expressions for the dominant terms in the LTE for each method.

6.4   The Runge–Kutta–Merson algorithm

Any reasonable algorithm for the numerical solution of an IVP should include an appropriate error control strategy. By this we mean that the algorithm should be capable of detecting where to use a small step length, and where a relatively large step may be tolerated in order to produce solutions which are correct to the user's accuracy requirements. At the same time efficiency must be borne in mind and, in the present context, this means that the number of evaluations of $f(t, y)$ must be kept to a minimum.

Ideally we should attempt to control the global error, so that

$$|y(t_{i+1}) - y_{i+1}| \le abserr + relerr|y(t_{i+1})|$$

where *abserr* and *relerr* are absolute and relative error tolerances respectively. However this is not always easy, or indeed possible, and the large majority of the software currently available attempts to control the local error at each step only, in which case the inequality

$$|z(t_{i+1}) - y_{i+1}| \leq abserr + relerr|y_{i+1}|$$

holds, where z represents the trajectory which passes through (t_i, y_i). Controlling the local error only will not always provide true global error control because of the wandering from one trajectory to another that occurs as the computation proceeds, and consequently the accuracy of numerical solutions returned by software should, whenever feasible, be assessed by reducing the error tolerances (say, by a factor of 10) and comparing the two sets of figures.

In order to control the local error, we need to be able to compute an error estimate. If this indicates a local error in excess of the user tolerance we reject the step and retry with a smaller step length. Efficiency considerations dictate that if the error estimate is considerably smaller than the user tolerance, we should contemplate increasing the step size at the next step. This assumes, of course, that the behaviour of the solution over the next interval will not differ substantially from that in the step just computed. Clearly we need to err on the side of caution; there is no point in increasing the step size if this is immediately followed by a step reduction.

As an example of the use of step control we consider the algorithm originally proposed by Merson [57] (see also Fox [31]) which is based on the use of Runge–Kutta formulae. If we define

$$
\begin{aligned}
s_1 &= f(t_i, y_i), \\
s_2 &= f(t_i + \tfrac{1}{3}h, y_i + \tfrac{1}{3}hs_1), \\
s_3 &= f(t_i + \tfrac{1}{3}h, y_i + \tfrac{1}{6}hs_1 + \tfrac{1}{6}hs_2), \\
s_4 &= f(t_i + \tfrac{1}{2}h, y_i + \tfrac{1}{8}hs_2 + \tfrac{3}{8}hs_3), \\
s_5 &= f(t_i + h, y_i + \tfrac{1}{2}hs_1 - \tfrac{3}{2}hs_3 + 2hs_4),
\end{aligned}
$$

then the two methods

$$
\begin{aligned}
y_{i+1}^{[1]} &= y_i + \tfrac{1}{6}h(s_1 + 4s_4 + s_5), \\
y_{i+1}^{[2]} &= y_i + \tfrac{1}{2}h(s_1 - 3s_3 + 4s_4),
\end{aligned}
$$

are fourth order. Under the assumption that $f(t, y)$ is linear in both t and y, that is, we may write

$$f(t, y) = At + By + C \qquad (6.21)$$

for some constants A, B and C, Merson showed that the LTEs for the two methods are given by

$$
\begin{aligned}
y_{i+1}^{[1]} &= y(t_{i+1}) - h^5 y^{(5)}(\eta_1)/720, \\
y_{i+1}^{[2]} &= y(t_{i+1}) - h^5 y^{(5)}(\eta_2)/120,
\end{aligned}
$$

Initialize:
    set step_size = (interval_end − interval_start)/100
    current_position = start
    y = initial_value
    $i = 0$

Repeat
    compute $y_{i+1}^{[1]}$ and error_estimate
    If (error_estimate > user_tolerance) Then
        halve step_size
    Else
        overwrite y with $y_{i+1}^{[1]}$, the accepted approximation
        current_position = current_position + step_size
        If (error_estimate is well below user tolerance) Then
            double step_size
        If (next step would go beyond interval_end) Then
            set step_size = interval_end − current_position
        $i = i + 1$
Until (current_position = interval_end)

Figure 6.8   Skeletal algorithm for Runge–Kutta–Merson method.

---

respectively for some $\eta_1, \eta_2 \in [t_i, t_{i+1}]$. Assuming that the fifth derivative of y is constant over the current interval of interest we may deduce that

$$y_{i+1}^{[1]} - y_{i+1}^{[2]} = 5h^5 y^{(5)}(\eta)/720,$$

and hence that

$$\begin{aligned}
\left| y_{i+1}^{[1]} - y(t_{i+1}) \right| &= \tfrac{1}{5} \left| y_{i+1}^{[1]} - y_{i+1}^{[2]} \right| \\
&= \tfrac{1}{30} h | -2s_1 + 9s_3 - 8s_4 + s_5 |.
\end{aligned} \tag{6.22}$$

For the general function we need to assume that the interval is small enough for $f(t, y)$ to be expressed in the form (6.21). We then compute $y_{i+1}^{[1]}$ (noting that the coefficient in the LTE for this method is smaller than that for $y_{i+1}^{[2]}$) and use (6.22) as an estimate of the error in this approximation. Since the two fourth-order methods employ the same function evaluations, the error estimate essentially comes free.

We comment on the ease of changing step size using a Runge–Kutta method in that if a step is rejected we immediately fall back to our last accepted solution and no extra computation is required. This should be compared with the methods to be discussed in Section 6.5 and Section 6.7.

In Figure 6.8 we present a skeletal algorithm for step control using Merson's method (or, indeed, any ODE solver). A simple implementation of the algorithm is given in Code 6.2. A step halve/double strategy is used if the error estimate is too big/small. Modern practical implementations of such step control strategies attempt to determine a more optimal ratio at each change. MERSON integrates a system of N equations from $t = $ A to $t = $ B in such a way that the computed local error estimate at each step is within the user defined absolute tolerance TOL. The initial conditions should be provided in the array Y which on successful exit contains the computed approximations at $t = $ B. The right-hand side of the system is defined by the user supplied routine FUNCT whose format is the same as in subroutine EULER in Code 6.1. The arrays S(IS, 5) (with IS $\geq$ N) and Y1 (of length at least N) are used for workspace. The parameter IFAIL is set to zero if the routine succeeds, and to one if inconsistent input values are detected. Note that on some systems a run-time error will occur at the point of declaration of Y if N $<$ 0, before this is trapped by the first IF statement. See Exercise 1.14. If the routine attempts to use too small a step size (the limit being specified by the internal variable SMLSTP) the routine will halt with IFAIL set to two.

```
      SUBROUTINE MERSON(Y,N,A,B,S,IS,Y1,FUNCT,TOL,IFAIL)
*
*.. IMPLEMENTS THE RUNGE-KUTTA-MERSON METHOD FOR THE
*.. APPROXIMATE SOLUTION OF A SYSTEM OF N INITIAL VALUE
*.. PROBLEMS. THE INITIAL CONDITIONS AT T = A ARE PROVIDED IN
*.. THE ARRAY Y AND AN ATTEMPT IS MADE TO INTEGRATE THE
*.. SYSTEM TO T = B SO THAT THE ABSOLUTE ERROR ESTIMATE AT
*.. EACH ACCEPTED POINT IS WITHIN TOL.
*
*.. THE USER SUPPLIED ROUTINE
*
*           SUBROUTINE FUNCT(T,Y,F,N)
*
*.. SHOULD RETURN F(T,Y)
*
*.. THE ARRAY S(IS,5) WHERE IS.GE.N IS USED FOR WORKSPACE
*.. AS IS Y1(N)
*
      INTEGER IFAIL,IS,N
      DOUBLE PRECISION A,B,S(IS,5),TOL,Y(N),Y1(N)
*
```

```
*.. SET VARIOUS CONSTANTS, EXPERIMENTAL FACTORS, ETC.
*.. SMLSTP SIGNIFIES THE SMALLEST ALLOWABLE STEPSIZE
*.. FRACT IS THE FRACTION OF THE INTERVAL USED AS AN
*.. INITIAL STEP
      DOUBLE PRECISION DOUBFC,EIGHT,FOUR,FRACT,ONE,PT2,SIX,
     +                 SMLSTP,THREE,TWO,ZERO

      PARAMETER(TWO = 2.0D0,THREE = 3.0D0,FOUR = 4.0D0,
     +          SIX = 6.0D0,EIGHT = 8.0D0,PT2 = 0.2D0,
     +          ZERO = 0.0D0,ONE = 1.0D0,FRACT = 100.0D0,
     +          DOUBFC = 64.0D0,SMLSTP = 1.0D-6)
*
      INTEGER I
      DOUBLE PRECISION ERROR,H,H2,H3,H6,H8,T,Y4
      LOGICAL INCR,LAST
      INTRINSIC ABS,SIGN
*
*.. CHECK INPUT PARAMETERS
      IF(N.LT.1 .OR. IS.LT.N .OR. TOL.LT.ZERO)THEN
        IFAIL = 1
        RETURN
      ENDIF
*
*.. RETURN IMMEDIATELY IF RANGE OF LENGTH ZERO
      IF(A.EQ.B) THEN
        RETURN
      ENDIF
*
*.. INITIALIZE
      T = A
      LAST = .FALSE.
      H = (B-A)/FRACT
*
*.. MAIN LOOP
10    CONTINUE
*
      H2 = H/TWO
      H3 = H/THREE
      H6 = H/SIX
      H8 = H/EIGHT
*
*.. S1 IS IN THE FIRST COLUMN OF S
      CALL FUNCT(T,Y,S(1,1),N)
      DO 20 I = 1,N
        Y1(I) = Y(I)+H3*S(I,1)
```

```
20    CONTINUE
*
*.. S2 IS IN THE SECOND COLUMN OF S
      CALL FUNCT(T+H3,Y1,S(1,2),N)
      DO 30 I = 1,N
        Y1(I) = Y(I)+H6*(S(I,1)+S(I,2))
30    CONTINUE
*
*.. S3 IS IN THE THIRD COLUMN OF S
      CALL FUNCT(T+H3,Y1,S(1,3),N)
      DO 40 I = 1,N
        Y1(I) = Y(I)+H8*(S(I,2)+THREE*S(I,3))
40    CONTINUE
*
*.. S4 IS IN THE FOURTH COLUMN OF S
      CALL FUNCT(T+H2,Y1,S(1,4),N)
      DO 50 I = 1,N
        Y1(I) = Y(I)+H2*(S(I,1)-THREE*S(I,3)+FOUR*S(I,4))
50    CONTINUE
*
*.. S5 IS IN THE FIFTH COLUMN OF S
      CALL FUNCT(T+H,Y1,S(1,5),N)
*
*.. CALCULATE ERROR AND VALUES AT T+H
      INCR = .TRUE.
      DO 60 I = 1,N
        Y4 = Y(I)+H2*(S(I,1)-THREE*S(I,3)+FOUR*S(I,4))
        Y1(I) = Y(I)+H6*(S(I,1)+FOUR*S(I,4)+S(I,5))
*
*.. NOW THE ERROR ESTIMATE
      ERROR = ABS(PT2*(Y4-Y1(I)))
      IF(ERROR.GT.TOL)THEN
*
*.. STEP TOO BIG HALVE IT AND RETRY
          IF(ABS(H2).LT.SMLSTP)THEN
            IFAIL = 2
            RETURN
          ENDIF
          H = H2
          LAST = .FALSE.
          GOTO 10
        ENDIF
*
* LOOK OUT FOR POSSIBLE INCREASE IN THE STEP SIZE
        IF(ERROR*DOUBFC.GT.TOL)THEN
```

```
         INCR = .FALSE.
      ENDIF
60    CONTINUE
*
*.. STEP HAS BEEN ACCEPTED-GET READY FOR THE NEXT STEP
*.. MOVE THE NEW VALUES INTO Y
      DO 70 I = 1,N
         Y(I) = Y1(I)
70    CONTINUE
      T = T+H
      IF(.NOT. LAST) THEN
         IF(INCR) THEN
            H = TWO*H
         ENDIF
*
*.. LAST STEP CHANGE H
         IF(SIGN(ONE,H)*(T+H).GT.B)THEN
            H = B-T
            LAST = .TRUE.
         ENDIF
         GOTO 10
      ENDIF
      END
```

Code 6.2   Subroutine MERSON.

The success of the Merson LTE estimator rests on the accuracy of the representation (6.21) and it is not difficult to devise problems which will fool the method. Shampine and Watts [73] are particularly critical to the point that they dismiss it from consideration as a possible scheme for local error control. Their main complaint is that the error estimates are too conservative, and hence any code which uses them is likely to be inefficient. This is not to say that such a program will be unreliable, just that the number of function calls made may be larger than necessary. Many other LTE estimators have been proposed, of which one of the best known is the Fehlberg 4-5 scheme (Fehlberg [28]), see Exercise 6.11. This scheme forms the basis of the integrator RKF45 (Shampine and Watts [74]), and of the global error estimating code GERK (Shampine and Watts [72]).

EXERCISE 6.11   An alternative strategy to Merson's method for step control using Runge–Kutta methods is to derive a first estimate, $y_{i+1}^{[1]}$, for $y(t_{i+1})$ using a pth-order method, and a second estimate, $y_{i+1}^{[2]}$, using a $(p+1)^{st}$-order method. Now, by increasing the order of the method we would hope to increase the accuracy of the approximation. Hence, if we take $y_{i+1}^{[2]}$ to be $y(t_{i+1})$, the quantity $\left| y_{i+1}^{[1]} - y_{i+1}^{[2]} \right|$ can be taken as an estimate of the magnitude of the error in $y_{i+1}^{[1]}$.

Fehlberg [28] has proposed a number of methods of this form, of which the best known is the fourth/fifth-order pair

$$y_{i+1}^{[1]} = y_i + \frac{25}{216}hs_1 + \frac{1408}{2565}hs_3 + \frac{2197}{4104}hs_4 - \frac{1}{5}hs_5,$$

$$y_{i+1}^{[2]} = y_i + \frac{16}{135}hs_1 + \frac{6656}{12825}hs_3 + \frac{28561}{56430}hs_4 - \frac{9}{50}hs_5 + \frac{2}{55}hs_6,$$

where

$$\begin{aligned}
s_1 &= f(t_i, y_i), \\
s_2 &= f\left(t_i + \tfrac{1}{4}h, y_i + \tfrac{1}{4}hs_1\right), \\
s_3 &= f\left(t_i + \tfrac{3}{8}h, y_i + \tfrac{3}{32}hs_1 + \tfrac{9}{32}hs_2\right), \\
s_4 &= f\left(t_i + \tfrac{12}{13}h, y_i + \tfrac{1932}{2197}hs_1 - \tfrac{7200}{2197}hs_2 + \tfrac{7296}{2197}hs_3\right), \\
s_5 &= f\left(t_i + h, y_i + \tfrac{439}{216}hs_1 - 8hs_2 + \tfrac{3680}{513}hs_3 - \tfrac{845}{4104}hs_4\right), \\
s_6 &= f\left(t_i + \tfrac{1}{2}h, y_i - \tfrac{8}{27}hs_1 + 2hs_2 - \tfrac{3544}{2565}hs_3 + \tfrac{1859}{4104}hs_4 - \tfrac{11}{40}hs_5\right).
\end{aligned}$$

Note that the s_rs are shared by the two methods and so the cost of the error estimate is just the function evaluation s_6. Write a routine, equivalent to MERSON (Code 6.2), which implements the Runge–Kutta–Fehlberg fourth/fifth-order method, and compare experimentally the performance of the two schemes.

EXERCISE 6.12 The second-order BVP

$$y'' = f(t, y, y'), \tag{6.23}$$

with

$$y(a) = \alpha, \qquad y(b) = \beta \tag{6.24}$$

can be converted to a coupled system of two first-order equations, for $y_1 \equiv y$ and $y_2 \equiv y'$, with the boundary conditions $y_1(a) = \alpha$, $y_1(b) = \beta$. If we guess a value μ, say, for $y_2(a)$, we have an initial value problem, for $y_1^{(\mu)}$ and $y_2^{(\mu)}$, say, which can be solved by any suitable integrator. The chances of the solution to this problem matching the known boundary condition at $t = b$ are remote, and hence further values should be chosen until this *shooting method* is on target. Effectively, we have a root-finding problem, since we are trying to find that value of μ which gives $F(\mu) \equiv y_1^{(\mu)} - \beta = 0$, and any appropriate technique may be used. Using the Runge–Kutta–Merson integrator MERSON of Code 6.2, write a program based on the Bus and Dekker algorithm (Section 2.3) for finding the value of $y'(a)$, where y satisfies the differential equation (6.23), with associated boundary conditions (6.24).

6.5 Linear multistep methods

We now investigate a second way of obtaining high-order methods for first-order IVPs which avoids the need to compute derivative values. So far we have always projected our approximation forward using just a single value of the independent variable, that is we use values computed at $t = t_i$ only to calculate the approximation at $t = t_{i+1}$. Methods which adopt this strategy are termed *one-step* for obvious reasons. The methods we next

NTIME	$y_{t=2}^{true} - y_{t=2}^{approx}$
16	0.208859×10^{-2}
64	0.121927×10^{-3}
256	0.749372×10^{-5}
1024	0.466428×10^{-6}
4096	0.291218×10^{-7}
16384	0.181965×10^{-8}
65536	0.113723×10^{-9}

Figure 6.9   Solution using a two-step linear multistep method.

consider use information at several previous points and are termed *linear multistep methods* (LMMs).

Consider the two-step scheme

$$y_{i+1} = y_i + \tfrac{1}{2}h(3f_i - f_{i-1}). \tag{6.25}$$

Using Taylor series expansions it is easy to show that

$$y(t_{i+1}) = y(t_i) + \tfrac{1}{2}h\big(3y'(t_i) - y'(t_{i-1})\big) + \tfrac{5}{12}h^3 y'''(t_i) + O\left(h^4\right),$$

which indicates that (6.25) is a second-order method.

Immediately we see that there is a computational problem associated with multistep methods, namely starting. The first application of (6.25) gives $y_2 = y_1 + \tfrac{1}{2}h(3f_1 - f_0)$ but we are provided only with the single starting value y_0 in the statement of the problem. We could generate y_1 using one of the one-step methods discussed earlier (Taylor series or Runge–Kutta), but we must ensure that the method used gives a sufficiently accurate value, that is, its order must be at least that of the LMM.

In Figure 6.9 we list results obtained on $[0, 2]$ using (6.25) to solve (6.11) using several values of h; in each case the exact value was used for y_1. The second-order behaviour of the method is clearly illustrated.

Having introduced a method we must, as usual, attempt to establish its stability characteristics, and again we do this in the context of the simple problem (6.5) with initial value $y(0) = 1$. Figure 6.10 shows the result of integrating to $t = 1$ using (6.25) with $h = 0.01$ for different values of λ. For $\lambda > 0$ the results exhibit the correct qualitative behaviour; that is, they grow exponentially with λ. For $\lambda < 0$, however, the approximation is adequate only for $\lambda > -100$. In order to explain this behaviour we consider the difference equation for the method which takes the form

$$y_{i+1} - (1 + \tfrac{3}{2}h\lambda)y_i + \tfrac{1}{2}h\lambda y_{i-1} = 0.$$

This has a general solution of the form

$$y_i = az_1^i + bz_2^i, \tag{6.26}$$

λ	$y_{t=1}^{approx}$	$y_{t=1}^{true} - y_{t=1}^{approx}$
-105	0.598509×10^2	0.598509×10^2
-95	0.888755×10^{-4}	0.888755×10^{-4}
-60	0.622917×10^{-22}	0.622829×10^{-22}
-10	0.474156×10^{-4}	0.201570×10^{-5}
-1	0.367894×10^0	0.152744×10^{-4}
1	0.271817×10^1	0.111502×10^{-3}
10	0.211935×10^5	0.832903×10^3
20	0.364902×10^9	0.120262×10^9
135	0.114003×10^{17}	0.114003×10^{17}
150	0.755564×10^{22}	0.755564×10^{22}

Figure 6.10   Effect of varying λ with fixed time step.

where z_1 and z_2 are the roots of the associated quadratic polynomial (the *characteristic polynomial*)

$$z^2 - (1 + \tfrac{3}{2}h\lambda)z + \tfrac{1}{2}h\lambda = 0. \tag{6.27}$$

The fact that (6.27) is a second-degree polynomial means that the underlying difference scheme is termed second order. This should not be confused with the order of the LMM, although the latter is second order as well in this case. Using the usual formula for determining the roots of a quadratic, and expanding the square root term in this formula using a binomial expansion, it can be shown that one of the roots of (6.27) may be written

$$z_1 = 1 + h\lambda + \tfrac{1}{2}(h\lambda)^2 - \tfrac{1}{4}(h\lambda)^3 + \cdots$$

which, up to $O\left(h^3\right)$ terms, is the Taylor series expansion for $e^{\lambda h}$ about 0. Hence the term in the general solution (6.26) corresponding to this root exhibits the correct qualitative behaviour. In consequence z_1 is termed the *principal root*. The second root, the *extraneous root*, is present because we have used a second-order difference approximation to a first-order equation, and its existence may influence the accuracy of our approximation. In fact it can be shown that if λ is real and negative, then no problems are encountered provided $h\lambda \geq -1$, as evidenced by the results shown in Figure 6.10 for which h was given the value 0.01. The LMM is termed *absolutely stable* for $h\lambda \geq -1$, $\lambda < 0$. For h outside this range the method is termed unstable. Indeed, all but some of the simplest of LMMs have restrictions on the size of $h\lambda$ for $\lambda < 0$ as a direct result of stability considerations.

Our stability analysis for the LMM (6.25) is valid only for a very simple problem. For practical purposes we need to generalize our concept of λ in (6.5) to cover more complex equations, and to embrace systems of initial value problems. If an LMM is used to solve a nonlinear problem it is possible to wander in and out of the region of stability. Ideally, we should embark on some sort of stability analysis each time a new problem is encountered so that we may obtain some insight into how a particular method is likely to perform. However, it is unreasonable to expect the average user of numerical software to set out on such an adventure, and so IVP codes must check for the possible occurrence of instability. We return to this in the next section.

In its most general form an r-step LMM may be written

$$\alpha_r y_{i+1} + \alpha_{r-1} y_i + \cdots + \alpha_0 y_{i-r+1}$$
$$= h(\beta_r f_{i+1} + \beta_{r-1} f_i + \cdots + \beta_0 f_{i-r+1}) \qquad (6.28)$$

where $\{\alpha_k, \beta_k \mid k = 0, 1, \ldots, r\}$ are constants and $\alpha_r = 1$ by convention. The method is explicit if $\beta_r = 0$, and implicit otherwise. Note that the method cannot be used to provide values for $y_1, y_2, \ldots, y_{r-1}$. To investigate the stability characteristics of such a method we again consider (6.5) and obtain the rth-order difference equation

$$(\alpha_r - h\lambda\beta_r)y_{i+1} + (\alpha_{r-1} - h\lambda\beta_{r-1})y_i + \cdots + (\alpha_0 - h\lambda\beta_0)y_{i-r+1} = 0,$$

which has a general solution of the form $y_i = \sum_{k=1}^{r} a_k z_k^i$, where the z_ks are the roots of the characteristic polynomial $\sum_{k=0}^{r}(\alpha_k - h\lambda\beta_k)z^k = 0$. We now have $r - 1$ extraneous roots, each of which is potentially capable of rendering an approximate solution worthless.

In a general LMM (6.28) there are $2r - 1$ free parameters available and we might hope to choose these so that, when each term in (6.28) is expanded by Taylor series about $t = t_i$, $y = y_i$, we match the expansion of $y(t_{i+1})$ about $t = t_i$, thus producing a method of order $2r$. However, if we are to control the behaviour of the 'spurious' terms associated with the extraneous roots, the maximum attainable order is $r + 2$ if r is even, and $r + 1$ if r is odd. Unfortunately, optimal-order even-step methods suffer from computational disadvantages and hence are rarely used.

Explicit LMMs of the form

$$y_{i+1} - y_i = h(\beta_{r-1} f_i + \beta_{r-2} f_{i-1} + \cdots + \beta_0 f_{i-r+1})$$

form a family of schemes known as *Adams–Bashforth* (AB) methods. We list in Figure 6.11 coefficients for $r = 1, 2, 3$ and 4, and observe that the majority of the coefficients are greater than one in modulus. This is a serious drawback since the effect will be a tendency to amplify rounding errors. The stability interval quoted in this table is for our model problem (6.5). The final column gives the coefficient of $h^{order+1}y^{(order+1)}$ in the LTE.

Implicit LMMs of the form

$$y_{i+1} - y_i = h(\beta_r f_{i+1} + \beta_{r-1} f_i + \cdots + \beta_0 f_{i-r+1})$$

are termed *Adams–Moulton* (AM) methods and characteristics of these methods for $r = 1, 2, 3$ and 4 are given in Figure 6.12.

r	β_0	β_1	β_2	β_3	Order	Stability interval	Coefficient in LTE
1	1				1	$(-2,0)$	$\frac{1}{2}$
2	$\frac{3}{2}$	$-\frac{1}{2}$			2	$(-1,0)$	$\frac{5}{12}$
3	$\frac{23}{12}$	$-\frac{16}{12}$	$\frac{5}{12}$		3	$(-\frac{6}{11},0)$	$\frac{3}{8}$
4	$\frac{55}{24}$	$-\frac{59}{24}$	$\frac{37}{24}$	$-\frac{9}{24}$	4	$(-\frac{3}{10},0)$	$\frac{251}{720}$

Figure 6.11   Adams–Bashforth formulae.

Comparing AB and AM methods of the same order, we see that the AM method is more appealing since the coefficients are all less than one in modulus, and the interval of stability is greater. Further, the coefficient of the LTE is smaller for an AM method than for the corresponding AB method. However, at each step we need to solve for y_{i+1} a nonlinear equation of the form

$$y_{i+1} - h\beta_r f(t_{i+1}, y_{i+1}) = g_i,$$

where

$$g_i = y_i + h\sum_{k=0}^{r-1} \beta_k f_{i-k+1},$$

and for systems of IVPs the situation is compounded since we will have a nonlinear system to solve. Any of the methods discussed in Chapter 2 for locating the root of an equation could, in principle, be used here, but a simpler approach is to use fixed-point iteration (q.v. Exercise 2.2). Starting with $y_{i+1}^{[0]}$, an initial estimate of y_{i+1}, we compute further iterates using

$$y_{i+1}^{[l]} = g_i + h\beta_r f\big(t_{i+1}, y_{i+1}^{[l-1]}\big), \qquad l = 1, 2, \ldots$$

until some suitable convergence criterion is satisfied. It may be shown (see Henrici [42]) that the iteration converges if $h < 1/|L\beta_r|$ where L is an upper

r	β_0	β_1	β_2	β_3	β_4	Order	Stability interval	Coefficient in LTE
1	$\frac{1}{2}$	$\frac{1}{2}$				2	$(-\infty,0)$	$-\frac{1}{12}$
2	$\frac{5}{12}$	$\frac{8}{12}$	$-\frac{1}{12}$			3	$(-6,0)$	$-\frac{1}{24}$
3	$\frac{9}{24}$	$\frac{19}{24}$	$-\frac{5}{24}$	$\frac{1}{24}$		4	$(-3,0)$	$-\frac{19}{720}$
4	$\frac{251}{720}$	$\frac{646}{720}$	$-\frac{264}{720}$	$\frac{106}{720}$	$-\frac{19}{720}$	5	$(-\frac{90}{49},0)$	$-\frac{3}{160}$

Figure 6.12   Adams–Moulton formulae.

limit on the magnitude of f_y. Hence we have an additional constraint on the step size chosen, although this is often not as severe as that imposed by stability considerations.

It is obviously computationally advantageous to choose $y_{i+1}^{[0]}$ as close to y_{i+1} as possible, since each iteration requires a function evaluation. The value at the previous point, y_i, immediately suggests itself as a candidate, but better still we could use an explicit LMM of the same order as the implicit LMM to provide the value used to start the iteration. When an explicit and implicit LMM are combined in this way they are said to form a *predictor–corrector* pair.

We now return to the problem of when to halt the iteration. The obvious approach is to continue the process until two successive iterates differ by less than some prescribed tolerance. If we do iterate, or correct, to convergence we may find that the number of function evaluations made is quite large. Further, at each iteration we have to check for convergence. However, we guarantee that the stability properties of the corrector are retained; the action of the predictor is solely to provide a starting value for iteration.

An alternative way of implementing the corrector iteration is to apply it a fixed number of times m, say, at each step. The final iterate, $y_{i+1}^{[m]}$, is then taken to be y_{i+1} and we move on to the next step. The role of the predictor is now more critical and the two methods must be treated as an intimately linked pair. By simplifying the iterative process we sacrifice some of the stability interval of the corrector formula, although with care we can ensure that the effect of this is not too great. To discuss the matter further it is convenient to identify the individual processes which occur at each step as

- P: an application of the predictor formula,

- E: an evaluation of f at $t = t_{i+1}$, $y = y_{i+1}^{[m]}$, where $y_{i+1}^{[m]}$ is the most recently computed estimate of y_{i+1}, and

- C: an application of the corrector formula.

Thus, a single iteration may be classed as PEC or PECE. At first it might appear that the final evaluation in PECE is wasted since it has no bearing on the value used for y_{i+1}, obtained after the use of the corrector. However, at each step we retain a value f_{i+1} for use at later stages. In the case of PEC this will be equal to $f(t_{i+1}, y_{i+1}^{[0]})$, but in PECE it will be $f(t_{i+1}, y_{i+1}^{[1]})$. In general the use of PECE mode gives a larger interval of stability than PEC, at the cost of a further function call. An additional application of the corrector (to give PECEC, or $P(EC)^2$ mode) is actually detrimental in the present context. As an example, for the fourth order AB-AM pair, the stability interval for PEC mode is $(-0.16, 0)$, for PECE mode is $(-1.25, 0)$, and for PECEC mode is $(-0.9, 0)$.

EXERCISE 6.13 The *midpoint* method $y_{i+1} = y_{i-1} + 2hf_i$ is second order. For the model problem (6.5) find the two roots of the characteristic polynomial of this method and investigate its stability properties. Determine also the stability region of the trapezoidal method

$$y_{i+1} = y_i + \tfrac{1}{2}h\left(f_i + f_{i+1}\right). \tag{6.29}$$

EXERCISE 6.14 LMM's may be derived by integrating the ODE and then employing a quadrature rule. Integrating (6.6) over $[t_{i-r+1}, t_{i+1}]$ gives

$$y(t_{i+1}) - y(t_{i-r+1}) = \int_{t_{i-r+1}}^{t_{i+1}} f(t, y)\, dt. \tag{6.30}$$

Write down the method corresponding to (i) $r = 1$ and the use of the trapezium rule (Section 3.1) to approximate the right-hand side of (6.30), (ii) $r = 2$ and the use of Simpson's rule (Exercise 3.6). Investigate experimentally the stability of the implicit fourth-order method (ii) when combined with the explicit fourth-order method

$$y_{i+1} = y_{i-3} + \tfrac{4}{3}h(2f_i - f_{i-1} + 2f_{i-2})$$

to form a predictor–corrector pair, correcting until some appropriate convergence criterion is achieved. (The combination is known as Milne's method.)

One way of deriving Simpson's rule is to integrate over $[t_{i-1}, t_{i+1}]$ the Lagrange interpolating polynomial (Section 3.2) based on the point set $\{t_{i-1}, t_i, t_{i+1}\}$. Indicate how this approach may be employed to derive the Adams–Bashforth and Adams–Moulton methods.

EXERCISE 6.15 The fourth-order AB and AM methods

$$\begin{aligned}
y_{i+1} &= y_i + \tfrac{1}{24}h(55f_i - 59f_{i-1} + 37f_{i-2} - 9f_{i-3}) \\
y_{i+1} &= y_i + \tfrac{1}{24}h(9f_{i+1} + 19f_i - 5f_{i-1} + f_{i-2})
\end{aligned}$$

can be combined to form a predictor–corrector pair. Write a program which implements this scheme using a fixed step length and (i) PEC, (ii) PECE, and (iii) PECEC mode. Carefully monitor the difference in the accuracy achieved by each mode on a range of problems.

EXERCISE 6.16 In the *method of undetermined coefficients* the function values in the r-step LMM (6.28) are expanded, using Taylor series, and the α_ks and β_ks chosen so that as many terms as possible agree with the Taylor series expansion for $y(t_{i+1})$. Show that the method $y_{i+1} = y_i + h(\gamma f_i + (1 - \gamma)f_{i+1})$ is first order (that is, the Taylor series agree up to $O\left(h^2\right)$), and write down the schemes corresponding to the choices $\gamma = 1, 0$ and $\tfrac{1}{2}$. Setting $\alpha_2 = 1$ and $\alpha_0 = c$, derive the equations which must be satisfied for a two-step method to be third order, and determine the solution. Write down the method corresponding to the choice $c = 1$ and show that it is fourth order.

6.6 The NAG Merson routines

Routines for the numerical solution of ordinary differential equations are contained in chapter D02 of the NAG library. Since we are only concerned with first-order initial value problems in this chapter of the book we restrict our discussions accordingly. The majority of the routines available can be classified as

- Merson's method routines,
- Adams' method routines,
- BDF routines.

Each routine is capable of solving a system of differential equations. We consider here the Merson routines; a pattern will then be set in which the remaining routines of interest may be discussed.

The Merson routines make use of the Runge–Kutta–Merson error control technique outlined in Section 6.4. There are five 'driver' routines available:

- D02BAF: integrates up to some user specified point;
- D02BBF: same as D02BAF but with intermediate output;
- D02BDF: same as D02BBF but returns a global error estimate and a stiffness check;
- D02BGF: same as D02BAF but integrates until a component of the solution vector attains a specified value;
- D02BHF: same as D02BAF but integrates until a function of the solution vector is zero.

Each routine makes a call to the 'comprehensive' integrator D02PAF which will be discussed in detail later.

In describing the purpose of D02BDF we have used a term which has not yet been introduced, namely, stiffness. Briefly, a problem is stiff if the solution contains both rapidly and slowly decaying components (e.g., e^{-100t} and $e^{-0.01t}$). Stiffness is a qualitative term, the rate of decay indicating the degree of stiffness. The methods considered so far do not deal efficiently with stiff systems; stability considerations dictate that a very small step length must be used throughout the entire range of integration, even though the components that give trouble rapidly tend to zero. We shall be content to work here with our vague definition of stiffness, although we note that the stiffness of a problem can be measured in terms of the eigenvalues of the Jacobian of f. A detailed discussion of the phenomenon of stiffness may be found in Gear [32] and Gear [33]. Special purpose algorithms exist for solving stiff systems; in the present context the BDF routines are of relevance and we consider their performance in Section 6.8.

A call to D02BAF takes the form

```
CALL DO2BAF(X,XEND,N,Y,TOL,FCN,W,IFAIL)
```

where N contains the number of equations in the system. On entry Y (of size at least N) must contain the initial conditions, that is, values of the vector-valued function y at the initial point of the range of interest, defined on entry by X. The vector-valued function f is defined using the parameter FCN, whose specification is

```
SUBROUTINE FCN(T,Y,F)
DOUBLE PRECISION T,Y(n),F(n)
```

The form of this routine is almost identical to that used in EULER (Code 6.1) and MERSON (Code 6.2). However, here n is not a parameter; it must be the actual value given to N in the call to DO2BAF. We recommend that the value used to dimension the arrays should be N in both FCN and the calling unit, and its value specified via a PARAMETER statement in both units to indicate its significance. Alternatively the value should be passed to FCN via a COMMON block. The workspace array W must be of size $(N, p(\geq 7))$.

On successful termination DO2BAF returns in Y an approximate solution at XEND. (Note that it is possible for XEND to be less than X, in which case integration proceeds in a negative direction.) The initial conditions are hence overwritten (as is X, which is assigned the value XEND) and care must be exercised if further calls to DO2BAF are to be made, for example, to check whether the required global accuracy has been obtained. The user supplied tolerance TOL is used to control the error incurred in a single Runge–Kutta–Merson step. However, there can be no guarantee that the computed values are correct to this accuracy. The recommendation is that the routine should be called with two values of TOL, say 10^{-k} and $10^{-(k+1)}$, and the two sets of solutions compared. The agreement between the two sets of figures will indicate the precision to which the results may be quoted with some degree of confidence. The value of TOL is normally unchanged on exit. However, if the interval [X,XEND] is regarded by DO2BAF as being so small that a small change in TOL is likely to have little effect on the computed solution, the sign of TOL is switched. This acts as a gentle warning that if the size of the interval is increased and the same value of TOL is used, the accuracy in the computed solution is likely to worsen.

There are four ways in which DO2BAF will terminate with an error:

- IFAIL = 1 is set if either TOL or N are non-positive on entry.

- IFAIL = 2 is set if the routine has reached a point at which no further progress may be made without violating the accuracy criterion. This could result simply from the choice of a very small value of TOL. It may also be symptomatic of some difficulty in the problem (such as a singularity). If the user feels that a realistic value for TOL has been chosen then he should attempt to analyse his problem in order to ascertain whether there are any points at which difficulties are likely to occur. In particular he should attempt to discover whether his problem is stiff. Fortunately the NAG library is able to help in this matter; D02BDF returns a value by which the stiffness of a problem may be gauged and we return to it shortly.

- IFAIL = 3 is set if the routine finds that the accuracy condition is too strict at the outset. Either TOL is too small or the system is stiff.

- IFAIL = 4 is set if a serious error occurs in D02PAF which is called by D02BAF. We consider the performance of D02PAF shortly, but note here the advice given in the NAG manual should this error be flagged. 'Check all subroutine calls and array dimensions. Seek expert help.' No matter how comprehensive NAG routines may be, there are certain circumstances that just cannot be catered for, or detected, so that specific advice may be given. This is of little comfort to the user, but fortunately such instances rarely occur.

If D02BAF halts with IFAIL = 2 or 4, the value of X indicates the point reached before the error occurred, and Y contains the approximate solution at this point.

D02BBF operates in a similar way to D02BAF; its call sequence is

```
    CALL D02BBF(X,XEND,N,Y,TOL,IRELAB,FCN,
   +             OUTPUT,W,IFAIL)
```

The only difference between this and the call sequence for D02BAF is the inclusion of two additional arguments, namely IRELAB and OUTPUT. IRELAB gives the user some control on the accuracy condition used. If est is the local error estimate at a particular point, the step is accepted if

$$est \leq \text{TOL} \times \max\{1.0, |Y(1)|, |Y(2)|, ..., |Y(N)|\} \quad \text{for} \quad \text{IRELAB} = 0,$$
$$est \leq \text{TOL} \quad \text{for} \quad \text{IRELAB} = 1,$$
$$est \leq \text{TOL} \times \max\{eps, |Y(1)|, |Y(2)|, ..., |Y(N)|\} \quad \text{for} \quad \text{IRELAB} = 2,$$

where eps is a small machine-dependent number equal to $dwarf/macheps$, where $dwarf$ is the smallest positive machine-representable real number (as given by X02ALF), and $macheps$ is the smallest positive real number for which $1.0 + macheps > 1.0$ (as given by X02AJF). On the VAX 11/780 the value of eps is 1.05×10^{-22}. It can be seen that the choice IRELAB = 1 corresponds to an absolute error test, IRELAB = 2 gives a relative error test, and IRELAB = 0 corresponds to a mixture of the two. (Note that a call of D02BAF corresponds to the choice IRELAB = 0.) If the value of est is very small an attempt will be made to increase the current step value when computing the next set of approximate values. If the error condition is not satisfied the step size is reduced and a further attempt made.

The user supplied subroutine OUTPUT permits the printing of approximate solutions at points interior to [X,XEND] (it corresponds to MONIT of EULER (Code 6.1)). Its specification is

```
SUBROUTINE OUTPUT(XSOL,Y)
DOUBLE PRECISION XSOL,Y(n)
```

As with FCN, the dimension of Y is not specified via the parameter list, and again we comment that care must be exercised. Input to OUTPUT consists of XSOL, a value of the independent variable at which solutions are available and Y, the value of the solution at this point. On exit XSOL should be set to the next point at which intermediate output is required. Hence, for example, if we require solutions at equally spaced points, distance 0.1 apart, we include the statement

```
XSOL = XSOL+0.1D0
```

before any return. (The initial call of OUTPUT made by D02BBF is with XSOL equal to the initial point of the range, the value specified by X on entry.) On no account should the values contained in Y be altered by this routine.

Essentially, OUTPUT defines an initial discretization of the range defined by [X,XEND]; as soon as a point is reached which is beyond the first point of this discretization the process halts and an estimate of the solution at XSOL is obtained using D02XAF (see below). This is repeated until XEND is reached. It is important, therefore, that the output value of XSOL is actually greater than the input value. If this is not so, D02BBF will continue to integrate towards XEND and return a value for IFAIL of 5. Note that D02BBF will not go beyond XEND, even if a value greater than XEND is returned via XSOL in OUTPUT.

The errors detected by D02BBF include those detected by D02BAF. D02BBF additionally checks the validity of the value given to IRELAB and, as we have already indicated, the values given to XSOL by OUTPUT. Further, D02BBF also calls D02XAF, and if a serious error occurs in this latter routine IFAIL is set to 7.

The call sequence for D02BDF is

```
CALL D02BDF(X,XEND,N,Y,TOL,IRELAB,FCN,STIFF,
+           YNORM,W,IW,M,OUTPUT,IFAIL)
```

and the parameters additional to the sequence for D02BBF are STIFF, YNORM, IW and M. The routine D02BDF is able to return via STIFF a value in the range [0,1] indicating the stiffness of the problem; the higher the value of STIFF, the more stiff the problem is. If, on entry, STIFF is positive a stiffness value is computed at the end of each integration step. At XEND a further stiffness estimate is computed and the value returned is the average of the two. If STIFF is negative on entry the second stiffness value at XEND is not computed; moreover the process used to determine the stiffness value, which is evolutionary in nature, is restarted every 1000 steps. If STIFF is zero on entry then no stiffness check is made. Precisely what the user should do with these stiffness estimates we leave until later when the purpose of the remaining parameters has been discussed. Details of how the stiffness estimates are determined may be found in Shampine [71].

The value of YNORM indicates on entry a limit on the size of the solution. If at any stage a component of the solution exceeds YNORM in magnitude the routine halts. The purpose of this parameter is, therefore, to give the user an opportunity to terminate the integration process if things are getting out of hand. YNORM should be given a non-negative value; if zero, it has no effect.

For DO2BDF the workspace array W is of a different size, and this means that care must be exercised if a decision is made to switch from, say DO2BAF to DO2BDF, or vice versa. Here W must be of size (N, IW) where

$$IW \geq 14 \quad \text{if} \quad STIFF > 0.0,$$
$$IW \geq 13 \quad \text{if} \quad STIFF < 0.0,$$
$$IW \geq 12 \quad \text{if} \quad STIFF = 0.0.$$

If the user thinks it likely that he will call DO2BDF with positive, negative and zero values for STIFF then the choice $IW \geq 14$ will cover every eventuality.

The form of the routine OUTPUT is also different for DO2BDF. Its specification is

```
SUBROUTINE OUTPUT(X,Y,W,STIFF)
DOUBLE PRECISION X,Y(n),W(n,iw)
```

where the values of n and iw must be the same as those given to N and IW respectively prior to the call of DO2BDF. On input the role of the first two parameters is the same as before. However, X must remain unchanged on exit; the points at which output is produced are controlled by the parameter M of DO2BDF, OUTPUT being called every M integration steps. (If M is negative no calls of OUTPUT are made.) At each Mth step the computed stiffness estimate is available and hence its value may be printed.

The interpretation of the stiffness estimates is a matter in which intuition plays an important role. However, the NAG documentation does lay down some guidelines as to how these values may be used. In particular, if STIFF is given a positive value on input, and the final value of STIFF exceeds 0.75, the use of a special purpose routine for stiff problems is recommended. The purpose of DO2BDF is essentially to provide the user with some feedback as to whether his problem is stiff or not. Unless the user knows in advance that his system is stiff some preliminary investigation using this routine should be performed. For non-stiff systems, continued use of a Runge–Kutta–Merson routine may be appropriate, although DO2BAF,

DO2BDF and DO2PAF are likely to prove less demanding on computing resources. Alternatively, use of one of the Adams routines may be more suitable, particularly if the right-hand sides of the differential equation are expensive to evaluate. If the user is convinced that his system is stiff, the use of a BDF routine may well prove more efficient. We return to this in Section 6.8 and in Section 6.9 where a comparison of the performance of the NAG ODE routines is given.

There are no less than eighteen error indicators that can be returned by DO2BDF. There is however a certain amount of replication. DO2BDF makes two calls to DO2PAF, and the values $IFAIL = 12, 13, \ldots, 18$ correspond to $IFAIL = 2, 3, \ldots, 8$ but the former apply to the second call of DO2PAF, whilst the latter correspond to the first call.

A call of DO2BGF takes the form

```
CALL DO2BGF(X,XEND,N,Y,TOL,HMAX,M,VAL,FCN,W,IFAIL)
```

In addition to the parameters for DO2BAF we have HMAX, M and VAL (but note that the second dimension of W needs to be at least 10). Recall that DO2BGF attempts to locate the point at which the Mth component of the solution equals some specified value (VAL). Starting at the initial point X, the routine integrates towards XEND until a point is reached at which $Y(M) - VAL$ changes sign. The conclusion is that a bracket for a zero of the Mth component has been found, and a refined estimate, returned in X, is determined using the root-finding technique implemented in the reverse communication routine CO5AZF (see Section 2.4). If no root is found then DO2BGF halts with $IFAIL = 4$. The role of HMAX is to ensure that the bracketing process does not miss a zero, should there be multiple roots, by limiting the step sizes in the integration process. The value of HMAX may be positive or negative; if zero, no special action is taken.

A call of DO2BHF takes the form

```
CALL DO2BHF(X,XEND,N,Y,TOL,IRELAB,HMAX,FCN,G,W,IFAIL)
```

By now, with the exception of G, the parameters in this call should require little introduction (although we note that, as with DO2BAF and DO2BBF, the second dimension of W need only be 7). Recall that this routine attempts to locate a point at which a function of the solution equals zero. This function is defined by G which has specification

```
FUNCTION G(T,Y)
DOUBLE PRECISION G,T,Y(n)
```

Again, we must be careful to ensure that Y is dimensioned correctly. The process by which the required point is located is essentially the same as the technique employed in DO2BGF. Starting at X, we integrate towards XEND and locate a bracket within which G changes sign. Using CO5AZF a refined estimate is obtained and returned via X. Again HMAX may be used to limit the size of the steps taken in the integration process.

We have already indicated that each driver routine calls the comprehensive integrator DO2PAF, for which the call sequence is

```
        CALL DO2PAF(X,XEND,N,Y,CIN,TOL,FCN,COMM,
    +               CONST,COUT,W,IW,IW1,IFAIL)
```

The number of parameters here, 14, although appearing a little large, may not put the novice off initially. However this is deceptive since CIN, COMM, CONST, and COUT are arrays of size 6, 5, 3 and 14 respectively, and each component has some effect on the way DO2PAF performs. For example, CIN(2) roughly corresponds to IRELAB of DO2BBF, DO2BDF and DO2BHF. Other parameters are used to control the strategy for step size selection; for example, indicating by how much the step size should be increased (subject to a maximum) if the error estimate is small, and by how much it should be decreased (subject to a minimum) if the error estimate is large. We note the comment given in the documentation for this routine that it is 'recommended only to experienced users', namely those who are both sufficiently conversant with the workings of the routine that they may choose appropriate values for the input parameters, and also know how to interpret the output parameters. Less able users will be satisfied with using a driver routine for which the many additional parameters to DO2PAF are given default values.

DO2PAF itself calls DO2YAF for which the call sequence is

```
        CALL DO2YAF(X,H,N,Y,FCN,W,IW1,IW2)
```

DO2YAF advances the integration from X to $X + H$ using a single step. The workspace array W must have IW($\geq$ N) rows and at least IW2 columns. Values for IW2 of 4, 6 and $\geq$ 7 are permitted. If IW2 = 4 then W contains no useful information on exit. If IW2 = 6 then, on exit, the Merson error estimates are contained in the fifth column of W, whilst the fourth column contains error estimates for Euler's method. If IW2 $\geq$ 7, these error estimates are returned in columns 6 and 7 respectively. On entry the first column of W must contain the values returned by FCN at the input point X. Again, it is unlikely that anyone other than an experienced user would be able to gain much benefit from using this routine.

Finally we consider the purpose of the two routines DO2XAF and DO2XBF. We observe that the Merson driver DO2BBF integrates up to a specified point. Only the value of the solution at that point is returned, although, as we have seen, intermediate solutions can be printed or saved by writing to files. It may be that the user is interested in values at points close to, but other than XEND, and these may be determined using a call of DO2XAF after a call of DO2PAF. We have already hinted that the use of DO2PAF requires specialized knowledge (or extreme dedication) and hence there is little point in our describing D02XAF in detail. However, we observe that in the call

```
CALL D02XAF(XSOL,X,COUT,N,Y,W,IW,SOL,IFAIL)
```

the parameters X, Y, COUT and W are input parameters and must contain values produced by a prior call of D02PAF. Care must therefore be taken in interfacing the two routines. D02XBF performs in a similar manner to D02XAF, except that only one component of the solution is considered. The call sequence is similar to that for D02XAF; there is an additional parameter, M, which indicates the component in question. Both routines return a solution at a specified point using quintic Hermite interpolation.

6.7   The NAG Adams routines

The Adams routines in the NAG library make use of the AB and AM formulae described in Section 6.5. With two exceptions (D02BDF and D02YAF) for each Merson routine there is a corresponding Adams routine; the four drivers are D02CAF, D02CBF, D02CGF and D02CHF, the comprehensive integrator is D02QAF, and the interpolation routines (which must be preceded by a call of D02QAF) are D02XGF and D02XHF.

The call sequences for the NAG Adams drivers are the same as for the corresponding Merson routines, although there are two minor discrepancies:

(1)   The number of columns for W must be at least 18 for D02CAF, D02CBF and D02CHF, and at least 22 for D02CGF.

(2)   For D02CBF and D02CGF the error test for IRELAB = 0 or 2 is slightly different. For example, if IRELAB = 2 the test

$$est \leq TOL \times \max\left(eps, \frac{1}{N}\sum_{i=1}^{N}Y^2(i)\right)$$

is made.

There is also a close correlation between the error messages for the Adams drivers and the corresponding Merson routines, although there are, again, a few minor differences.

The Runge–Kutta–Merson routines use step control in an attempt to satisfy the user's accuracy criterion. Since Runge–Kutta schemes are one-step a change in the local step size may be made with ease. The Adams routines also use step control, but since the underlying methods are multistep, a change in the step size requires more care. For example, suppose that we are using the second order AB and AM methods ($r = 2$ in Figure 6.11 and $r = 1$ in Figure 6.12) as a predictor–corrector pair, and

Figure 6.13   Original step sizes.

suppose that we have reached the sixth integration point having used a step size of 0.001 throughout (see Figure 6.13). If, having computed an error estimate, a decision is made to increase the step length, a doubling strategy can be implemented with ease. In computing a value for y_7, an estimate of $y(a + 6 \times 0.001 + 0.002)$, we omit the point t_5; in the predictor formula f_{i-1} is f_4, not f_5 (see Figure 6.14). Matters are not so straightforward if a step reduction is required. The new value of t_6 will be $t_5 + 0.0005$ and in order to produce an estimate of y and/or f at this point the predictor will require a function value at a point $\hat{t}_4$, say, mid-way between t_4 and t_5 (see Figure 6.15). There are various ways in which this mid-value may be obtained, the most obvious being interpolation.

It is clear that the higher the order of the method, the greater the number of 'back values' which will have to be determined if a step reduction is required. Note that some of these values may already be available if a step increase was only recently performed. Further, if a step increase is deemed appropriate we must ensure that we are sufficiently far enough along the range of integration for this to be possible, but not so far that we exceed the upper limit of the range. Of course, we may not be able to hit the final point exactly, and hence interpolation here may be required. It should be clear by now that careful book-keeping is required, although the step doubling and halving algorithm is quite simple in principle.

Our second-order predictor–corrector method may be quite adequate for some problems, but for others may impose a severe constraint on the step length used for sufficiently accurate solutions to be computed. The use of a higher-order method may be more appropriate; however, there is no guarantee that this may be so, and, in addition, the higher the order of the method, the more care that must be taken with the step control process. This suggests that a variable-order/variable-step approach may yield a more efficient, if complicated, algorithm. When the accuracy condition is not satisfied we could reduce the step size and/or increase the order of the method being employed; an equivalent strategy could also be used

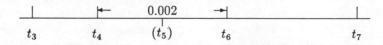

Figure 6.14   Effect of doubling the step size.

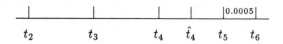

Figure 6.15   Effect of halving the step size.

whenever the tolerance is satisfied comfortably. Note that in the former case action must be taken; in the latter we have the choice of whether to stick with the current step length and method, or to change one of them. However, there is little point in making a change if we are brought back to where we started because the accuracy at the next point is unacceptable. Hence a switch should only be made if the error estimate is well within the constraints imposed by the user.

The comprehensive integrator D02QAF which is called by the four Adams driver routines implements such a variable-step/variable-order algorithm. A full description of the way D02QAF works may be found in Hindmarsh [44]. An error estimate is obtained by comparing methods of different order (cf. Exercise 6.11).

For non-stiff problems the use of an Adams routine is likely to prove less costly than the use of a corresponding Merson routine. The reason for this is quite easy to see. At each Merson step five function evaluations have to be made. At each Adams step, however, the number of function calls made is equal to the number of times the corrector is applied whatever the order used. Even if iteration to convergence were performed, it is unlikely that more than five cycles would be required.

EXERCISE 6.17   If the fourth-order AB and AM methods (Exercise 6.15) are to be used in a piece of software implementing some step control strategy, the two values $y_{i-\frac{1}{2}}$ and $y_{i-\frac{3}{2}}$, or $f_{i-\frac{1}{2}}$ and $f_{i-\frac{3}{2}}$, will need to be formed if a step reduction is necessary at the $(i+1)^{st}$ step, and the order of accuracy of these estimates must be at least that of all other approximations. Using f_i, f_{i-1}, f_{i-2}, f_{i-3} and f_{i-4} and the Newton forward difference form of the interpolating polynomial, (5.23), derive fourth-order approximations to $f_{i-\frac{1}{2}}$ and $f_{i-\frac{3}{2}}$. From Figure 6.11 and Figure 6.12, the LTEs of the predictor and corrector are $(251/720)h^5 y^{(v)}(\eta_1)$ and $-(19/720)h^5 y^{(v)}(\eta_2)$. Assuming the fifth derivative of y to be constant and $y_{pred} + \text{LTE}_{pred} = y_{corr} + \text{LTE}_{corr}$, determine an error estimate which can be used for step control.

6.8 Methods for stiff problems

If we attempt to use a Merson or Adams method to solve a stiff system the constraints on the step size will be such that a large number of steps, and hence function calls, will need to be made if any reasonable accuracy

r	β_r	α_{r-1}	α_{r-2}	α_{r-3}	α_{r-4}	α_{r-5}	α_{r-6}	Order
1	1	-1						1
2	$\frac{2}{3}$	$-\frac{4}{3}$	$\frac{1}{3}$					2
3	$\frac{6}{11}$	$-\frac{18}{11}$	$\frac{9}{11}$	$-\frac{2}{11}$				3
4	$\frac{12}{25}$	$-\frac{48}{25}$	$\frac{36}{25}$	$-\frac{16}{25}$	$\frac{3}{25}$			4
5	$\frac{60}{137}$	$-\frac{300}{137}$	$\frac{300}{137}$	$-\frac{200}{137}$	$\frac{75}{137}$	$-\frac{12}{137}$		5
6	$\frac{60}{147}$	$-\frac{360}{147}$	$\frac{450}{147}$	$-\frac{400}{147}$	$\frac{225}{147}$	$-\frac{72}{147}$	$\frac{10}{147}$	6

Figure 6.16   Backward difference formulae.

is to be achieved at the upper end of the range of integration. That is, the stability intervals of these methods are prohibitively small. In an attempt to improve this situation, Gear [32] and [33] proposes a sequence of r-step LMMs with the β_is all zero except for β_r (which means that these methods are implicit). (Again, by convention we take $\alpha_r = 1$.) In Figure 6.16 we list coefficients corresponding to $r = 1, 2, \ldots, 6$; Gear showed that if r is within this range, his *Backward Difference Formula* (BDF) methods possess desirable stability characteristics.

When $r = 1$ the method is

$$y_{i+1} = y_i + hf(t_{i+1}, y_{i+1}) \tag{6.31}$$

and is known as the *backward Euler method*. Its derivation is quite simple; as with Euler's method we start with a Taylor series expansion. We have $y(t_i) = y(t_{i+1}) - hy'(t_{i+1}) + R$, where R is the remainder term. The backward Euler method is obtained by ignoring R, which is thus the LTE for the method. The form of the LTE for this and the 'forward' Euler method is the same; that is, the two methods are both first-order, and the coefficient of h^2y'' is $\frac{1}{2}$ in each case. There seems little benefit in using the backward form of the method, particularly as some iterative method has to be used to solve the nonlinear equation for y_{i+1} at each step. However, a stability analysis for both methods reveals that the backward Euler scheme has superior stability characteristics.

In Section 6.2 we analysed the performance of Euler's method when applied to the model problem (6.5). Repeating the exercise for the backward Euler method, we have $y_{i+1} = y_i + h\lambda y_{i+1}$, or $y_{i+1} = y_i/(1 - h\lambda)$, and hence, if the initial condition is $y(0) = 1$, $y_{i+1} = 1/(1 - h\lambda)^{i+1}$. If λ is negative the approximate solutions will decrease with increasing i for any step length h, that is the interval of stability for the method is $(-\infty, 0)$. This should be compared with the stability interval for the 'forward' Euler method which is $(-2, 0)$. For a stiff problem in which λ is of the order of 100, or 1000, say, the backward Euler method will prove to be the more efficient of the two.

For nonlinear problems we still have the problem of iteration to consider. Recall from Section 6.5 that for an LMM, simple iteration will converge provided h is small enough, the bound on h for a single equation being inversely proportional to the magnitude of f_y. For stiff problems this restriction on h can be severe. However, the aim of BDF methods is to relax constraints on the step length, and hence some alternative iterative scheme must be employed. Newton's method (see Section 2.6) would appear to be a serious candidate; given a good initial estimate, convergence is guaranteed, irrespective of the size of h. However, to implement Newton's method for a system we must form the Jacobian matrix $\partial f/\partial y$. For the novice, formal differentiation of f can be a formidable task, and methods have been devised which estimate the Jacobian as the integration proceeds. The quadratic order of convergence is reduced, but the ensuing algorithm is likely to prove more amenable when implemented in a piece of numerical software (q.v. Section 2.10).

The driver routines D02EAF, D02EBF, D02EGF and D02EHF/D02EJF are available and there is a one-to-one correspondence between the basic function of these routines and the like-named Merson and Adams routines. The comprehensive integrator is D02QBF. The parameter list for D02EAF is almost the same as that for D02BAF and D02CAF. The only difference is the inclusion of an additional parameter, IW($\geq 18 + $N), which is used to indicate the size of the second dimension of the workspace array W. The error indicators are virtually the same as for D02BAF/D02CAF. Comparing D02EBF with D02BBF, the additional parameters are MPED, PEDERV and IW. PEDERV can be used to define the Jacobian matrix $J_{ij} = \partial f_i/\partial y_j$ which is used in the Newton iterations at each step. The specification of PEDERV is

```
SUBROUTINE PEDERV(T,Y,PW)
DOUBLE PRECISION T,Y(n),PW(n,n)
```

Note again the need to dimension the arrays correctly. Given the input value T, and the approximate solutions, Y, at this point, the value of the Jacobian matrix at T should be returned via PW. If the user is able to supply the Jacobian in this way, MPED should be set equal to 1 before entry to D02EBF. Users who are unable, or unwilling, to form the y-partial derivatives of f may indicate this by setting MPED equal to 0, in which case the Jacobian is estimated internally. The routines D02EGF and D02EHF require little introduction; only the latter allows the user to specify the form of the Jacobian. A further routine, D02EJF, acts as D02EHF, but with intermediate output.

Unlike the Merson and Adams routines the BDF routines are built from a whole subchapter of comprehensive routines. Subchapter D02N contains an extremely flexible set of routines for solving stiff systems of ODEs. These routines offer facilities

- to solve more complicated problems (for example, implicit systems of ODEs and coupled differential and algebraic equations);

- to customize the linear algebra routines for increased efficiency of solution (for example, to take account of sparseness or banded structure in the Jacobian);

- to use reverse communication for function and Jacobian evaluations;

- to select a wide range of options for a particular integrator.

The solution of a single problem requires calls to a number of routines and use of this subchapter is not to be undertaken lightly. D02N routines should only be considered if the far simpler to use BDF drivers have failed to provide a satisfactory solution.

We observed earlier (Section 2.10) that C05ZAF is available for checking that a Jacobian evaluation routine is consistent with a function evaluation routine, and an example of its use is given in Code 2.11. To be of use here, the routine FUN of Code 2.10 requires modification to take account of the form of FCN and PDERV. A suitable format is given in Code 6.3.

```
      SUBROUTINE FUN(N,T,Y,FVEC,FJAC,LDFJAC,IFLAG)
      INTEGER IFLAG,LDFJAC,N
      DOUBLE PRECISION FJAC(LDFJAC,N),FVEC(N),T,Y(N)
*
      IF(IFLAG.EQ.1)THEN
*
*.. FUNCTION EVALUATION
         CALL FCN(T,Y,FVEC)
      ELSE IF(IFLAG.EQ.2)THEN
*
*.. JACOBIAN EVALUATION
         CALL PEDERV(T,Y,FJAC)
      ENDIF
      END
```

Code 6.3 Jacobian checking program.

6.9 Using the NAG ODE solvers

As with any mathematical problem being considered for solution by a NAG routine, it is important that the user make some preliminary investigation in order to gain some indication as to whether any serious problems are likely to occur. In the present context this means looking to see whether the

NAG ODE solvers are likely to hit a singularity or discontinuity, whether the problem is stiff, etc. A stiffness check can, of course, be made by calling DO2BDF first and, unless the user is confident that his problem is not stiff, this would appear to be a good place to start. If the stiffness value returned by DO2BDF is large, then the user has little choice but to use the BDF routine which best suits his needs. If the user is able to evaluate the Jacobian matrix, it is likely that a call to DO2EBF with this specified will prove more efficient than a call in which the Jacobian matrix is estimated internally. No hard and fast rule can be given; the relative performance depends very much on the cost of evaluating the partial derivatives.

If the stiffness value returned by DO2BDF is reasonably small there is little point in using a BDF routine; the cost of the Newton iterations would outweigh any possible gain there might be. The user is therefore left with the choice of a Merson or Adams routine. Generally, the Adams routine will be the more efficient since it is likely to use fewer function calls. However, again we must qualify this by saying that the relative performance of the two routines will depend very much on how easily **f** may be evaluated. If the cost of each function call is not too large, then the fairly simple algorithm on which the Merson routine is based is likely to prove more efficient than a rather complicated variable-step/variable-order algorithm, with its need to fill in back values whenever a step reduction is required.

Having established the routine which best suits his needs, the user should now attempt several runs and monitor the relative performance of each. Only then will he be in a position to quote results with any degree of confidence.

We conclude by considering the equations

(1)

$$
\left.\begin{array}{ll}
y_1' & = -0.04y_1 + 10^4 y_2 y_3, \\
y_2' & = 0.04y_1 - 10^4 y_2 y_3 - 3 \times 10^7 y_2^2, \\
y_3' & = 3 \times 10^7 y_2^2, \\
y_1(0) = 1.0, \quad y_2(0.0) = 0.0, \quad y_3(0.0) = 0.0,
\end{array}\right\} x \in [0,1],
$$

(2)

$$
\left.\begin{array}{ll}
y_1' & = \tan(y_3), \\
y_2' & = -0.032\tan(y_3)/y_2 - 0.02y_2\sec(y_3), \\
y_3' & = -0.032/y_2^2, \\
y_1(0) = 0.0, \quad y_2(0) = 0.5, \quad y_3(0) = \pi/5
\end{array}\right\} x \in [0,100].
$$

Problem (1) is a stiff system and is used in the example programs for the BDF routines; problem (2) is a non-stiff system, used in the example programs for the Merson and Adams routines.

Problems (1) and (2) were solved using DO2BBF, DO2CBF and DO2EBF. In each case values of TOL = 10^{-6} and IRELAB = 0 were taken, and OUTPUT

Routine	Problem (1)		Problem (2)	
	Function calls	Cpu time	Function calls	Cpu time
D02BDF	13905	0.322	299	0.069
D02BBF	3416	0.139	159	0.147
D02CBF	5149	0.456	83	0.152
D02EBF	103	0.088	138	0.173

Figure 6.17   Comparison of NAG ODE solvers.

incremented XSOL by 1.0 at each call. Further, in the call to D02EBF MPED was set equal to 0, which meant that the Jacobian was computed internally. The approximate values returned by the three routines were similar but the relative cost of each routine was markedly different. We summarize in Figure 6.17 the number of function calls made by each routine, and the total time, in arbitrary units, required for execution of the complete program. Also included are results for calls to D02BDF, with STIFF given a positive value on input; on output its value was 0.9107 for problem (1), and 0.0 for problem (2). For problem (1) we see that D02BDF is very expensive indeed, but since it solves the problem twice with different tolerances using two calls to the comprehensive integrator D02PAF, we should not be too surprised by this. The Adams routine D02CBF makes far less function calls but the overall time is even larger, indicating that the step control algorithm is using up considerable resources. The straightforward Merson routine D02BBF does better still, but nowhere near as well as the BDF routine D02EBF whose performance is striking. D02EBF does not perform so well, however, on the non-stiff problem. If we consider function calls only, the Adams routine clearly out-performs its competitors. However, function calls are reasonably cheap and D02BBF does slightly better, in terms of overall time, using nearly twice as many function calls. Remarkably, D02BDF is over twice as fast despite the relatively large number of calls of FCN. It would appear that the interpolation process used to provide solutions at points specified by OUTPUT is slowing down D02BBF, D02CBF and D02EBF considerably. In fact, increasing the increment in OUTPUT to such an extent that no intermediate output is produced reduces the timings for D02BBF, D02CBF and D02EBF to 0.068, 0.083 and 0.088 respectively.

For a detailed comparison of the performance of software for ODEs see Hull et al. [45], and Enright et al. [27].

6.10 Summary

The NAG chapter for ordinary differential equations is D02, where problems are classified as initial value, boundary value and Sturm–Liouville. Here we are concerned only with methods for initial value problems, although some of the boundary value codes make use of similar techniques. Three basic methods are employed; a fixed-order Runge–Kutta–Merson method, a variable-order Adams method, and a variable-order BDF method. Each works in a step-by-step fashion and, by adjusting the step length, each attempts to keep the local error within bounds specified by the user. (The Adams and BDF routines may additionally adjust the order of the method.) This does not necessarily mean that the final result will be correct to this tolerance, and the user must be prepared to make a number of calls, possibly of several routines, before accepting any value returned. The BDF routines are particularly well suited to the solution of stiff problems. For non-stiff problems the choice between the Merson and Adams methods will largely be governed by the cost of evaluations of the functions appearing in the differential equation; if these are relatively cheap, the use of a Merson routine is likely to prove the more efficient.

For the Merson and Adams methods a comprehensive integration routine is available. These are D02PAF (Merson) (Section 6.6) and D02QAF (Adams) (Section 6.7). For stiff systems the D02N subchapter contains an extensive collection of comprehensive integrators (Section 6.8). All these routines give the user full control over step selection, etc., and as a consequence the parameter list of each is extensive and likely to prove daunting to the novice. The remaining routines call these comprehensive integrators but assume default values for certain of the parameters. For integration over a range $[a, b]$ the Merson routine D02BAF is available and returns an estimate of the solution to the differential equation at b. If, in addition, values at intermediate values within the range are required, D02BBF should be used. D02BGF integrates from the initial point until a component of the solution attains a specified value. D02BHF performs in a similar manner but attempts to locate a point at which a function of the solution is zero (Section 6.6). For each Merson driver, there is a corresponding Adams routine (D02CAF, D02CBF, D02CGF, D02CHF) (Section 6.7) and a BDF routine (D02EAF, D02EBF, D02EGF, D02EHF, D02EJF) (Section 6.8). Additionally, D02YAF uses the Merson method to integrate over a single step specified by the user, and D02BDF integrates over a range, but returns a global error estimate and a value indicating the stiffness of the problem (Section 6.6). In the absence of any other information, a preliminary run with this latter routine is strongly recommended to determine whether or not a BDF routine should be used subsequently.

Although the Merson comprehensive routine D02PAF integrates over a range, the process may be interrupted at, for example, each step. (The routine can, therefore, be used in a similar way to the reverse communication root finding routines (Section 2.4).) Following such a temporary halt, information is returned by D02PAF which allows an interpolated value for all components of the differential equation to be determined by a subsequent call of D02XAF. If only one component is of interest D02XBF should be used (Section 6.6). In a similar manner a call of the comprehensive Adams integrator D02QAF may be followed by D02XGF (all components) or D02XHF (one component) (Section 6.7).

The need for extensive experimentation with the NAG ODE solvers cannot be overstressed. No one routine should be taken in isolation and treated as a black box. An initial call of D02BDF should be followed by calls of an appropriate Merson, Adams and BDF routine, and the results analysed carefully. Further runs may be necessary before any method is isolated as being the one to use for a particular problem, and even then several runs with different accuracy tolerances may be necessary to check for stability and convergence. Without such experimentation, undue faith should not be placed in any values returned.

Appendix A
Solutions to Exercises

1.2: Design, implementation, coding, debugging, machine time, testing, maintenance, disc usage, tape usage.

1.3: (a) absolute difference $|x_{n+1} - x_n| < \epsilon$, (b) relative difference $|x_{n+1} - x_n|/|x_{n+1}| < \epsilon$, (c) attempt to avoid fluke convergence $|x_{n+1} - x_n| < |x_n - x_{n-1}| < \epsilon$. Relative error control gives a specified number of significant digits, regardless of the size of the iterates. Absolute error control is more useful for iterates $O(1)$ in magnitude.

1.4:

(a)
```
    *..   SEE WHETHER INTEGER OVERFLOW DETECTED
          I = 1
    10    I = 2*I
          PRINT*,I
          GOTO 10
```

(b) If both operands have the same sign bit but the sign bit of the result is different, an overflow has occurred. In all other cases, the operation has not overflowed. Subtraction is just a special case of addition.

(c) Numerator = largest negative integer, denominator = -1.

1.5: (a) Largest = either sign, mantissa and exponent all set to one = $\pm(2 - 2^{-24}) \times 2^{128} = \pm 0.68 \times 10^{39}$. (b) Machine epsilon = $2^{-23} = 0.119 \times 10^{-6}$ (if rounding is towards $-\infty$ then 2^{-24} has the same effect). (c) $1/2^{-126}$ is representable.

1.6: $\epsilon_n = (n!/M!)\epsilon$ where $e_n = I_n - \tilde{I}_n$ and $\epsilon = I_M$.

1.7:

(a) The program will not terminate because $(0.1)_{10}$ cannot be expressed exactly in binary; there is therefore a small representation error in the assignment `H = 0.1D0`. H added to itself 10 times will not equal `1.0D0`, which can be stored exactly. The comparison in the statement labelled 10 will thus always fail.

(b) Most I/O systems massage their output; the first write statement will often produce 0.1D0 and 0.1D1. The second will generate a small value representing the rounding error.

1.8: 1/N is an integer divide; H is thus set to zero.

1.9: Y

1.10: In the following / represents a newline:

```
1   7/   2   8/   3   9/   4 10/   5 11/   6 12/
1   5   9/   2   6 10/   3   7 11/   4   8 12/
```

1.11: Assigns 3×3 submatrix in top left corner of 10×10 matrix. The call results in the first 9 elements of the first column being considered as a 3×3 matrix (the last 6 of these would have been unassigned although many compilers will assume they have been set to zero).

1.12: First fragment $2 \times 100 \times 100 + 100$, second $4 \times 100 \times 100 + 100$. The second can be reduced to $3 \times 100 \times 100 + 100$ by assigning X(J) to a temporary, simple variable before the innermost loop.

1.14: Some, but not all, systems will report a run-time error.

1.16: An example argument list and error checking code:

```
        SUBROUTINE MATPRD(C,IC,A,IA,B,IB,M,N,P,IFAIL)
        INTEGER IA,IB,IC,M,N,P,IFAIL
        DOUBLE PRECISION A(IA,N),B(IB,P),C(IC,P)
*
*.. FORM THE M BY P MATRIX C AS THE PRODUCT OF THE M BY N
*.. MATRIX A AND THE N BY P MATRIX B.
*..
*.. CHECK THAT THE DIMENSIONING PARAMETERS ARE CONSISTENT
        IF(M.GT.IA .OR. N.GT.IB .OR. M.GT.IC)THEN
           IFAIL = 1
           RETURN
        ENDIF
*
*.. CHECK PARAMETERS ARE POSITIVE
        IF(M.LT.1 .OR. N.LT.1 .OR. P.LT.1 .OR. IA.LT.1 .OR.
     +     IB.LT.1 .OR. IC.LT.1)THEN
           IFAIL = 2
           RETURN
        ENDIF
```

1.17: If $a = \max(x, y)$ and $b = \min(x, y)$, $b^2/a^2 \leq 1$. Thus the square root argument is always ≤ 2. Worst case is when $x = y = maxreal/\sqrt{2}$, which is computed as $maxreal/\sqrt{2} \times \sqrt{2}$, rounding error permitting. If performed in a naive way we obtain an overflow when squaring either x or y.

1.18: Overhead of the do-loop i.e., counter incrementation and testing. Beware the code optimizer which notices that the add only needs to be done once and moves the assignment out of the loop. Other effects – machine loading (if multiuser), use of pipelines, vector or array processors.

1.23: Parameter statements allow the definition of program constants, e.g., maximum dimensions of arrays, to be isolated within a program unit. Unfortunately such constants are only defined locally and thus have to be included in each program unit they are required. Thus a change to any parameter value may involve changes to a number of parameter statements to ensure consistency. A Toolpack tool is available to check for the consistency of parameter statements (see Appendix C).

1.27: The convergence becomes linear when the starting value is a long way from the solution. There are problems with computing $\sqrt{0}$ as both f and f' are zero at zero.

1.28: $x_1 = \frac{1}{2}(x_0 + a/x_0) = g(x_0)$. Minimum of $g(x)$ occurs when $g'(x) = \frac{1}{2}(1 - a/x^2) = 0$, i.e., $x = \sqrt{a}$ for $x > 0$. Hence $x_n \geq \sqrt{a}$, $\forall n \geq 1$. Thus $\forall x_0 \in (0, \infty)$, $x_1 \geq \sqrt{a}$. $x_{n+1} - \sqrt{a} = \frac{1}{2}(x_n + a/x_n) - \sqrt{a} = \frac{1}{2}(x_n^2 + a - 2\sqrt{a}x_n) = (x_n - \sqrt{a})^2/(2x_n)$. Rearranging, $(x_{n+1} - \sqrt{a})/(x_n - \sqrt{a}) = \frac{1}{2}(1 - \sqrt{a}/x_n)$. From above $x_n \geq \sqrt{a}$ and thus $\frac{1}{2}(1 - \sqrt{a}/x_n)$ lies in $(0, \frac{1}{2}]$, hence $x_{n+1} - \sqrt{a} \leq x_n - \sqrt{a}$ and the result follows. Equality only occurs if $x_n = \sqrt{a}$ for some n, otherwise there is strict inequality and the sequence converges monotonically from above. For this equation the Newton iteration converges for all positive starting values.

1.29: Let $c^3 = a$ then $f(x) = x^3 - c^3$ and the Newton iteration is $x_{n+1} = (2x_n + c^3/x_n^2)/3 = g(x_n)$. $g(x)$ has a minimum at c for $x_0 > 0$, i.e., $\forall x_0 \in (0, \infty)$, $x_1 \geq c$. $x_{n+1} - c = (x_n - c)^2(2x_n + c)/(3x_n^2)$ gives $(x_{n+1} - c)/(x_n - c) = \frac{1}{3}((x_n - c)/x_n)((2x_n + c)/x_n)$. For $n > 0$, $x_n \geq c$, hence $(1 - c/x_n) \in (0, 1]$ and $(2 + c/x_n) \in (2, 3]$ gives $(x_{n+1} - c)/(x_n - c) \leq 1$, and the result follows as in Exercise 1.28.

1.30: $|a| \geq |b|$ implies $a \oplus b = \sqrt{a^2 + b^2} \leq \sqrt{2a^2} = \sqrt{2}|a|$. $|b| \geq 0$ implies $a \oplus b \geq \sqrt{a^2} = |a|$. Thus $a \oplus b \in [|a|, \sqrt{2}|a|]$.

Assuming that the function $pythag(a, b)$ returns $a \oplus b$ the following is a pseudocode implementation of a Euclidean norm function

```
double precision function euclid(x,n)
double precision x(n),s,pythag
s = 0
for i = 1,n
    s = pythag(s,x(i))
endfor
euclid = s
end
```

For an extended discussion of this exercise see Bentley [8].

1.31: The rate of convergence is cubic. A full description of this algorithm is given by Moler and Morrison [59]. Scale $\sqrt{a^2 + b^2}$ as $c\sqrt{1 + d^2}$ where $c = \max(|a|, |b|)$ and $d = \min(a^2/c^2, b^2/c^2)$.

1.32: An ill-conditioned linear system of order 2 can be interpreted geometrically as a pair of almost parallel straight lines in the x-y plane. Their meeting point is the solution. A small change to any of the defining coefficients moves the intersection point a disproportionate distance. When $\epsilon = 0$ in (1.7) the lines are coincident.

1.33: (a) $(-3, 1)$. (b) $(-5 - \epsilon, 2)$. No, the definition of ill-conditioning requires that the change be small relative to the defining coefficients.

1.34: Write e^{-10} as $1/e^{10}$ or $(e^{-1})^{10}$ and then use (1.10) with $x = 10$ or $x = -1$.

1.35: Number of operations for adding two square matrices of order n is n^2. Time $= 200n^2$ nanoseconds. Maximum $n = 38730$. Number of operations for product of two square matrices of order n is n^3 multiplications and $n^3 - n^2$ additions. Ignoring the n^2 term time taken is $500n^3$ nanoseconds. Maximum $n = 843$. Other considerations: paging, ability to declare and store very large matrices.

1.37: $\|A\|_p = \sup_{\mathbf{x} \neq 0} \|A\mathbf{x}\|_p / \|\mathbf{x}\|_p$ implies $\|A\|_p \geq \|A\mathbf{x}\|_p / \|\mathbf{x}\|_p$, $\forall \mathbf{x} \neq 0$ since the supremum is the least upper bound and the result follows.

1.38: $\|\mathbf{a}\|_2 = \sqrt{\sum_i a_i^2}$. (a) Since a_i is real $a_i^2 \geq 0$, $\forall i$ thus $\|\mathbf{a}\|_2 \geq 0$, $\forall \mathbf{a}$. (b) $\|\mathbf{a}\|_2 = 0$ iff $a_i = 0$, $\forall i$. (c) $\|c\mathbf{a}\|_2 = \sqrt{\sum_i c^2 a_i^2} = |c|\sqrt{\sum_i a_i^2} = |c| \|\mathbf{a}\|_2$, $\forall$ scalar c. (d) $\|\mathbf{a} + \mathbf{b}\|_2^2 = (\mathbf{a} + \mathbf{b})^T(\mathbf{a} + \mathbf{b}) = \mathbf{a}^T\mathbf{a} + 2\mathbf{a}^T\mathbf{b} + \mathbf{b}^T\mathbf{b} \leq \|\mathbf{a}\|_2^2 + 2\|\mathbf{a}\|_2\|\mathbf{b}\|_2 + \|\mathbf{b}\|_2^2 = (\|\mathbf{a}\|_2^2 + \|\mathbf{b}\|_2^2)^2$, where the inequality follows from the Cauchy–Schwarz inequality. $\|\mathbf{a}\|_\infty = \max_i\{|a_i|\}$.

(a) $|a_i| \geq 0$, $\forall i$ implies $\|a\|_\infty \geq 0$, $\forall a$. (b) $\|a\|_\infty = 0$ iff $a_i = 0$, $\forall i$. (c) $\|ca\|_\infty = \max_i\{|ca_i|\} = \max_i\{|c|\,|a_i|\} = |c|\max_i\{|a_i|\} = |c|\,\|a\|_\infty$. (d) $\|a+b\|_\infty = \max_i\{|a_i + b_i|\} \leq \max_i\{|a_i|\} + \max_i\{|b_i|\} = \|a\|_\infty + \|b\|_\infty$. Computationally $\|\cdot\|_2$ requires n multiplies, n adds and a square root (see also the exercises at the end of Section 1.14). $\|\cdot\|_\infty$ requires n ABS calls and $n-1$ comparisons (calls to MAX).

1.39: $\|a\|_1 = \sum_i\{|a_i|\} \geq \max_i\{|a_i|\} = \|a\|_\infty$. $\|a\|_1 = \sum_i\{|a_i|\} \leq \sum_i \max_i\{|a_i|\} = n\|a\|_\infty$. Thus $\|a\|_\infty \leq \|a\|_1 \leq n\|a\|_\infty$.

1.40: (a) $\|A\|_F \geq 0$ since $a_{ij}^2 \geq 0$, $\forall i,j$. (b) $\|A\|_F = 0$ iff all the $a_{ij} = 0$. (c) $\|cA\|_F = \sqrt{\sum_i\sum_j(ca_{ij})^2} = c\sqrt{\sum_i\sum_j a_{ij}^2} = c\|A\|_F$. (d) $\|A+B\|_F^2 = \sum_i\sum_j(a_{ij}+b_{ij})^2 = \sum_i\sum_j a_{ij}^2 + \sum_i\sum_j b_{ij}^2 + 2\sum_i\sum_j a_{ij}b_{ij} \leq \|A\|_F^2 + \|B\|_F^2 + 2\|A\|_F\|B\|_F$. (e) Let a_i be the ith row of A and b_j be the jth column of B. Then $A^T = (a_1, a_2, \ldots, a_n)$ and $Ax = (a_1^T x, a_2^T x, \ldots, a_n^T x)$. Thus $\|Ax\|_2^2 = \sum_i(a_i^T x)^2$ and by the Cauchy–Schwarz inequality (see Exercise 1.38) $(a_i^T x)^2 \leq \|a_i\|_2^2\|x\|_2^2$ giving $\|Ax\|_2^2 \leq \|x\|_2^2 \sum_i \|a_i\|_2^2 = \|x\|_2^2\|A\|_F^2$. $\|A.B\|_F^2 = \|(Ab_1, Ab_2, \ldots, Ab_n)\|_F^2 = \sum_j \|Ab_j\|_2^2 \leq \|A\|_F^2 \sum_j \|b_j\|_2^2 = \|A\|_F^2\|B\|_F^2$.

1.41: $\|H_n\|_1 = \sum_{k=1}^n 1/k = S_n$ where S_n is the nth Harmonic number (see Exercise 1.36). $\|H_3^{-1}\|_1 = 408$. The exact values of $\|H_n\|\,\|H_n^{-1}\|$ are 748, 28375, 9.4×10^5, 2.9×10^7, 9.9×10^8, 3.4×10^{11}, 1.1×10^{12}, 3.5×10^{14}, for $n = 3, 4, \ldots, 10$. Values obtained by actually computing inverses are liable to be less accurate.

2.1: Roots are at the intersection of the straight line $1.6x - 10$ and the curve 1.2^x. Since $1.2^x > 0$, $\forall x$ and $1.6x - 10 < 0$ for $x < 6.25$ there is no root < 6.25. Also 1.2^x increases exponentially with x, hence once $1.2^x > 1.6x - 10$ for some large enough x there will be no further intersections. A careful sketch in the range 6.25 to 20.0 reveals roots in $[10, 11]$ and $[13, 14]$.

2.2: Consider $g(x) = 1 - \frac{2}{3}x^3$, $g'(x) = -2x^2$. This will converge if the root and the initial guess are both in $(-1/\sqrt{2}, 1/\sqrt{2})$. Unfortunately this interval does not contain the root. Next consider $g(x) = 1.5/(1.5 + x^2)$, $g'(x) = -3x/(1.5 + x^2)^2$. Now $|g'(x)|$ has a maximum value at $\pm 1/\sqrt{2}$ of 0.53, hence the iteration will converge for all starting values. When x_0 is chosen large, x_1 will be close to zero (since $1.5/(1.5 + x^2) \approx 0$) and the iteration will converge to 0.735. Only a single root is ever found as the remaining roots are complex.

2.3: $x\tan(x) - 1$ has problems around $x = n\pi/2$; $\tan(x) - 1/x$ has problems around $x = n\pi/2$ and 0; $x - 1/\tan(x)$ and $x - \arctan(1/x)$ both have problems around $x = 0$.

2.5: Because there is no sign change either side of the zero. If $f(x)$ has a root, α, of multiplicity r then we may write $f(x) = (x - \alpha)^r g(x)$, then $f'(x) = (x - \alpha)^{r-1}\big(rg(x) + (x - \alpha)g'(x)\big)$ and $f(x)/f'(x) = (x - \alpha)g(x)/\big(rg(x) + (x - \alpha)g'(x)\big)$ has a simple root at α.

2.6: Near the root the iteration converges from one side, the interval length does not tend to zero.

2.8: When x_k is close to a root we have $x_k \approx x_{k-1}$ and $f_k \approx f_{k-1} \approx 0$. With (2.3) and the formula given in the exercise the iteration values are generated from combinations of subexpressions which may be badly affected by rounding errors (e.g., $f_k/(f_k - f_{k-1})$). Formula (2.4) is constructed so that the iteration is of the form $x_k + h_k$ where h_k tends to zero as x_k converges to the root.

2.10: The number of negative real roots of $p_n(x)$ is at most equal to the number of sign changes of $p_n(-x)$. $p_5(x) = 8x^5 - 6x^4 - 3x^2 + 5x - 4$ has 3 sign changes and since $3 - pos$ must be a nonnegative even integer there must be either 3 or 1 positive roots. $p_5(-x) = -8x^5 - 6x^4 - 3x^2 - 5x - 4$ has no sign changes and hence there are no negative roots. Thus there are either 3 positive real roots and a pair of complex roots or 1 positive real root and 2 pairs of complex roots. (The roots are $+1$, $0.565 \pm 0.526i$, $-0.690 \pm 0.602i$.)

2.11: (a) $g(x) = x - f(x)/f'(x)$, $g'(x) = f(x)f''(x)/\big(f'(x)\big)^2$. For a simple root at α, $f(\alpha) = 0$ and $f'(\alpha) \neq 0$ and hence $g'(\alpha) = 0$. Using the result of Exercise 2.2 we require $|ff''/f'^2| < 1$ in an interval containing x_0 and α. (b) $g(x) = x - rp/p'$, $g'(x) = (1 - r) + rpp''/p'^2$. $p(x) = (x - \alpha)^r q$, $p'(x) = (x - \alpha)^{r-1}\big(rq + (x - \alpha)q'\big)$, $p''(x) = (x - \alpha)^{r-2}\big(r(r - 1)q + 2r(x - \alpha)q' + (x - \alpha)^2 q''\big)$. Whence $rpp''/p'^2|_{x=\alpha} = r - 1$ and the result follows. A solution for the case where f is a general function is given in Ralston and Rabinowitz [67], p. 354.

2.12: $f(x + h) = f(x) + hf'(x) +$ higher-order terms; set $\alpha = x + h$ where $f(\alpha) = 0$ then $f(\alpha) \approx f(x) + (\alpha - x)f'(x)$ giving $\alpha = x - f(x)/f'(x)$.

2.15: See Figure 2.2

2.16: Roots are $\{i\}_{i=1}^{10}$. The equation is ill-conditioned, see Wilkinson [82], pp. 38–43 for further details.

2.17: Because each function, f_i, is only dependent on at most three dependent variables it is possible to save function evaluations by incrementing a number of the x_is at once. Thus we may simultaneously compute approximations to columns 1, 4, 7, ... of the Jacobian from the difference

$f(x + \sum_k he_{1+3k}) - f(x)$ where $k = 0, 1, \ldots$ and e_i is the ith column of the unit matrix. Similarly the other columns can be computed from two further function evaluations of the form $f(x + \sum_k he_{j+3k})$ for $j = 2, 3$. For details of the extension to general banded and sparse Jacobians see Curtis et al. [22].

3.1: At $x = a$ and $x = \infty$, $z = -1$ and 1 respectively. Rearranging, we obtain the transformation function $\phi(z) = (1 + z)/(1 - z) + a$ and the result follows since $\int_a^\infty f(x)\,dx = \int_{-1}^1 f(\phi(z))\,d\phi(z)/dz$. It is possible for the transformed integrand to be infinite at the left-hand end-point.

3.2: Define a continuously differentiable function $f(t)$ which maps $[a, \infty)$ onto $[b, c]$ such that $f(b) = a$ and $f(c) = \infty$. So using $x = f(t)$ and $dx = f'(t)\,dt$ the result follows. If $f(t) = t/(1 - t)$ then $f(0) = 0$ and $f(1) = \infty$ then $b = 0$ and $c = 1$. $\phi(x) = (1+x)^{-m}$ gives $\phi\{f(t)\} = (1-t)^m$ and $f'(t) = (1 - t)^{-2}$ whence $\int_0^\infty (1 + x)^{-m}\,dx = \int_0^1 (1 - t)^{m-2}\,dt$.

3.3: (a) From (3.5) with $x_{i+1} - x_i = h$, thus $R = \frac{1}{2}h(f_1 + 2f_2 + \cdots + 2f_{n-1} + f_n)$. (b) Using $h = 1$; $I_0 = \int_0^1 dx = 1$, $R = \frac{1}{2}(1 + 1) = 1$; $I_1 = \int_0^1 x\,dx = \frac{1}{2}$, $R = \frac{1}{2}(0 + 1) = \frac{1}{2}$; $I_2 = \int_0^1 x^2\,dx = \frac{1}{3}$, $R = \frac{1}{2}(0 + 1) = \frac{1}{2}$; we deduce that the degree of precision is one. (c) If the function values used at one step are $\{f_1, f_2, \ldots, f_n\}$ where $f_1 = f(a)$, $f_n = f(b)$, $f_i = f(x_i)$ and $x_{i+1} - x_i = h$ then at the next step we would use $\{f_i\}_{i=1}^n$ and $\{f_{i+\frac{1}{2}}\}_{i=1}^{n-1}$ where $f_{i+\frac{1}{2}} = f(x_i + \frac{1}{2}h)$. The first approximation is then given by $R_1 = h\sum_{i=1}^{\prime\prime n} f_i$ and the second by $R_2 = \frac{1}{2}h\sum_{i=1}^{\prime\prime n} f_i + \frac{1}{2}h\sum_{i=1}^{n-1} f_{i+\frac{1}{2}} = \frac{1}{2}R_1 + \frac{1}{2}h\sum_{i=1}^{n-1} f_{i+\frac{1}{2}}$.

3.4: $F(x_{i+1}) = F(x_i + h) = F(x_i) + hF'(x_i) + \frac{h^2}{2!}F''(x_i) + \cdots$. But $F' = f$, $F'' = f'$ etc. and $F(x_i) = 0$. Thus $F(x_{i+1}) = hf_i + \frac{h^2}{2!}f_i' + \frac{h^3}{3!}f_i'' + \cdots$. Using the Taylor series expansion for $f(x_{i+1})$ the trapezium rule for $[x_i, x_{i+1}]$ gives $\frac{1}{2}h(f_i + f_{i+1}) = \frac{1}{2}h(f_i + f_i + hf_i' + \frac{h^2}{2!}f_i'' + \cdots) = hf_i + \frac{h^2}{2!}f_i' + \frac{h^3}{4}f_i'' + \cdots$. Hence the leading error term is $(\frac{h^3}{3!} - \frac{h^3}{4})f_i'' = -\frac{1}{12}h^3 f_i''$.

3.5: As h is successively halved the error in the computed integral should gradually decrease until the effects of rounding error become noticeable. As a result reducing h further will increase the error in the computed integral. The value of h producing the optimal computed error will thus be dependent on the underlying floating-point arithmetic.

3.6: $p_{2,i-1} = \frac{1}{2}(x - x_{i-1})(x - x_i)f_{i+1}/h^2 - (x - x_{i-1})(x - x_{i+1})f_i/h^2 + \frac{1}{2}(x - x_i)(x - x_{i+1})f_{i-1}/h^2$. Integration gives the required result. On halving the step size these two summations are combined to form the first term of the next approximation and the sum of the function values at the newly introduced points forms the second term. Approximating $I_k = \int_0^1 x^k\,dx$,

$k = 0, 1, 2, 3, 4$ using the basic rule $S_k = \frac{1}{6}(f_1 + 4f_2 + f_3)$, we have $f_1 = 1(k = 0)$, $f_1 = 0(k > 0)$, $f_2 = (\frac{1}{2})^k$ and $f_3 = 1$. Then $I_k = 1, \frac{1}{2}, \frac{1}{3}, \frac{1}{4}, \frac{1}{5}$ and $S_k = 1, \frac{1}{2}, \frac{1}{3}, \frac{1}{4}, \frac{5}{24}$, showing that the rule does not exceed degree of precision 3. Generalize for the integration range $[a, b]$. The composite Simpson's rule is of the form $\frac{1}{3}h(f_1 + 4f_2 + 2f_3 + 4f_4 + \cdots + 4f_{2n} + f_{2n+1}) = \frac{1}{3}h(2\sum''^{n+1}_{i=1} f_{2i-1} + 4\sum^n_{i=1} f_{2i})$.

3.7: For $f(x) = 1$ we obtain $\int_{-\pi/2}^{\pi/2} \sin(x)\,dx = 0 = w_1 + w_2$; similarly for $f(x) = x$, x^2 and x^3 we obtain three further equations $w_1 x_1 + w_2 x_2 = 2$; $w_1 x_1^2 + w_2 x_2^2 = 0$; $w_1 x_1^3 + w_2 x_2^3 = 3\pi^2/2 - 12$. Solving these gives $x_1 = -x_2 = 1.184146656$ and $w_1 = -w_2 = 0.84448999195$.

3.8: $H_4(x) = 16x^4 - 48x^2 + 12$. Rearrange the Rodriguez formula to obtain the expression for $g^{(n)}$. Differentiating, $g^{(n+1)}(x) = (-1)^n e^{-x^2}(-2xH_n(x) + H_n'(x))$. But $g^{(n+1)}(x) = (-1)^{n+1}e^{-x^2}H_{n+1}(x)$, and the result follows. Starting with $H_0(x) = 1$ it follows by induction that the coefficient of x^n in H_n is 2^n. Now, let $H_m(x)$ and $H_n(x)$ be Hermite polynomials of degree m and n respectively, with $m < n$. Then $I = \int_{-\infty}^{\infty} e^{-x^2} H_n(x)H_m(x)dx = (-1)^n \int_{-\infty}^{\infty} g^{(n)}(x)H_m(x)dx$. Using integration by parts we obtain $I = (-1)^n g^{(n-1)}(x)H_m(x) \big|_{-\infty}^{\infty} + (-1)^{n-1}\int_{-\infty}^{\infty} g^{(n-1)}(x)H_m'(x)dx$ and the first term vanishes. Continuing, we deduce $I = \int_{-\infty}^{\infty} g(x)H_m^{(n)}(x)dx$. But for $m < n$, $H_m^{(n)}(x) = 0$ and so the polynomials are orthogonal. Further, if $n = m$, $I = 2^n n! \int_{-\infty}^{\infty} g(x)dx = 2^n n!\sqrt{\pi}$, so divide the Hermite polynomials by the square root of this to obtain an orthonormal set.

3.9: G_{2^i} requires 2^i function evaluations. $G_1, G_2, G_4,\ldots$ thus need 1, 2, 4,... evaluations, giving totals of 1, 3, 7,... function evaluations which are all different as Gauss rules using even numbers of points have no evaluations in common. Patterson rules use 1, 3, 7, 15,... points where each contains all the points of its predecessor. The degree of precision of the first six Gauss rules is 1, 3, 7, 15, 31, 63 and the Patterson rules 1, 5, 11, 23, 47, 95.

3.12: $P(x) = erfc(-x/\sqrt{2})/2$, $Q(x) = erfc(x/\sqrt{2})/2$.

3.13: The numerator and denominator both tend to zero as x tends to unity.

3.14: $p_3(x_{i+1}) = \frac{1}{2}(f_{i+1} + f_{i+2} + (x_{i+1} - x_{i+2})(f_{i+2} - f_{i+1})/(x_{i+2} - x_{i+1})) = f_{i+1}$ and similarly for $p_3(x_{i+2})$.

$$p_3(x_i) = \frac{1}{2}\big(f_{i+1} + f_{i+2} + (2x_i - x_{i+1} - x_{i+2})f[x_{i+1}x_{i+2}] + (x_i - x_{i+1})(x_i - x_{i+2})(f[x_{i+1}x_{i+2}x_{i+3}] + f[x_ix_{i+2}x_{i+3}])\big)$$

$$+ (x_i - x_{i+1})(x_i - x_{i+2})(x_i - x_{i+3})f[x_i x_{i+1} x_{i+2} x_{i+3}]$$
$$= \tfrac{1}{2}\big(f_{i+1} + f_{i+2} + (2x_i - x_{i+1} - x_{i+2})f[x_{i+1} x_{i+2}]$$
$$+ 2(x_i - x_{i+1})(f[x_i x_{i+1}] - f[x_{i+1} x_{i+2}]))\big)$$
$$= \tfrac{1}{2}\big(f_{i+1} + f_{i+2} + (x_{i+1} - x_{i+2})f[x_{i+1} x_{i+2}]$$
$$+ 2(x_i - x_{i+1})f[x_i x_{i+1}]\big)$$
$$= f_i$$

and similarly for $p_3(x_{i+3})$.

4.1: Let $A = L^{(1)} L^{(2)}$, where A, $L^{(1)}$ and $L^{(2)}$ have elements $\{a_{ij}\}$, $\{l_{ij}^{(1)}\}$ and $\{l_{ij}^{(2)}\}$ respectively, and $L^{(1)}$ and $L^{(2)}$ are lower triangular. Then $a_{ij} = \sum_{k=1}^{n} l_{ik}^{(1)} l_{kj}^{(2)}$. But $l_{kj}^{(2)} = 0$ for $k < j$, and $l_{ik}^{(1)} = 0$ for $k > i$. Hence A is lower triangular. No workspace arrays are required if A is assembled in the order $a_{n1}, a_{n2}, \ldots, a_{n,n-1}, a_{n-1,1}, \ldots$.

4.2: Although two vectors may have some non-zero entries in common positions, there may also be some non-common entries, and the formed sum must allow for this.

4.6: At the first stage $a_{ij}^{(2)} = a_{ij} - a_{i1} a_{1j}/a_{11}$ and so max $|a_{ij}^{(2)}| \leq$ max $|a_{ij}| +$ $\max(|a_{i1}||a_{ij}|/|a_{11}|)$. But if row interchanges have taken place, $|a_{i1}|/|a_{11}| \leq 1$, and hence $\max_{i,j} |a_{ij}^{(2)}| \leq 2 \max_{i,j} |a_{ij}|$.

4.7: The elimination of the coefficient matrix needs to be done only once.

4.9: (a) I_n forms the diagonal of M_k. Then in $\mathbf{m}_k \mathbf{e}_k^T$, the first $k + 1$ rows are all zero, and the remaining rows have zeros everywhere except in the kth column, where the i, k^{th} entry is m_{ik}. (b) $(I_n + \mathbf{m}_k \mathbf{e}_k^T)(I_n - \mathbf{m}_k \mathbf{e}_k^T) = I_n - \mathbf{m}_k \mathbf{e}_k^T \mathbf{m}_k \mathbf{e}_k^T$. But $\mathbf{e}_k^T \mathbf{m}_k = 0$ and the result follows. M_k^{-1} is the same as M_k, but with the minus signs in the kth column changed to plus signs. (c) Show that if $\tilde{M} = M_{i+1}^{-1} M_{i+2}^{-1} \cdots M_{n-1}^{-1}$ is lower triangular with $\tilde{m}_{jj} = 1, j = 1, 2, \ldots, n$ and $\tilde{m}_{jk} = m_{jk}, k > i, \quad j > i + 1$, then $M_i^{-1} \tilde{M}$ is the same as $\tilde{M}$ but with column i replaced by $(0, 0, \ldots, 0, 1, m_{i+1,i}, m_{i+2,i}, \ldots, m_{ni})^T$. Now use induction. (d) Let P_1 and P_2 be two permutation matrices. Then $P_2 P_1$ is the same as P_1 but with certain rows interchanged. The result is just a permutation of I_n.

4.12: The Crout factorization for a general matrix is given in detail in (4.8) and (4.9). For the symmetric, positive definite case there is no need to pivot. Using the symmetry of the coefficient matrix the first column of L is $l_{j1} = a_{j1} = a_{1j}$ and the first row of U is $u_{1j} = a_{1j}/l_{11}$ for $j = 2, 3, \ldots, n$; hence $u_{1j} = l_{j1}/l_{11}$. Once column i of L has been computed all the information is available to compute row i of U. Assume that when computing u_{ij} that $u_{rs} = l_{sr}/l_{rr}$ for $r < i$ and for $s < j$ and $r = i$, then (all summations are for $k = 1$ to i) $a_{ij} = \sum l_{ik}u_{kj}$, $i \leq j$ and $a_{ji} = \sum l_{jk}u_{ki}$, $i \geq j$. By symmetry $\sum l_{ik}u_{kj} = \sum l_{jk}u_{ki}$ and with the above assumption $u_{ij} = l_{ji}/l_{ii}$ and we have an inductive proof. We may thus write $A = \hat{L}D\hat{L}^T$ where $D = \text{diag}\{l_{11}, l_{22}, \ldots, l_{nn}\}$ and $\hat{L} = LD^{-1}$. We thus have a three-stage algorithm for solving $Ax = b$, i.e., $\hat{L}z = b$, $Dy = z$ and $\hat{L}^T x = y$, where the same vector may be used to store x, y and z. There is no need to store D explicitly. No square roots are required in either the backward or forward substitutions.

4.13: A is a symmetric, positive definite matrix of bandwidth k, i.e., $a_{ij} = 0$ for $|i - j| > k$. A may be factored into LL^T where $l_{ij} = 0$ for $|i - j| > k$, $l_{ij} = \left(a_{ij} - \sum_{r=i-k}^{i-1} l_{ir}l_{jr}\right)/l_{jj}$, $(j = i - k, i - k + 1, \ldots, i - 1)$, and $l_{ii} = \left(a_{ii} - \sum_{r=i-k}^{i-1} l_{ir}^2\right)^{\frac{1}{2}}$. Whence we have $y_i = \left(b_i - \sum_{r=i-k}^{i-1} l_{ir}y_r\right)/l_{ii}$, $(i = 1, 2, \ldots, n)$ and $x_i = \left(y_i - \sum_{r=i+1}^{i+k} l_{ri}x_r\right)/l_{ii}$, $(i = n, n - 1, \ldots, 1)$.

4.14: Storing by diagonals is more efficient on storage for small bandwidths but makes the accessing of the coefficients slightly more complicated. The amount of array space required is n^2 regardless of the bandwidth. The second storage method requires $(2k + 1)n$ elements. The cross-over point is thus when $k = (n - 1)/2$. Note that a full matrix has bandwidth n.

4.15: A row of the LU factors may contain up to $2k + 1$ non-zero elements due to the interchange of rows required by pivoting.

4.17: Store the coefficient matrix and the factors by rows rather than by columns, i.e., $A(2, N)$, $LLT(2, N)$. This allows for array slicing and the deletion of the parameter NROWS.

4.18: To form L^{-1} we solve $Lx_i = e_i$ for $i = 1, 2, \ldots, n$. There are n solutions to be determined, each of which involves $O\left(n^2\right)$ operations in the forward substitution process, giving a total of $O\left(n^3\right)$. The operation count for forming U^{-1} follows similarly.

4.19: $a_{ij}^{(2)} = a_{ij} - a_{i1}a_{1j}/a_{11} = \bar{a}_{ji} - \bar{a}_{1i}\bar{a}_{j1}/\bar{a}_{11} = \bar{a}_{ji}^{(2)}$. Now if $x^H = (\bar{x}_2, \bar{x}_3, \ldots, \bar{x}_n)$ is any non-zero $(n - 1)$-vector

$$x^H A^{(2)} x = \begin{pmatrix} 0 & x^H \end{pmatrix} \begin{pmatrix} 0 & 0 \\ 0 & A^{(2)} \end{pmatrix} \begin{pmatrix} 0 \\ x \end{pmatrix}$$

$$= \begin{pmatrix} 0 & x^H \end{pmatrix} \begin{pmatrix} 0 & 0 \\ 0 & A^{(2)} \end{pmatrix} \begin{pmatrix} y \\ x \end{pmatrix}$$

$$= \begin{pmatrix} 0 & \mathbf{x}^H \end{pmatrix} \begin{pmatrix} 0 & & & & \\ -\frac{a_{21}}{a_{11}} & 1 & & & \\ -\frac{a_{31}}{a_{11}} & 0 & 1 & & \\ \vdots & \vdots & \vdots & \ddots & \\ -\frac{a_{n1}}{a_{11}} & \cdots & \cdots & \cdots & 1 \end{pmatrix} A \begin{pmatrix} y \\ \mathbf{x} \end{pmatrix}$$

$$= \begin{pmatrix} -\sum_{i=2}^n \frac{a_{i1}}{a_{11}} \overline{x}_i & \mathbf{x}^H \end{pmatrix} A \begin{pmatrix} y \\ \mathbf{x} \end{pmatrix},$$

for any scalar y. By setting $y = -\left(\sum_{i=2}^n \overline{a}_{i1}/a_{11}\right)x_i$ and using the fact that A is positive definite, we may deduce that $\mathbf{x}^H A^{(2)} \mathbf{x} > 0$, that is, $A^{(2)}$ is positive definite. Now, in an Hermitian positive definite matrix the element of maximum modulus is on the diagonal. Suppose otherwise and that a_{pq} is the largest element with $p \neq q$. Pre- and postmultiply A by the vector with non-zeros in positions p and q, and zeros everywhere else. Then the matrix

$$\begin{pmatrix} a_{pp} & a_{pq} \\ a_{qp} & a_{qq} \end{pmatrix}$$

must be positive definite and have a positive determinant, that is, $a_{pp}a_{qq} > a_{qp}a_{pq} = \overline{a}_{pq}a_{pq} = |a_{pq}|^2$. It follows that we must have $a_{pp} > |a_{pq}|$ or $a_{qq} > |a_{pq}|$, which contradicts the assumption that a_{pq} is a element of maximum modulus. We now deduce $a_{ii}^{(2)} = a_{ii} - a_{i1}a_{1i}/a_{11} = a_{ii} - a_{i1}\overline{a}_{i1}/a_{11} \leq a_{ii}$, and hence $|a_{ij}^{(2)}| \leq \max_{i,j} |a_{ij}|$.

5.1: Assume all summations are from $k = 1$ to m. We require $\partial S/\partial a_0 = \partial S/\partial a_1 = 0$ for a minimum. $\partial S/\partial a_0 = \sum -2(f_k - a_0 - a_1 x_k) = 0$ gives $a_0 \sum 1 + a_1 \sum x_k = \sum f_k$ and $\partial S/\partial a_1 = \sum -2x_k(f_k - a_0 - a_1 x_k) = 0$ gives $a_0 \sum x_k + a_1 \sum x_k^2 = \sum x_k f_k$. The solution of $\begin{bmatrix} a_{11} & a_{12} \\ a_{21} & a_{22} \end{bmatrix} \begin{bmatrix} z_1 \\ z_2 \end{bmatrix} = \begin{bmatrix} b_1 \\ b_2 \end{bmatrix}$ gives $z_1 = (a_{22}b_1 - a_{12}b_2)/\Delta$, $z_2 = (a_{11}b_2 - a_{21}b_1)/\Delta$ where $\Delta = a_{11}a_{22} - a_{21}a_{12}$. Substitution gives the required result.

5.2: If $pv^\gamma = c$, then $\ln(p) + \gamma\ln(v) = \ln(c)$. Hence fit $\ln(p)$ to $\ln(v)$ using a straight line whose slope is $-\gamma$ and p-intercept is $\ln(c)$. Now use the result of Exercise 5.1 to find $\gamma = 1.4223$ and $c = 0.9972$.

5.3: Assume all summations are from $k = 1$ to m. The weighted least squares approximation is of the form $S = \sum \left(w_k(f_k - p(x_k))\right)^2$ where $p(x) = a_0 + a_1 x + \cdots + a_n x^n$. To minimize S we require $\partial S/\partial a_i = 0$, for $i = 0, 1, \ldots, n$. $\partial S/\partial a_i = \sum 2w_k\left(f_k - p(x_k)\right)(-x_k^i)$, that is, $\sum w_k f_k x_k^i = a_0 \sum w_k x_k^i + a_1 \sum w_k x_k^{i+1} + \cdots + a_n \sum w_k x_k^{i+n}$. This leads to a system of $n + 1$ equations of the form $A\mathbf{a} = \mathbf{b}$ where $A_{ij} = \sum w_k x_k^{i+j-2}$, $(i, j = 1, 2,$

$\ldots, n)$, $\mathbf{a} = (a_0, a_1, \ldots, a_n)^T$ and $b_i = \sum w_k x_k^{i-1} f_k$, $(i = 1, 2, \ldots, n)$. For large values of n this system becomes highly ill-conditioned if the points are evenly distributed in $[0, 1]$ (see Ralston and Rabinowitz [67], p. 252).

5.4: In the text we have shown that p_0 and p_1 are orthogonal w.r.t. the inner product. Assume that the orthogonality relationship is true for $j = 0$, $1, \ldots, r$: we wish to show that $< p_j, p_{r+1} > = \sum w_k p_j(x_k) p_{r+1}(x_k) = 0$ (all sums are from $k = 1$ to m) for $j = 0, 1, \ldots, r$. Using the three-term recurrence relation we have $p_{r+1}(x) = \lambda_r(x - \alpha_{r+1}) p_r(x) - \beta_r p_{r-1}(x)$ and hence $< p_j, p_{r+1} > = \lambda_r < x p_j, p_r > - \lambda_r \alpha_{r+1} < p_j, p_r > - \beta_r < p_j, p_{r-1} >$. For $j = 0, 1, \ldots, r - 2$ the induction hypothesis ensures that the last two terms are zero. For these values of j we have that $x p_j(x)$ is a polynomial of degree at most $r - 1$ and is therefore expressible as a linear combination of the $p_j(x)$, $j = 0, 1, \ldots, r - 1$, which, by the induction hypothesis, means that the first term is also zero. Thus for $j = 0, 1, \ldots, r - 2$ we have $< p_j, p_{r+1} > = 0$ for any value of α_{r+1}, β_r. For $j = r - 1$, $< p_j, p_r > = 0$ hence $< p_{r-1}, p_{r+1} > = \lambda_r < x p_{r-1}, p_r > - \beta_r < p_{r-1}, p_{r-1} >$. From the recurrence relation we have

$$< p_r, p_r > = \lambda_{r-1} < x p_r, p_{r-1} > - \lambda_{r-1} \alpha_r < p_r, p_{r-1} > - \beta_{r-1} < p_r, p_{r-2} >$$

giving $< p_r, p_r > = \lambda_{r-1} < x p_r, p_{r-1} >$ and substituting for β_r we obtain $< p_{r-1}, p_{r+1} > = 0$. For $j = r$ the third term vanishes to give $< p_r, p_{r+1} > = \lambda_r < x p_r, p_r > - \lambda_r \alpha_{r+1} < p_r, p_r > = 0$ by the definition of α_{r+1}, and the induction is complete.

5.5: $< f - p, f - p > = < f, f > - 2 < f, p > + < p, p >$. Assume all sums are for $i = 0$ to n. Then $< f, p > = < f, \sum c_i p_i > = \sum c_i < f, p_i > = \sum c_i^2$ $< p_i, p_i >$ from (5.5). Also $< p, p > = < \sum c_i p_i, \sum c_i p_i > = \sum c_i^2 < p_i, p_i >$ and the result follows.

5.6: $T_0(x) = \cos(0) = 1$, $T_1(x) = \cos(\cos^{-1} x) = x$. The identity $\cos(2A) = 2\cos^2(A) - 1$ gives $T_2(x) = 2x^2 - 1$. The maximum value of $\cos(\theta)$ is 1 for all θ, and hence $|T_i(x)| \leq 1$. $T_i(-1) = \cos(i \cos^{-1}(-1)) = \cos(i\pi) = (-1)^i$. $T_i(1) = \cos(0) = 1$.

5.7: The integrals of T_0 and T_1 are x and $x^2/2$ respectively. Use the change of variable $x = \cos(\theta)$ to show that

$$\int T_i(x) dx = -\int \cos(i\theta) \sin(\theta) d\theta$$

$$= -\tfrac{1}{2} \int \left(\sin\big((i+1)\theta\big) - \sin\big((i-1)\theta\big) \right) d\theta$$

and the result follows. Now, if $p(x) = \sum_{i=0}'^n a_i T_i(x)$, use the results $T_i(-1) = (-1)^i$ and $T_i(1) = 1$ to show that

$$\int_{-1}^1 p(x) dx = a_0 - \sum_{i=2}^n \left(\frac{1 + (-1)^i}{i^2 - 1} \right) a_i.$$

5.9: Use the identity $\cos(A) + \cos(B) = 2\cos\left(\frac{A+B}{2}\right)\cos\left(\frac{A-B}{2}\right)$ with $A = i\cos^{-1}x$ and $B = j\cos^{-1}x$. See Broucke [12] for Fortran routines for multiplication, etc. of Chebyshev series.

5.10: Use the three-term recurrence relation (5.6) to generate T_2, T_3 and T_4, and then express x^i as a linear combination of T_i, T_{i-1}, ..., T_0. Rewriting p in terms of the Chebyshev polynomials, the coefficient of T_4 is $1/192$. Recalling that the maximum value of T_i is 1 (Exercise 5.6), the effect of ignoring this term results in a bound on the further error of about 0.0052, compared with $1/24 \approx 0.042$ if the x^4 term is omitted from the original expansion.

5.11: For the first part, use the identity

$$\sin(A) - \sin(B) = 2\cos\left(\frac{A+B}{2}\right)\sin\left(\frac{A-B}{2}\right)$$

with $A = \sin\left((k-1+\frac{1}{2})x\right)$ and $B = \sin\left((k-1-\frac{1}{2})x\right)$. Substituting for $\cos\left((k-1)x\right)$, the terms in the summation sign cancel, except the first and last. The second result follows from the first using the identity $\sin(A+B) = \sin(A)\cos(B)+\cos(A)\sin(B)$, with $A = (m-1)x$ and $B = x/2$. To prove the orthogonality property, form $<T_i, T_j> = \sum''^m_{k=1} \cos\left(\frac{i(k-1)\pi}{m-1}\right)\cos\left(\frac{j(k-1)\pi}{m-1}\right)$ Then use the identity $\cos(A) + \cos(B) = 2\cos\left(\frac{A+B}{2}\right)\cos\left(\frac{A-B}{2}\right)$ with $A = (i+j)(k-1)\pi/(m-1)$ and $B = (i-j)(k-1)\pi/(m-1)$ to give $<T_i, T_j> = \frac{1}{2}\sum''^m_{k=1} \cos\left((i+j)(k-1)\pi/(m-1)\right) + \cos\left((i-j)(k-1)\pi/(m-1)\right)$. For $i \neq j$ apply the second result above to show that all terms in the sum are zero. For $i = j = 0$ or $m-1$ both terms in the summation are equal to one, and so $<T_i, T_i> = m-1$. For $i = j \neq 0$ or $m-1$ the second term is one again; apply the second result above to show that the first term is zero, and hence that $<T_i, T_i> = (m-1)/2$. For $i \neq j$ apply the second result above to both terms, deduce that they are zero, and hence that $<T_i, T_j> = 0$.

5.12: Use (5.15) to write $N_{i,2}$ in terms of $N_{i,0}$, $N_{i+1,0}$ and $N_{i+2,0}$. Deduce that

$$N_{i,2} = \frac{1}{2h^2}\begin{cases} (x - z_i)^2, & x \in [z_i, z_{i+1}]; \\ (x - z_i)(z_{i+2} - x) + (z_{i+3} - x)(x - z_{i+1}), & x \in [z_{i+1}, z_{i+2}]; \\ (z_{i+3} - x)^2, & x \in [z_{i+2}, z_{i+3}]. \end{cases}$$

When sketching $N_{i,2}$ note that $N_{i,2}(x) = 0, x \notin (z_i, z_{i+3})$, $N_{i,2}(z_{i+1}) = N_{i,2}(z_{i+2}) = \frac{1}{2}$, $N_{i,2}(z_{i+\frac{1}{2}}) = N_{2,i}(z_{i+\frac{5}{2}}) = \frac{1}{8}$, and $N_{2,i}(z_{i+\frac{3}{2}}) = \frac{3}{4}$.

5.13: Let $d_i^{(j)}$ be the coefficients in the B-spline representation of the jth derivative of a function, with $d_i^{(0)} = d_i$, and $p(x) = \sum d_i N_{k,i}(x)$. Whence we have $p^{(j)}(x) = k(k-1)\cdots(k-j+1)\sum d_i^{(j)} N_{k-j,i}(x)$ $d_i^{(j)} = (d_i^{(j-1)} - d_{i-1}^{(j-1)})/(z_{i+k+1+j} - z_i)$ (see de Boor [24]).

5.14: On $[z_i, z_{i+1}]$ only $N_{i,1}$ and $N_{i+1,1}$ are non-zero, and

$$N_{i,1}(x) + N_{i+1,1}(x) = (x - z_i)/h_i + (z_{i+1} - x)/h_i = 1$$

5.15: $d^3 B_i(x)/dx^3 = 3/(2h^3)$, $z_{i-2} < x < z_{i-1}$; $-9/(2h^3)$, $z_{i-1} < x < z_i$; $9/(2h^3)$, $z_i < x < z_{i+1}$; $-3/(2h^3)$, $z_{i+1} < x < z_{i+2}$; 0, $x \notin [z_{i-2}, z_{i+2}]$. This is discontinuous at the knot points.

5.16: The pattern of non-zero entries in the coefficient matrix is

$$
\begin{array}{ccccc}
x & x & x & 0 & 0 \\
x & x & x & x & 0 \\
x & x & x & x & 0 \\
0 & x & x & x & x \\
0 & x & x & x & x \\
0 & x & x & x & x \\
0 & 0 & x & x & x \\
\end{array}
$$

5.17: Dropping the superscript (pq) on $Q^{(pq)}$, forming QQ^T preserves the ones on the diagonal, $(QQ^T)_{pp} = (QQ^T)_{qq} = \sin^2\theta + \cos^2\theta = 1$, $(QQ^T)_{pq} = (QQ^T)_{qp} = \sin\theta\cos\theta - \sin\theta\cos\theta = 0$. Thus $QQ^T = I_m$ and Q is orthonormal. Premultiplying X by Q replicates all rows of X except for rows p and q, whose elements are replaced by $\tilde{X}_{pi} = sX_{pi} + cX_{qi}$, $\tilde{X}_{qi} = -cX_{pi} + sX_{qi}$, $i = 1, 2, \ldots, n$, where $s = \sin\theta = X_{pp}/\Delta$, $c = \cos\theta = X_{qp}/\Delta$, and $\Delta = \sqrt{X_{pp}^2 + X_{qp}^2}$. In particular $\tilde{X}_{qp} = -cX_{pp} + sX_{qp} = 0$.

5.18: The routine gives a function estimate of 0.313568, an error of 0.588.

n	Errors in interpolation			
	0.0125	0.5	1.25	1.985
11	0.175×10^{-3}	0.136×10^{-4}	0.349×10^{-5}	-0.141×10^{-3}
21	0.137×10^{-8}	0.211×10^{-11}	0.756×10^{-14}	-0.103×10^{-10}
31	0.657×10^{-4}	-0.368×10^{-7}	-0.288×10^{-12}	-0.272×10^{-9}
41	-0.362	0.658×10^{-4}	0.168×10^{-10}	0.111×10^{-6}
51	0.153×10^6	0.482	0.132×10^{-8}	0.743×10^{-5}

Never expect too much from an interpolation routine!

6.1: Define $y_i \equiv y^{(i-1)}$. Then the system becomes

$$
\begin{aligned}
y_1' &= y_2 \\
y_2' &= y_3 \\
y_3' &= y_4 \\
y_4' &= 3601y_3 - 3600y_1 + 1800t^2
\end{aligned}
$$

subject to $y_1(0) = y_2(0) = y_3(0) = y_4(0) = 0$.

6.2: Define $y_1 \equiv y$, $y_2 \equiv y'$, $y_3 \equiv y''$, $y_4 \equiv x$, $y_5 \equiv x'$, $y_6 \equiv x''$, $y_7 \equiv x'''$. Then the required system is

$$
\begin{aligned}
y_1' &= y_2 \\
y_2' &= y_3 \\
y_3' &= (y_5 y_2 - y_6)/\sin(t) \\
y_4' &= y_5 \\
y_5' &= y_6 \\
y_6' &= y_7 \\
y_7' &= (e^{t y_4 y_1} - \sin(y_4)y_3 - y_6 - y_6 y_2)/y_5
\end{aligned}
$$

6.3: The perturbation ϵ in the initial condition on $y(0)$ changes the solution by an amount $\frac{1}{2}\epsilon(e^{2t} + e^{-2t})$. The solution to the original problem exhibits exponential decay; the perturbed problem exhibits exponential growth. The problem is therefore badly conditioned. Solving the problem numerically we would expect rounding, and other, errors rapidly to become significant. (Recall (Section 1.3) that 0.1 is not exactly representable in binary.)

6.4: Euler's method gives $y_{i+1} = y_i + hy_i = (1+h)y_i$, and so $y_i = (1+h)^i$. But $h = x/i$ and, assuming convergence of the method, the required result follows.

6.9:

$$
\begin{aligned}
\theta''' &= -(g/l)\cos(\theta)\theta' \\
\theta^{(4)} &= -(g/l)\cos(\theta)\theta'' + (g/l)\sin(\theta)(\theta')^2 \\
\theta^{(5)} &= -(g/l)\cos(\theta)\theta''' + 3(g/l)\sin(\theta)\theta'\theta'' + (g/l)\cos(\theta)(\theta')^3.
\end{aligned}
$$

Substituting into the Taylor series expansions we find $\theta(0.1) \approx 0.205$, $\theta(0.2) \approx 0.192$, $\theta(0.3) \approx 0.169$, $\theta'(0.1) \approx -0.873$, $\theta'(0.2) \approx -0.179$, $\theta'(0.3) \approx -0.279$. Here we only use the Taylor series expansions about $t = 0$; in Exercise 6.7 when computing an approximation at t_{i+1} we use expansions about t_i. The advantage of the former method is that we have no error accumulation, as in Euler's method, and we do not have to determine values of the higher derivatives at 0.1 and 0.2; the disadvantage is that the expansions used become less accurate as t increases.

6.10: (6.20) gives the general second-order two-stage Runge–Kutta method as $y_{i+1} = y_i + h(c_1 f(t_i, y_i) + c_2 f(t_i + a_2 h, y_i + a_2 h s_1))$ where $c_1 + c_2 = 1$ and $a_2 c_2 = \frac{1}{2}$. Thus we obtain the triples (a_2, c_1, c_2); $(1, \frac{1}{2}, \frac{1}{2})$ – known as the *improved Euler method*, $(\frac{1}{2}, 0, 1)$ – known as the *modified Euler method*, $(\frac{1}{3}, -\frac{1}{2}, \frac{3}{2})$ and $(\frac{1}{4}, -1, 2)$. Defining $F = f_t + f f_y$ and $G = f_{tt} + 2f f_{ty} + f^2 f_{yy}$ and with the assumption that $y_i = y(t_i)$ we obtain $T_{i+1} = y(t_{i+1}) - y_{i+1} = h(f + \frac{1}{2}hF + \frac{1}{6}h^2(Ff_y + G)) - h((c_1 + c_2)f + hc_2 a_2 F + \frac{1}{2}h^2 c_2 a_2^2 G) + O(h^4)$ where the right-hand side is evaluated at $t = t_i$, $y = y(t_i)$. Since the method must be of order two we may use (6.20) to simplify this to $T_{i+1} = h^3\left(\frac{1}{6}Ff_y + (\frac{1}{6} - \frac{1}{4}a_2)G\right) + O\left(h^4\right)$. The multipliers of G are $-\frac{1}{12}$, $\frac{1}{24}$, $\frac{1}{12}$, $\frac{5}{48}$ for $a_2 = 1, \frac{1}{2}, \frac{1}{3}, \frac{1}{4}$ respectively.

6.13: The characteristic polynomial of the midpoint method is $z^2 - 2h\lambda z - 1$ which has roots $\lambda h \pm \sqrt{\lambda^2 h^2 + 1}$. For $\lambda < 0$ the root $\lambda h - \sqrt{\lambda^2 h^2 + 1}$ is greater than one in absolute value and so the method is absolutely unstable. The characteristic polynomial of the trapezoidal method is $z(1 - h\lambda/2) - (1 + h\lambda/2)$ which has root $(1 + h\lambda/2)/(1 - h\lambda/2)$. The method is absolutely stable for all $h\lambda < 0$, and so the region of absolute stability is $(-\infty, 0)$.

6.14: (a) gives the trapezoidal method (6.29); (b) gives $y_{i+1} = y_{i-1} + \frac{h}{3}\left(f_{i+1} + 4f_i + f_{i-1}\right)$. Milne's method is unstable. To derive an Adams method, we integrate the ODE over $[t_i, t_{i+1}]$. f is replaced by an interpolating polynomial based on $\{t_i, t_{i-1}, \ldots, t_{i-r+1}\}$ for an AB method, and $\{t_{i+1}, t_i, \ldots, t_{i-r+1}\}$ for an AM method.

6.16: $y_{i+1} - y_i - h\left(\gamma y_i' + (1 - \gamma)y_{i+1}'\right) = h^2(1 - \gamma - \frac{1}{2})y_i'' + O(h^3)$, so the method is first order, unless $\gamma = \frac{1}{2}$, in which case the method is at least second order. If $\gamma = 1$ we get Euler's method; if $\gamma = 0$ we get the backward Euler method (6.31), and if $\gamma = \frac{1}{2}$ we get the trapezoidal method (6.29) which has LTE $-\frac{1}{12}h^3 y'''$. For the two-step method to be third order we require

$$c + \alpha_1 + 1 = 0$$
$$\alpha_1 + 2 - \beta_0 - \beta_1 - \beta_2 = 0$$
$$\tfrac{1}{2}(\alpha_1 + 4 - 2\beta_1 - 4\beta_2) = 0$$
$$\tfrac{1}{6}(\alpha_1 + 8 - 3\beta_1 - 12\beta_2) = 0.$$

This system has solution $\alpha_1 = -(1 + c)$, $\beta_0 = -\frac{1}{12}(1 + 5c)$, $\beta_1 = \frac{8}{12}(1 - c)$, $\beta_2 = \frac{1}{12}(5 + c)$, with LTE $-\frac{1}{24}(1 + c)h^4 y^{(4)} + O(h^5)$. If $c = 1$ we obtain the Simpson's rule method (Exercise 6.15), which is at least fourth order. In fact, the LTE for the method is $-\frac{1}{90}h^5 y^{(v)}$.

6.17: $f_{i-\frac{1}{2}} = \frac{1}{128}(35f_i + 140f_{i-1} - 70f_{i-2} + 28f_{i-3} - 5f_{i-4})$; $f_{i-\frac{3}{2}} = \frac{1}{64}(-f_i + 24f_{i-1} + 54f_{i-2} - 16f_{i-3} + 3f_{i-4})$. If $y_{i+1}^{(P)}$ is the predictor estimate and $y_{i+1}^{(C)}$ is the corrector estimate, the quantity $\frac{19}{270}|y_{i+1}^{(P)} - y_{i+1}^{(C)}|$ can be used as a measure of the LTE in $y_{i+1}^{(C)}$.

Appendix B
NAG Graphics

NAG offers a suite of routines for producing graphical output which sits on top of the numerical subroutine library. Like the main library this consists of a set of user callable routines each of which performs a specific graphical task, for example, drawing axes, labelling axes, plotting curves, etc. The advantages of using this library are that programs may be ported with a minimum of effort between different output devices at the same site as well as to other sites supporting the library.

The library consists of two levels of routines; the lower level, known as the graphical interface, is dependent upon the output device and, although some of the routines may be user callable (e.g., low level line drawing routines) the majority are used only by the higher level routines.

Two different sets of coordinates are used by the graphics routines; first there are the user coordinates $(xmin, ymin)$, $(xmax, ymax)$ defining the rectangular region of the user's data which is of interest, known as the user's window. Note that this need not define the actual extremes of the data plotted (see Figure B.1). Each plotting device has a predefined default plotting region or frame; this could be the screen of a graphics display terminal or a predefined area of paper on a hard copy plotter. The second set of coordinates $(umin, vmin)$, $(umax, vmax)$ define a rectangular window onto this plotting area; the entire default area has these coordinates set as $(0, 0)$, $(1, 1)$. By defining appropriate windows the user may plot a number of graphs on a single frame or preserve some aspect ratio in the plot (e.g., a height to width ratio — see Figure B.2).

To generate graphical output using the NAG graphical supplement library:

(1) Initialize the plotter dependent package.

> ```
> CALL XXXXXX
> ```

selects a particular device and may initialize various plotter dependent routine values etc. Not only is this call site dependent but it may also need to be changed for different devices; the name used above is thus fictitious.

(2) Initialize the NAG graphics system.

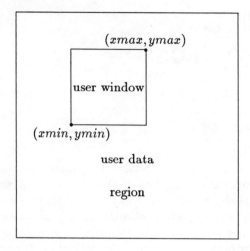

Figure B.1   An example of a user window.

```
CALL J06WAF
```

initializes the NAG graphical interface routines and the call must follow step (1). Note: steps (1) and (2) must only be performed once during a particular program execution.

(3) Define the data window.

```
CALL J06WBF(XMIN,XMAX,YMIN,YMAX,MARGIN)
```

defines the range of x and y values that the user wishes to plot. The call creates a rectangular window with its bottom left-hand corner at $(\texttt{XMIN},\texttt{YMIN})$ and its top right-hand corner at $(\texttt{XMAX},\texttt{YMAX})$. The integer parameter $\texttt{MARGIN}$ may be set to either 0 or 1. If set to zero the whole region is used for plotting; with $\texttt{MARGIN} = 1$ a border is left around the plotting area to allow for titles etc.

(4) To define the window into the plotting area

```
CALL J06WCF(UMIN,UMAX,VMIN,VMAX)
```

declares that the portion of the plotting area to be used is the rectangle with diagonal corners at $(\texttt{UMIN},\texttt{VMIN})$ and $(\texttt{UMAX},\texttt{VMAX})$. Note that $0 \leq \texttt{UMIN} < \texttt{UMAX} \leq 1$ and $0 \leq \texttt{VMIN} < \texttt{VMAX} \leq 1$.

(5) Call user level routines to generate the required graphical output. Full details of the available routines may be found in the NAG Graphics manual. The routine documentation layout is exactly the same as for the main subroutine library. Routines are available for drawing

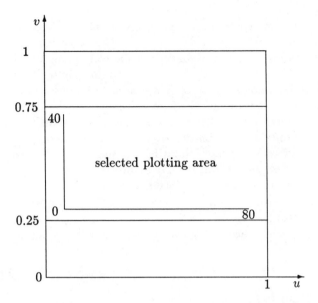

Figure B.2   Window onto plotting area.

axes, producing legends and titles, simple curve plotting both from data and from user defined functions, contour plotting, histograms, pie charts, etc.

(6) Advance to a new frame.

       `CALL J06WDF`

creates a new plotting area (e.g., clears the screen of a graphics display or advances the paper on a hard copy plotter). The user may now return to step (3).

(7) Terminate graphical output.

       `CALL J06WZF`

must be the last call to a NAG graphics routine. If omitted it is possible that part of the plot may be lost. Note that it is not necessary to advance to a new frame before terminating output.

Errors are generally not fatal, for example, on detecting illegal arguments the routine call is ignored and an error message is output on the error message channel (the default channel may be changed by a call to `X04AAF` see Section 1.9).

The mechanism for running a program containing NAG graphics routines will be installation dependent. Usually the execution of such a program results in the creation of an output file containing an encoded representation of the plot. Either this file is sent directly to the device, i.e., the output is device dependent, or it will need to be processed by another program before it is sent, in which case it is said to be device independent. Device independent output has the advantage that a single run of the program results in the definition of the picture which may be viewed on a number of different devices without having the expense of rerunning the program. For example the output may be previewed on a graphics display before being sent to a hard copy plotter.

We conclude with a sample program which generates a plot of the function *erfc* from $x = -2$ to $x = 2$. The resultant picture should bear some passing resemblance to Figure 3.5, Section 3.8.

```
      PROGRAM MAIN
*
*.. PROGRAM TO PLOT THE COMPLEMENTARY ERROR FUNCTION
*.. FROM X=-2 TO X=2 USING ITS FUNCTION DEFINITION.
      DOUBLE PRECISION ONE,TWO,ZERO
      CHARACTER*(*) TITLE
      PARAMETER (TITLE='PLOT OF ERFC(X)')
      PARAMETER (ZERO = 0.0D0,ONE = 1.0D0,TWO = 2.0D0)
      DOUBLE PRECISION FUN,UMIN,UMAX,VMIN,VMAX,XMIN,
     +                 XMAX,YMIN,YMAX
      INTEGER MARGIN
      EXTERNAL FUN,XXXXXX,J06WAF,J06WCF,J06WBF,J06EAF,J06EAF,
     +         J06AAF,J06AHF,J06WZF
*
*.. CALL THE POSSIBLY DEVICE DEPENDENT INITIALIZATION
*.. ROUTINE. NOTE THIS IS A FICTICIOUS NAME.
      CALL XXXXXX
*
*.. INITIALIZE THE NAG GRAPHICAL INTERFACE.
      CALL J06WAF
*
*.. DEFINE A WINDOW INTO THE PLOTTING AREA
*.. IN THIS CASE WE USE ALL THE AVAILABLE SPACE.
      UMIN = ZERO
      UMAX = ONE
      VMIN = ZERO
      VMAX = ONE
      CALL J06WCF(UMIN,UMAX,VMIN,VMAX)
```

```
*
*.. DEFINE THE USER COORDINATE WINDOW
      XMIN = -TWO
      XMAX = TWO
      YMIN = ZERO
      YMAX = TWO
      MARGIN = 1
      CALL J06WBF(XMIN,XMAX,YMIN,YMAX,MARGIN)
*
*.. PLOT FUNCTION DEFINED BY FUN(X) AS A SMOOTH
*.. CURVE.
      CALL J06EAF(FUN,XMIN,XMAX)
*
*.. DRAW AXES
      CALL J06AAF
*
*.. ADD A TITLE
      CALL J06AHF(TITLE)
*
*.. TERMINATE NAG GRAPHICS OUTPUT
      CALL J06WZF
      END

      DOUBLE PRECISION FUNCTION FUN(X)
      INTEGER IFAIL
      DOUBLE PRECISION S15ADF,X
      EXTERNAL S15ADF
*
      IFAIL = 0
      FUN = S15ADF(X,IFAIL)
      END
```

Code B.1   NAG graphics program for plotting $erfc(x)$.

Appendix C
Toolpack

The Toolpack suite of Fortran software tools offers programmers a wide range of useful facilities to assist in the production, maintenance and testing of Fortran 77 software. The facilities available include the following:

(1) A formatter, or pretty printer, which will renumber statement labels, indent blocks of statements and generally improve the readability of the code. In addition this tool may be tuned by the user, via an options editor, to generate almost any required 'house style'.

(2) A declaration standardizer which rebuilds the declarative parts of Fortran program units. This ensures that all the variables used are explicitly declared and names which are declared and not used can be removed.

(3) A precision transformer which converts Fortran code from single to double precision and vice versa.

(4) A do-loop unroller which produces code that may be more easily optimized by compilers for vector machines (e.g., Crays).

(5) A standard verifier which checks a complete program for conformance to a large subset of the ANSI Fortran 77 standard (see ANSI [5]). This tool is an invaluable aid to code portability.

The source code for the full suite of tools is in the public domain and is available for a modest handling charge from NAG. In its basic form the software is not easy to implement and the tools tend to run rather slowly. For this reason a number of tailored versions are available for several common machine/operating system combinations (for example, VAX/VMS, VAX/UNIX, SUN/UNIX, IBM/MVS). The number of these implementations is growing and full details of their availability may be obtained from NAG.

Bibliography

[1] ABRAMOWITZ, M., AND STEGUN, I. A. *Handbook of Mathematical Functions.* Dover, New York, 1965.

[2] AMIT, D. *Field Theory, the Renormalization Group and Critical Phenomena.* McGraw-Hill, New York, 1978.

[3] ANONYMOUS. Index by subject to algorithms. *Commun. ACM 7* (1964), 146–148.

[4] ANSI. *Programming Language Fortran X3.9-1966.* American National Standards Institute, New York, 1966.

[5] ANSI. *Programming Language Fortran X3.9-1978.* American National Standards Institute, New York, 1979.

[6] ATKINSON, L. V., AND HARLEY, P. J. *An Introduction to Numerical Methods with Pascal.* Addison-Wesley, London, 1983.

[7] BALFOUR, A., AND MARWICK, D. *Programming in Standard Fortran 77.* Heinemann, London, 1979.

[8] BENTLEY, J. Programming pearls — birth of a cruncher. *Commun. ACM 29* (1986), 1155–1161.

[9] BERNTSEN, J. *A Test of some well known One-Dimensional General Purpose Automatic Quadrature Routines.* Tech. Rep. 20, University of Bergen, Bergen, Norway, 1986.

[10] BORWEIN, J. M., AND BORWEIN, P. B. The arithmetic-geometric mean and fast computation of elementary functions. *SIAM Rev. 26* (1984), 351–366.

[11] BRENT, R. P. *Algorithms for Minimization without Derivatives.* Prentice-Hall, Englewood Cliffs, New Jersey, 1973.

[12] BROUCKE, R. Algorithm 446: Ten subroutines for the manipulation of Chebyshev series [c1]. *Commun. ACM 16* (1973), 254–256.

[13] BUS, J. C. P., AND DEKKER, T. J. Two efficient algorithms with guaranteed convergence for finding a zero of a function. *ACM Trans. Math. Softw. 1* (1975), 330–345.

[14] CARLSON, B. C. *Special Functions of Applied Mathematics.* Academic Press, London, 1977.

[15] CLENSHAW, C. W. Curve fitting with a digital computer. *Comput. J. 2* (1960), 170–173.

[16] CLENSHAW, C. W. *Mathematical Tables Volume 5. Chebyshev Series for Mathematical Functions.* H.M.S.O., London, 1962.

[17] CLENSHAW, C. W., AND HAYES, J. G. Curve and surface fitting. *J. Inst. Math. Applics. 1* (1965), 164–183.

[18] CODY, JR., W. J., AND WAITE, W. *Software Manual for the Elementary Functions.* Prentice-Hall, Englewood Cliffs, New Jersey, 1980.

[19] CONTE, S. D., AND DE BOOR, C. *Elementary Numerical Analysis — An Algorithmic Approach.* McGraw-Hill, New York, 1980.

[20] COPSON, E. T. *An Introduction to the Theory of Functions of a Complex Variable.* Oxford University Press, Oxford, 1935.

[21] COX, M. G. An algorithm for spline interpolation. *J. Inst. Math. Applics. 15* (1975), 95–108.

[22] CURTIS, A. R., POWELL, M. J. D., AND REID, J. K. On the estimation of sparse Jacobian matrices. *J. Inst. Math. Applics. 13* (1974), 117–119.

[23] DAVIS, P. J., AND RABINOWITZ, P. *Methods of Numerical Integration,* second ed. Academic Press, Orlando, Florida, 1984.

[24] DE BOOR, C. On calculating with B-splines. *J. Approx. Theory 6* (1972), 50–62.

[25] DEKKER, T. J. Finding a zero by means of successive linear interpolation. In *Constructive Aspects of the Fundamental Theorem of Algebra,* B. Dejon and P. Henrici, Eds., Wiley-Interscience, London, 1969, pp. 37–48.

[26] DONGARRA, J., BUNCH, J. R., MOLER, C. B., AND STEWART, G. W. *LINPACK Users' Guide.* SIAM Publications, Philadelphia, 1979.

[27] ENRIGHT, W. H., HULL, T. E., AND LINDBERG, B. Comparing numerical methods for stiff systems of ordinary differential equations. *BIT 15* (1975), 10–48.

[28] FEHLBERG, E. Klassische Runge-Kutta Formeln vierter und niedrigerer Ordnung mit Schrittweiten-Kontrolle und ihre Anwendung auf Warmeleitungs-probleme. *Computing 6* (1970), 61–71.

[29] FORSYTHE, G., AND MOLER, C. B. *Computer Solution of Linear Algebraic Systems.* Prentice-Hall, Englewood Cliffs, New Jersey, 1967.

[30] FORSYTHE, G. E. Generation and use of orthogonal polynomials for data fitting with a digital computer. *J. Soc. Ind. Appl. Math. 5* (1957), 74–88.

[31] FOX, L. *Numerical Solution of Ordinary and Partial Differential Equations.* Pergamon, New York, 1962.

[32] GEAR, C. W. The automatic integration of stiff ordinary differential equations. In *Information Processing '68*, Morrell, Ed., North Holland, Amsterdam, 1969, pp. 187–193.

[33] GEAR, C. W. *Numerical Initial Value Problems in Ordinary Differential Equations.* Prentice-Hall, Englewood Cliffs, New Jersey, 1971.

[34] GENTLEMAN, W. M. Least squares computation by Givens transformations without square roots. *J. Inst. Math. Applics. 12* (1973), 329–336.

[35] GEORGE, J. A., AND LIU, J. W. *Computer Solution of Large Sparse Positive Definite Systems.* Prentice-Hall, Englewood Cliffs, New Jersey, 1981.

[36] GILL, P. E., AND MILLER, G. F. An algorithm for the integration of unequally spaced points. *Comput. J. 15* (1972), 80–83.

[37] GOLUB, G. H., AND WELSCH, J. H. Calculation of Gauss quadrature rules. *Math. Comput. 23* (1969), 221–230.

[38] GONET, G. H. *Handbook of Algorithms and Data Structures.* Addison-Wesley, Reading, Massachusetts, 1984.

[39] GRANT, J. A., AND HITCHINS, G. D. Two algorithms for the solution of polynomial equations to limiting machine precision. *Comput. J. 18* (1975), 258–264.

[40] HAGEMAN, L. A., AND YOUNG, D. M. *Applied Iterative Methods.* Academic Press, New York, 1981.

[41] HALL, G., AND WATT, J. M. *Modern numerical methods for ordinary differential equations.* Clarendon Press, Oxford, 1976.

[42] HENRICI, P. *Discrete Variable Methods in Ordinary Differential Equations.* John Wiley and Sons, London, 1962.

[43] HENRICI, P. *Elements of Numerical Analysis.* John Wiley and Sons, New York, 1964.

[44] HINDMARSH, A. C. *Ordinary Differential Equation Solver.* Tech. Rep. UCID-30001, Rev 3, Lawrence Livermore Laboratory, 1974.

[45] HULL, T. E., ENRIGHT, W. H., FELLEN, B. M., AND SEDGWICK, A. E. Comparing numerical methods for ordinary differential equations. *SIAM J. Numer. Anal. 9* (1972), 603–637.

[46] JOHNSON, L. W., AND RIESS, R. D. *Numerical Analysis*, second ed. Addison-Wesley, Reading, Massachusetts, 1982.

[47] KELLER, H. B. *Numerical Methods for Two-point Boundary Value Problems*. Blaisdell, Waltham, Massachusetts, 1968.

[48] KNUTH, D. E. *The Art of Computer Programming Vol 1 : Fundamental Algorithms*, second ed. Addison-Wesley, London, 1973.

[49] KNUTH, D. E. *The Art of Computer Programming Vol 2 : Seminumerical Algorithms*, second ed. Addison-Wesley, London, 1981.

[50] KNUTH, D. E. *The Art of Computer Programming Vol 3 : Searching and Sorting*. Addison-Wesley, London, 1973.

[51] KROGH, F. T. Efficient algorithms for polynomial interpolation and numerical differentiation. *Math. Comput. 24* (1970), 185–190.

[52] KRONROD, A. S. *Nodes and Weights of Quadrature Formulas*. Tech. Rep., Consultants Bureau, New York, 1965.

[53] LAWSON, C. L., AND HANSON, R. J. *Solving Least Squares Problems*. Prentice-Hall, Englewood Cliffs, New Jersey, 1974.

[54] LAWSON, C. L., HANSON, R. J., KINCAID, D. R., AND KROGH, F. T. Basic linear algebra subprograms for Fortran usage. *ACM Trans. Math. Softw. 5* (1979), 308–371.

[55] MALCOLM, M. A. Algorithms to reveal properties of floating point arithmetic. *Commun. ACM 15* (1972), 949–951.

[56] MALCOLM, M. A., AND SIMPSON, R. B. Local versus global strategies for adaptive quadrature. *A.C.M. Trans. Math. Softw. 1* (1975), 129–146.

[57] MERSON, R. H. An operational method for the study of integration processes. In *Proceedings of a Symposium on Data Processing* (Salisbury, S.Australia, 1957), Weapons Research Establishment.

[58] METCALF, M. *Fortran Optimization*. Academic Press, London, 1982.

[59] MOLER, C., AND MORRISON, D. Replacing square roots by Pythagorean sums. *IBM J. Res. Develop. 27* (1983), 577–581.

[60] NONWEILER, T. R. F. *Computational Mathematics*. Ellis Horwood, Chichester, 1984.

[61] PATTERSON, T. N. The optimum addition of points to quadrature formulae. *Math. Comput. 22* (1968), 847–856.

[62] PETERS, G., AND WILKINSON, J. H. Practical problems arising in the solution of polynomial equations. *J. Inst. Math. Applics. 8* (1971), 16–35.

[63] PIESSENS, R., DE DONCKER-KAPENGA, E., UBERHUBER, C. W., AND KAHANER, D. K. *QUADPACK - A Subroutine Package for Automatic Integration.* Springer-Verlag, New York, 1983.

[64] POWELL, M. J. D. *Approximation Theory and Methods.* Cambridge University Press, Cambridge, 1981.

[65] POWELL, M. J. D. A hybrid method for nonlinear equations. In *Numerical Methods for Nonlinear Algebraic Equations* (London, 1970), P. Rabinowitz, Ed., Gordon and Breach, pp. 87–114.

[66] PRESS, W. H., FLANNERY, B. P., TEUKOLSKY, S. A., AND VETTERLING, W. T. *Numerical Recipes. The Art of Scientific Computing.* Cambridge University Press, Cambridge, 1986.

[67] RALSTON, A., AND RABINOWITZ, P. *A First Course in Numerical Analysis*, second ed. McGraw-Hill, New York, 1978.

[68] RICE, J. R. An educational adaptive quadrature algorithm. *SIGNUM Newsletter 8* (1973), 27–41.

[69] RICE, J. R. A metalgorithm for adaptive quadrature. *JACM 22* (1975), 61–82.

[70] SCHONFELDER, J. L. Whither Fortran. *University Computing 8* (1986), 67–76.

[71] SHAMPINE, L. F. Stiffness and nonstiff differential equation solvers II: detecting stiffness with Runge-Kutta methods. *ACM Trans. Math. Softw. 3* (1977), 44–53.

[72] SHAMPINE, L. F., AND WATTS, H. A. Algorithm 504 GERK: global error estimation for ordinary differential equations [D]. *ACM Trans. Math. Softw. 2* (1976), 200–203.

[73] SHAMPINE, L. F., AND WATTS, H. A. Comparing error estimators for Runge-Kutta methods. *Math. Comput. 25* (1971), 445–455.

[74] SHAMPINE, L. F., AND WATTS, H. A. Global error estimation for ordinary differential equations. *ACM Trans. Math. Softw. 2* (1976), 172–186.

[75] SINCOVEC, R. F., AND MADSEN, N. K. Software for nonlinear partial differential equations. *ACM Trans. Math. Softw. 1* (1975), 232–260.

[76] SKEEL, R. D. Iterative refinement implies numerical stability for Gaussian elimination. *Math. Comput. 35* (1980), 817–832.

[77] STEFFENSON, J. F. *Interpolation*. Chelsea, New York, 1950.

[78] STEWART, G. W. *Introduction to Matrix Computations*. Academic Press, New York, 1973.

[79] STRANG, G. *Linear Algebra and Its Applications*. Academic Press, New York, 1980.

[80] STROUD, A. H., AND SECREST, D. H. *Gaussian Quadrature Formulas*. Prentice-Hall, Englewood Cliffs, New Jersey, 1966.

[81] SWIFT, A., AND LINDFIELD, G. R. Comparison of a continuation method for the numerical solution of a single nonlinear equation. *Comput. J. 21* (1978), 359–362.

[82] WILKINSON, J. H. *Rounding Errors in Algebraic Processes*. H.M.S.O., London, 1964.

[83] WILKINSON, J. H., AND REINSCH, C. *Handbook for Automatic Computation Volume II - Linear Algebra*. Springer-Verlag, New York, 1971.

[84] WYNN, P. On a device for calculating the $e_m(s_n)$ transformation. *M.T.A.C. 10* (1956), 91–96.

NAG Routine Index

Index